Nitrogen-Containing Macromolecules in the Bio- and Geosphere

Nitrogen-Containing Macromolecules in the Bio- and Geosphere

B. Artur Stankiewicz, EDITOR
University of Bristol

Pim F. van Bergen, EDITOR
University of Bristol

Developed from a symposium sponsored by the Division
of Geochemistry at the 214th National Meeting
of the American Chemical Society,
Las Vegas, Nevada,
September 7–11, 1997

American Chemical Society, Washington, DC

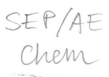

Library of Congress Cataloging-in-Publication Data

Nitrogen-containing macromolecules in the bio- and geosphere / B. Artur Stankiewicz, editor, Pim F. van Bergen, editor.

p. cm.—(ACS symposium series ; 707)

"Developed from a symposium sponsored by the Division of Geochemistry at the 214th National Meeting of the American Chemical Society, Las Vegas, Nevada, September 7–11, 1997."

Includes bibliographical references and index.

ISBN 0–8412–3582–1

1. Geochemistry—Congresses. 2. Macromolecules—Congresses. 3. Nitrogen cycle—Congresses. 4. Biosphere—Congresses.

I. Stankiewicz, B. Artur, 1965– . II. Van Bergen, Pim F., 1965– . III. American Chemical Society. Division of Geochemistry. IV. American Chemical Society. Meeting (214th : 1997 : Las Vegas, Nev.) V. Series.

QE514.N58   1998
551.9—dc21   98–25956
                                                                     CIP

Foreword

THE ACS SYMPOSIUM SERIES was first published in 1974 to provide a mechanism for publishing symposia quickly in book form. The purpose of the series is to publish timely, comprehensive books developed from ACS sponsored symposia based on current scientific research. Occasionally, books are developed from symposia sponsored by other organizations when the topic is of keen interest to the chemistry audience.

Before agreeing to publish a book, the proposed table of contents is reviewed for appropriate and comprehensive coverage and for interest to the audience. Some papers may be excluded in order to better focus the book; others may be added to provide comprehensiveness. When appropriate, overview or introductory chapters are added. Drafts of chapters are peer-reviewed prior to final acceptance or rejection, and manuscripts are prepared in camera-ready format.

As a rule, only original research papers and original review papers are included in the volumes. Verbatim reproductions of previously published papers are not accepted.

ACS BOOKS DEPARTMENT

Contents

CHITIN: THE "FORGOTTEN" MACROMOLECULE

SOURCES OF ORGANIC NITROGEN IN
SEDIMENTARY ORGANIC MATTER

ROLE AND IMPORTANCE OF NITROGEN IN
CROPS, WASTE, AND SOIL

Preface

Nitrogen is the fourth most abundant element in the bio- and geosphere. The most significant sources of nitrogen in the biosphere are in N-containing biomolecules such as proteins, amino sugars, nucleic acids, and pigments, for example cholorophylls; the importance of these molecules in lifecycles is self-evident. Most of these nitrogen compounds are biodegraded or mineralized upon the death of an organism, but a minute amount is retained in the geosphere. Much is known about the chemical transformations undergone by carbon, hydrogen, and oxygen during diagenesis, but the fate of nitrogen is less clearly understood. The past few years have seen a substantial increase in research on the fate of nitrogen in both biological and geological cycles. However, the main emphasis of this research has been on the nitrogen cycle, focusing on C/N ratios and simple nitrogen forms such as ammonia, nitrate, and nitrogen oxide, rather than on (bio)macromolecular forms.

Nitrogen preserved in the geosphere can sometimes be related to biological macromolecular precursors, for example, proteins, amino sugars, and nucleic acids. Although there have been studies on protein moieties in many disciplines of the life sciences, research into the geochemistry of proteins has been rather scattered. Another important class of biomacromolecules in this context is the amino sugars and in particular chitin, which is often cross-linked with proteins in animal tissues. Although chitin has been studied extensively in the past 20 years, mostly because it has significant commercial application, the geochemical processes leading to the degradation and transformation of this macromolecule during diagenesis are not well known. With respect to the preservation of nucleic acids in ancient materials, this is still a lively and controversial topic with the main problems relating to the amplification of DNA. Apart from the fact that nitrogen-containing biomacromolecules may be incorporated into so-called geopolymers, substantial amounts of the organic nitrogen present in the soils and sediments occur as 'unidentified' forms. Despite the importance of these forms of nitrogen, the origin and exact molecular composition of the unidentified structural moieties, which are believed to be primarily macromolecular and chemically complex, are still poorly understood.

The symposium, upon which this book is based, was organized to provide a forum for scientists studying N-containing macromolecules. It was the first opportunity to review the current state-of-the-art research in this field and subsequently compile a volume that draws upon research from a wide range of disciplines all of which focus on (bio)macromolecular nitrogen. In general, the book addresses the key questions pertaining to the stability of N-containing biomacromolecules, and the transformations that they undergo upon diagenesis, and provides the most recent ideas on the fate of these molecules in the bio- and geosphere. It presents the latest research trends and advances on the fate of proteins and/or amino acids in the natural environment which is especially important as amino acids contribute to the plant and animal necromass in all sedimentary systems. This book also provides the reader with comprehensive overviews including the distribution, biodegradation, and fate of the amino sugar, chitin, in the bio- and geosphere. The extensive overview on DNA research presented here, sheds new light on the problems involving amplification of DNA in ancient samples and shows the recent advances in this field. Finally, the volume gives an extensive summary of the fate of so-called 'unidentified' organic nitrogen in soils and sedimentary rocks.

This book also includes several invited overview chapters written by experts in their respective fields. The focus of this volume on the macromolecular forms and the biogeochemical processes is unique such that the theme has never been published before in such a comprehensive form. The papers collated here cover a wide range of topics and are of interest to soil scientists, geochemists, biochemists, geologists, biologists, and archaeologists.

Acknowledgments

This volume and the symposium, on which this book is based, are the result of great support, advise, and help of many individuals and organizations. We thank the ACS Division of Geochemistry, Inc., that sponsored the symposium and its chairmen George Luther III and Ken Anderson for advise and help. We extend our gratitude to all speakers who presented their work during the symposium as well as to all contributors to this book. We especially thank all the referees who donated their precious time to review the papers, thus ensuring the highest scientific standard possible. Acknowledgment is made to the Donors of The Petroleum Research Fund, administered by the ACS, for partial support (ACS-PRF 32472-SE) of this symposium, to which ensured several foreign speakers were able to attend the meeting. B. Artur Stankiewicz and Pim F. van Bergen were supported by NERC grants, GST/02/1027 to Derek E. G. Briggs and Richard P. Evershed, and GR3/9578 to Richard P. Evershed respectively while editing this volume, and while they were employed by University of

Bristol, United Kingdom. B. Artur Stankiewicz is especially grateful for all the support and friendship of Derek E.G. Briggs while in Bristol.

B. ARTUR STANKIEWICZ
Shell E&P Technology Co.
3737 Bellaire Blvd.
Houston, TX 77025

PIM F. VAN BERGEN
Organic Geochemistry Group
Faculty of Earth Sciences
Utrecht University
P.O. Box 80021
3508 TA Utrecht, The Netherlands

Chapter 1

Nitrogen and N-Containing Macromolecules in the Bio- and Geosphere: An Introduction

B. Artur Stankiewicz[1-3] and Pim F. van Bergen[2,4]

[1]Biogeochemistry Centre, Department of Geology, University of Bristol, Wills Memorial Building, Queens Road, Bristol BS8 1RJ, United Kingdom
[2]Organic Geochemistry Unit, School of Chemistry, University of Bristol, Cantock's Close, Bristol BS8 1TS, United Kingdom

A brief overview of our current understanding of nitrogen-containing macromolecules and their fate in the bio- and geosphere is presented in the context of their significance in the N cycle. The major biological macromolecular sources, such as amino sugars, proteins and nucleic acids, as well as abiogenic forms of N, are briefly reviewed. The analytical techniques and methods used in studies of nitrogen in the bio- and geosphere are also summarized.

The Global Nitrogen Cycle

The nitrogen cycle is second only to carbon in its potential to drive global change (*1*) and is a key element in controlling the diversity, dynamics and functioning of many marine, freshwater and terrestrial ecosystems (*2-4*). The global N cycle consists of various pools of nitrogen in the atmosphere, hydrosphere, biosphere and geosphere. However, most of this nitrogen is unavailable to organisms.

Prior to human activities the nitrogen available to organisms was derived mainly from biological fixation and lightning (*1, 5*). However, in recent years availability has changed dramatically due to human activities (*6, 7*). In fact, more N is fixed annually due to the impact of mankind, than that fixed through natural processes (*6*). Most important are industrial and agricultural nitrogen fixation and N-fixation through fossil-fuel combustion. Furthermore, biomass burning and conversion/land clearing of soils accelerate the mobilization of nitrogen (*3, 6*).

Nitrogen cycling is strongly dependent on climatic and environmental factors such as water availability and pH, salinity, aeration, forestation, type of vegetation and temperature. This has led to an increasing number of studies of the N cycle related to small ecosystems such as arid areas (*8, 9*), forests (*8, 10*) and tundra (*11*). Understanding 'small' N cycling is essential to advance our knowledge of the global

[3]Current address: Shell E&P Technology Company, 3737 Bellaire Boulevard, Houston, TX 77025
[4]Current address: Organic Geochemistry Group, Faculty of Earth Sciences, Utrecht University, P.O. Box 80021, 3508 TA Utrecht, The Netherlands

cycle. However, in sharp contrast to the carbon cycle, we are still far from a quantitative understanding of the global nitrogen cycle.

The nitrogen cycle is dependent on several reactions occurring in the environment (8). Biological fixation of N takes place mainly through enzymatic reactions and occurs only in certain organisms which include prokaryotes, cyanobacteria and some symbiotic plants such as legumes. Moreover, ammonia (NH_3) is released in the process of mineralization following death and decomposition of living organisms. The next step involves assimilation of NH_3 by organisms, which primarily involves plants as animals can not assimilate this form of N. This process involves glutamine and glutamate enzymes. Another important process in nitrogen cycle involves conversion of ammonium to nitrate (NO_3^-) or nitrification. This is carried out mainly by nitrifying bacteria belonging to the genera *Nitrobacter* and *Nitrosomonas*. Nitrate can be utilized by plants, fungi, eubacteria and archaebacteria, but not animals. Nitrate can be further converted to nitrite (NO_2^-) and utilized by various organisms in the process of denitrification which produces N_2O and N_2. Denitrification occurs under anaerobic conditions in almost all known environments. Finally, despite the fact that assimilation of NH_3 is more favorable than that of NO_3^-, the latter can be directly converted to organic nitrogen by many micro-organisms and plants (8). Abiological processes of nitrogen fixation within the biosphere and geosphere are discussed later.

Distribution of Nitrogen

The Earth's nitrogen is present in four major pools: lithosphere (10^{23} g), atmospheric (3.9×10^{21} g), terrestrial (4.8×10^{17} g) and aquatic (2.3×10^{19} g) (8, 12). Although the bulk of nitrogen is accumulated in rocks, sediments and organic deposits (lithosphere) its availability is severely restricted (12, 13). This overall deficiency, not only from the lithosphere, makes nitrogen one of the most important limiting nutrients (13).

Figure 1 is a schematic representation of the various different nitrogen pools. More than 99.9 % of nitrogen is present as gaseous N_2, the rest being present as so-called 'combined' nitrogen (Fig. 1). N_2 constitutes in both the atmospheric and aquatic pools the bulk of the nitrogen. Within the atmosphere it is over 99.99% (13) whereas in the aquatic setting, dissolved dimolecular nitrogen (N_2) contributes for over 95% (13). Within the 'combined' pool (Fig. 1), nitrogen is distributed between organic (57%) and inorganic forms (43%) (8, 13). Organic nitrogen in terrestrial systems occurs primarily in dead organic matter as a component of soil, *ca* 96% of the total. The remaining 4% of terrestrial organic N occurs in living biomass (8, 12) of which, 94-99%, is present in plants, and only 1-6% derived from animals and micro-organisms, although this proportion is somewhat dependent on the specific ecosystem (8, 12). Similar to the terrestrial pool, organic nitrogen in the aquatic pool occurs primarily as dissolved and particulate organic matter (DOM and POM; Fig. 1). However, at least part of POM may contain living biomass. In both the terrestrial and aquatic pools of inorganic N substantial amounts of ammonia occur in anoxic sediments and are adsorbed onto surfaces of clay minerals. It should be emphasized that the values presented here are average values based on a number of different

TOTAL AVAILABLE NITROGEN

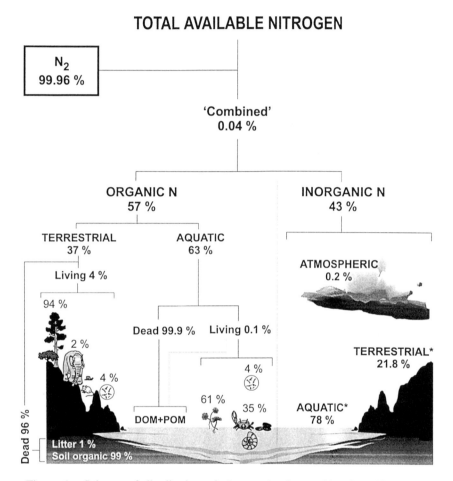

Figure 1. Scheme of distribution of nitrogen in the world. (Modified from references *8, 12-14*). The N_2 (99.96 %) represents the dimolecular nitrogen present in atmospheric, terrestrial and aquatic pools. Note: DOM-dissolved organic matter; POM-particulate organic matter (partly composed of living biomass); * - includes ammonia fixed in clay minerals (terrestrial) and in anoxic sediments (aquatic).

studies (*1, 3, 6, 8, 12-14*). For additional and more detailed information about the topic of global nitrogen cycling and fluxes the reader is referred to various specialist publications (*1, 3, 6, 8, 12-14*).

Organic Forms of Nitrogen

Apart from the inorganic forms of nitrogen approximately half of the pool of 'combined' nitrogen (Fig. 1) occurs as organic forms. Significantly, many publications addressing nitrogen fixation and cycling are based mainly on studies of elemental N or simple forms like NH_3 or NO_x (*1, 3, 8*). Such studies therefore concentrate primarily on N prior to its incorporation into, or after its release from, organic forms. These latter forms are often directly related to various (bio)macromolecular structures (e.g., DNA, proteins, amino sugars). However, in stark contrast to the bulk of literature on 'simple' N, relatively little is known about the fate of these N-containing macromolecules in the biosphere and geosphere. Moreover, the fact that much organic nitrogen relates to one of these biomacromolecules indicates that a clearer insight into the latter may aid in general studies of the nitrogen cycle in all ecological niches.

Biological Forms. Nitrogen can be found in living organisms as a gas (in cells), and/or in oxidized or reduced form (*8*). Nitrogen in the reduced form is the major component of the three most important biological macromolecular structures: (i) proteins/polypeptides, (ii) DNA and RNA, and (iii) polymers of amino sugars. The preservation potential in sedimentary strata is considered very low for DNA and low for amino sugars (*15, 16*). Although, the original source of nitrogen in sediments is rarely evident, due to decomposition processes, a macromolecular origin appears likely. Apart from the three main groups, other minor biological contributors to organic nitrogen are smaller molecules such as porphyrins (including chlorophylls and hemoglobin), and various N-containing secondary plant metabolites (e.g. alkaloids and sphingolipids).

Proteins and polypeptides. Proteins and polypeptides are important macromolecules in the life cycle and are composed of series of amino acids. They have several fundamental biological functions, for example as structural material (mostly fibrous structures, e.g., keratins, collagen or myosin). They also can serve as biological catalysts (e.g., enzymes) or regulators (globular forms). They are often associated with other macromolecules and large molecular complexes (e.g., chitin-protein in arthropods exoskeletons, glycoproteins in plant cell walls). Since proteins are fundamental molecules in living organisms, they are an important source of biological N. However, they are assumed to have a low preservation potential and are known to be prone to decomposition (usually depolymerized by enzymatic hydrolysis) to peptides and further to single amino acids. In this way they become easily available to micro-organisms and thus contribute to the N cycle. The fate of proteins in the bio- and geosphere and the latest ideas on their geochemical cycle(s) are discussed in subsequent chapters of this volume (*17-20*).

DNA and RNA. In addition to proteins, DNA (deoxyribonucleic acid) and RNA (ribonucleic acid) are fundamental to life on Earth. Chains of these acids are composed of millions of repeating monomers or nucleotides which can reach a molecular weight of more than one thousand million. These nucleotides are composed of a phosphate, sugar and heterocyclic base unit (i.e. purines: adenine, guanine, and pyrimidines: uracil, cytosine, thymine). Their major function is storage and replication of genetic information, as well as serving as the blueprint for protein production. DNA is known to be easily decomposed in the biosphere and is rarely found in ancient materials (*21-23*). The last decade witnessed many controversial studies regarding survival of this macromolecule in the geosphere (*24-27*), but recent reports indicate a limit for the survival of reliable DNA at *ca* 100 Ka (*21, 28*). A detailed discussion on the fate of DNA in ancient materials is presented later in this volume (*23*).

Amino sugar polymers. These substances are often associated with other biopolymers and form complexes known as mucopolysaccharides (polysaccharide+protein, e.g. mureins in eubacterial cell walls, glycoproteins in plant cell walls), lipopolysaccharides (polysaccharide+lipids, in mycobacterial waxes) or other combinations (*29, 30*). Chitin is the most important N-containing polysaccharide. It was first described in 1811 (*31*) and later named by Odier (*32*) from the greek word χιτων (tunic, envelope). It is also the most resistant to degradation of all N-containing biomacromolecules (*15*). The fate of chitin and its biotransformation have been studied, especially in the marine ecosystem (*33*). In contrast, the chemical changes occurring during the transformation from bio- to geosphere are still poorly understood. Although chitin has been reported in sediments as old as 25 Ma (*34, 35*), the detail degradation and transformation processes involved are not known. Because of this it is difficult to estimate how much of the nitrogen present in sedimentary basins is of chitin/amino sugar origin. The fate of chitin in the biosphere, and its degradation in marine ecosystems and in sedimentary strata up to 60 Ma, are discussed in subsequent chapters, providing a comprehensive summary of current knowledge on the biogeochemical fate of this important type of macromolecule (*30, 33, 35, 36*).

Other N-containing molecules. Porphyrins and various N-containing secondary plant metabolites are relatively 'small', naturally occurring molecules, which appear to be minor contributors to the nitrogen in the bio- and geosphere. Porphyrins have been found preserved in sediments millions of years old (*37-42*). However, it should be emphasized that they occur mainly as 'free' compounds and rarely as part of macromolecular structures. They are composed of a tetrapyrrole heterocycle and occur as metal complexes in materials such as chlorophyll (plants) and hemoglobin (animals). Although not necessarily very abundant, these small molecules, including the plant metabolites, may become part of macromolecular nitrogen-containing complexes if the original molecule had reactive functionalities such as double bonds, free hydroxyl or carboxyl groups. These functionalities could be involved in reactions with other organic materials or reactive compounds leading to their incorporation into larger macromolecular structures. These in turn could become part of kerogen, solvent-insoluble sedimentary organic matter (*43*).

Abiological Forms. One of the problems related to understanding the global nitrogen cycle, and the chemical transformations leading to nitrogen-containing macromolecular complexes in the geosphere, is the formation of abiological forms of organic nitrogen. In terrestrial ecosystems, apart from the known forms of organic nitrogen such as amino acids, proteins, hexoamines and nucleic acids, sometimes up to 50% of the organic nitrogen is present as 'unknown' or 'unidentified' forms (*44, 45*). This unknown nitrogen is much more difficult to mineralize and although potentially available for biological processes, becomes immobilized. The consequences of nitrogen immobilization are far reaching as it reduces soil fertility (*46*) and causes changes in the extent of litter decomposition, which in turn directly affects carbon dioxide emission to the atmosphere (*47*).

To date, a number of different formation pathways have been suggested for these forms of 'unidentified' organic nitrogen. These two most commonly proposed are (i) reactions of amino acids or ammonia with phenols or quinones (*12, 44, 47*) and (ii) condensation of amino acids with saccharides through the so-called Maillard reaction (*44, 48*). For additional formation pathways see also Knicker and Kögel-Knaber (*68*). The first reaction pathway has often been considered of only minor significance, particularly, with respect to lignin related phenols (*44*). However, more recent work suggests that the reactions of amino compounds with lignin phenols may be of considerable importance, especially in the context of soil fertility (*46*) and the decomposition of litter (*47*). In view of the presence of abundant polysaccharides and amino acids among the available organic substrates present in both terrestrial and aquatic settings the Maillard reaction might be expected to occur readily (*44*). This process plays a part in the formation of macromolecular nitrogen complexes during organic matter decomposition leading to the formation of insoluble organic matter in soils and sediments (*48*). However, to date unequivocal evidence for the Maillard reaction has been reported from only one set of samples (*20, 49*). Apart from these two scenarios, most recent studies on the preservation of proteins have revealed evidence of increasing cross-linking of proteins/polypeptides leading to the formation of very large macromolecular structures which are no longer susceptible to standard protein analyses (*19*). Such complexes may well be another, important, source of unknown N in soils and sediments. In addition to the possible contribution of proteins, nucleic acids and amino sugars, this unknown source would be a significant contributor to the nitrogen observed in kerogens and coals.

Techniques for the study of macromolecular N in bio- and geosphere.

Studies of macromolecular N in bio- and geomaterials are directly related to developments in analytical techniques. While traditional approaches such as elemental analysis and calculations of C/N ratios are still routinely used (*12, 50*), various spectroscopic and biochemical methods have become more popular in the last decade. These latter methods can provide detailed molecular information on the chemical structure(s) of N-containing molecules. It is possible to determine the type of intact or partially degraded macromolecule, or predict the original source of nitrogen on the basis of partial molecular information (especially in the geosphere). It should be emphasized that most techniques target particular compound classes eliminating the possibility of investigating other macromolecules simultaneously (e.g., polymerase chain reaction (PCR), enzymatic assays, immunochemical methods).

Although many of the analytical approaches can yield important insights, to date there is no single technique which provides a simple detailed picture of the material studied. High performance liquid chromatography (HPLC) is one of the most popular methods in qualitative and quantitative determination of amino acids (51-53). In combination with mass spectrometry (HPLC/MS) this technique can yield information on the presence of DNA nucleotides in ancient materials (54). Two-dimensional electrophoresis offers the possibility of studying proteins without hydrolysis to single amino acids, and offers the highest resolution for this material (19, 55). Other techniques for protein analyses include radio-immunoassay and Western blot (56). An array of colorimetric and enzymatic assays has been used for studies of not only proteinaceous materials, but also amino sugars (57, 58). Since its introduction (59) PCR has become a major technique for the study of DNA (21, 23, 60, 61).

Solid state ^{13}C and ^{15}N nuclear magnetic resonance (NMR), can elucidate the nature of protein and chitin cross-links in arthropod cuticle (62, 63), but requires relatively large amounts of material to obtain reliable spectra. Recently, solids NMR has gained in popularity and has been applied to studies of organic nitrogen in soils, humics and kerogens (64-70). Fourier transform infrared spectroscopy (FTIR) is used in investigations of proteins and amino sugars and may yield structural information (71-76). However, this method has limitations when both macromolecules are present in the sample, due to overlapping of characteristic bands (e.g., amide I and II). Gas chromatography (GC), in combination with MS, is often used in studies of bio- and geomaterials and can yield important structural information (77-81). But it requires elaborate preparation or derivatization steps prior to analyses involving degradation of the macromolecular structure. In recent years, combinations of analytical pyrolysis with GC, MS or GC/MS have become widely used in determining the molecular structure of various bio- and geopolymers (50, 82-90). Although it is an ideal technique for the study of recalcitrant polymeric organic matter, even when more than one macromolecule is present, it involves thermal decomposition of chemical structures and provides only partial and sometimes biased molecular information (91). Although not providing structural information, stable isotope δ^{13}C, δ^{18}O, δ^{15}N, δD analyses (IRMS) may yield specific insights into the source of organic matter under study (92-100).

Future directions

One of the important problems related to understanding the global nitrogen cycle and the chemical formation of nitrogen-containing macromolecular complexes in the geosphere, is the source of abiological forms of organic nitrogen. Information on these unknown forms of organic nitrogen affects nitrogen fluxes to and from the other 'compartments' and, when present in terrestrial ecosystems, can affect soil fertility. Detailed characterization is needed, but research in this field is still severely hampered by the absence of a single simple method that provides detailed molecular information on these moieties. Apart from characterizing these materials there is a greater need to determine the important chemical processes that are involved in the formation of N-containing moieties in ancient and fossil material. In particular, isotope analyses of selectively released N-containing moieties may provide insights into the origin of the nitrogen in macromolecular structures in the geosphere, and give answers as to whether the three main N-containing biomacromolecules are really the major sources. Finally, is there really fossil DNA, and if so is there a limit as to how long can it can survive?

Acknowledgments

We would like to thank all those whose through their discussion have helped us in developing and carrying this project. Both, BAS and PVB were supported by NERC grants GST/02/1027 to D. E. G. Briggs and R. P. Evershed, and GR3/9578 to R. P. Evershed respectively while editing this volume.

References

1.   Vitousek, P. M.; Matson, P. A. *Biogeochemistry of Global Change*; Chapman & Hall: New York London, 1993; pp 193-208.
2.   Lerman, A.; Mackenzie, F. T.; Ver, L. M. *Chem. Geol.* **1993**, *107*, 389-392.
3.   Vitousek, P. M.; Aber, J. D.; Howarth, R. W.; Likens, G. E.; Matson, P. A.; Schindler, D. W.; Schlesinger, W. H.; Tilman, D. G. *Ecol. Applic.* **1997**, *7*, 737-750.
4.   Clark, F. E. In *Terrestrial Nitrogen Cycles*; Clark, F. E.; Rosswall, T., Eds.; Ecological Bulletins 33; Swedish Natural Science Research Council: Stockholm, Sweden, 1981; pp 13-24.
5.   Schlesinger, W. H. *Biogeochemistry An Analysis of Global Change*; Academic Press: San Diego, CA, 1991, pp 322-335.
6.   Vitousek, P. M. *Ecology* **1994**, 75, 1861-1876.
7.   Cornell, S.; Rendell, A.; Jickells, T. *Nature* **1995**, *376*, 243-246.
8.   Sprent, J. I. *The Ecology of Nitrogen Cycle*; Cambridge Studies in Ecology; Cambridge University Press: Cambridge, UK, 1987; pp 1-151.
9.   Skujinš, F. E. In *Terrestrial Nitrogen Cycles*; Clark, F. E.; Rosswall, T., Eds.; Ecological Bulletins 33; Swedish Natural Science Research Council: Stockholm, Sweden, 1981; pp 477-492.
10. Gosz, J. R. In *Terrestrial Nitrogen Cycles*; Clark, F. E.; Rosswall, T., Eds.; Ecological Bulletins 33; Swedish Natural Science Research Council: Stockholm, Sweden, 1981; pp 405-426.
11. van Cleve, K.; Alexander, V. In *Terrestrial Nitrogen Cycles*; Clark, F. E.; Rosswall, T., Eds.; Ecological Bulletins 33; Swedish Natural Science Research Council: Stockholm, Sweden, 1981; pp 375-404.

12. *Soil Microbiology and Biochemistry*; Paul, E. A.; Clark, F. E., Eds.; 2nd edition; Academic Press: San Diego, 1996.
13. *The Major Biogeochemical Cycles and Their Interactions*; Bolin, B.; Cook, R. B., Eds.; John Wiley & Sons: Chichester, 1983.
14. Rosswall, T. In *The Major Biogeochemical Cycles and Their Interactions*; Bolin, B.; Cook, R. B., Eds.; John Wiley & Sons: Chichester, 1983; pp 46-50.
15. Tegelaar, E. W.; de Leeuw, J. W.; Derenne, S.; Largeau, C. *Geochim. Cosmochim. Acta*, **1989**, *53*, 3103-3106.
16. de Leeuw, J. W.; Largeau, C. In *Organic Geochemistry Principles and Application*; Engel, M. H.; Macko, S. A., Eds.; Topics in Geobiology 11; Plenum Press: New York and London, 1993; pp 23-72.
17. Bada, J. This volume.
18. Collins, M. J.; Walton, D.; King, A. This volume.
19. Nguyen, R. T.; Harvey, H. R. This volume.
20. Bland, H. A.; van Bergen, P. F.; Carter, J. F.; Evershed, R. P. This volume.
21. Lindahl, T. *Nature* **1993**, *362*, 709-715.
22. Pääbo, S. *Proc. Natl. Acad. Sci.* **1989**, *86*, 1939-1943.
23. Poinar, H. N. This volume.
24. Higuchi, R.; Wilson, A. C. *Federation Proc.* **1984**, *43*, 1557.
25. Cano, R. J.; Poinar, H. N.; Pieniezak, N. S.; Poinar Jr., G. O. *Nature* **1993**, *363*, 536-538.
26. DeSalle, R.; Gatesy, J.; Wheeler, W.; Grimaldi, D. *Science* **1992**, *257*, 1933.
27. Austin, J. J.; Ross, A.; Smith, A.; Fortey, R.; Thomas, R. *Proc. Royal Soc. London B* **1997**,.*264*, 467-474.
28. Krings, M.; Stone, A.; Schmitz, R. W.; Krainitzki, H.; Stoneking, M.; Pääbo, S. *Cell* **1997**, *90*, 19-30.
29. Kent, P. W.; Whitehouse, M. W. *Biochemistry of Aminosugars*; Butterworths Scientific Publications: London, 1955; pp 1-311.
30. Muzzarelli, R. A. A.; Muzzarelli, C. This volume.
31. Braconnot, H. *Ann. Chi. Phys.* **1811**, *79*, 265-304.
32. Odier, A. *Mém. Soc. Hist. Nat. Paris* **1823**, *1*, 29-42.
33. Poulicek, M.; Gail, F.; Goffinet, G. This volume.
34. Stankiewicz, B. A.; Briggs, D. E. G.; Evershed, R. P.; Flannery, M. B.; Wuttke, M. *Science* **1997**, *276*, 1541-1543.
35. Stankiewicz, B. A.; Briggs, D. E. G.; Evershed, R. P.; Miller, R.; Bierstedt, A. This volume.
36. Schimmelmann, A. Wintsch, R. P., Lewan, M. D., DeNiro, M. J. This volume.
37. Sinninghe Damsté, J. S.; Eglinton T. I.; de Leeuw J. W. *Geochim. Cosmochim. Acta* **1992**, *56*, 1743-1751.
38. Louda, J. W.; Baker, E. W. *ACS Symp. Ser.* **1986**, *305*, 107-126.
39. Keely, B. J.; Maxwell, J. R. *Org. Geochem.*, **1991**, *17*, 663-669.
40. Eckardt, C. B.; Keely, B. J.; Waring J. R.; Chicarelli, M. I.; Maxwell, J. R. *Phil. Trans. R. Soc. Lond. B* **1991**, *333*, 339-348.
41. Chicarelli, M. I., Hayes, J. M.; Popp, B. N.; Eckardt, C. B.; Maxwell, J. R. *Geochim. Cosmochim. Acta* **1993**, *57*, 1307-1311.

42. Marriott, P. J.; Gill, J. P.; Evershed, R. P.; Eglinton,G.; Maxwell, J. R. *Chromatographia* **1982**, *16*, 304-308.

43. Durand B. *Kerogen-insoluble organic matter from sedimentary rock.* Technip. Paris, 1980.

44. Anderson, H. A.; Bick, W.; Hepburn, A.; Stewart, M. In *Humic Substances II*; Hayes, M. H. B.; MacCarthy, P.; Malcolm, R. L.; Swift, R. S., Eds.; John Wiley & Sons: New York, 1989; pp 223-253.

45. Schulten, H.-R.; Sorge-Lewin, C.; Schnitzer, M. *Biol. Fertil. Soils* **1997**, *24*, 249-254.

46. Olk, D. C.; Cassman, K. G.; Randall, E. W.; Kinchesh, P.; Sanger, L. J.; Anderson, J. M. *Euro. J. Soil Sci.* **1996**, *47*, 293-303.

47. Coûteaux, M.-M.; Bottner, P.; Berg, B. *TREE.* **1995**, *10*, 63-66.

48. Ikan, R.; Rubinsztain, Y.; Nissenbaum, A.; Kaplan, I. R. In *The Maillard reaction: Consequences for the clinical and life sciences*; R. Ikan, Ed.; John Wiley & Sons: Chichester, 1996; 1-25.

49. Evershed, R. P.; Bland, H. A.; van Bergen, P. F.; Carter, J. F.; Horton, M. C.; Rowley-Conwy, P. A. *Science* **1997**, *278*, 432-433.

50. Schulten, H.-R.; Sorge, C.; Schnitzer, M. *Biol. Fertil. Soils* **1995**, *20*, 174-184.

51. Elmore, D. T. *Peptides and proteins*; Cambridge University Press; London, UK, 1968.

52. Lea, P. J.; Wallsgrove, R. M.; Miflin, B.J. In *Chemistry and Biochemistry of Amino Acids*; Barrett, G. C., Ed.; Chapman and Hall: London, 1985; pp 196-226.

53. Hunt, S. In *Chemistry and Biochemistry of Amino Acids*; Barrett, G. C., Ed.; Chapman and Hall: London, 1985; pp 415-425.

54. O'Donoghue, K.; Brown, T. A.; Carter, J. F.; Evershed, R. P. *Rapid Commun. Mass Spectrom.* **1996**, *10*, 495-500.

55. Dunbar, B. S. *Two-dimensional electrophoresis and immunological techniques;* Plenum Press: New York, 1987; pp. 372.

56. Tuross, N.; Stathoplos, L. *Meth. Enzym.* **1993**, *224*, 121-129.

57. Hackman, R. H.; Goldberg, M. *Anal. Biochem.* **1981**, *110*, 277-280.

58. Jeuniaux, C., *Bull. Soc. Chem. Biol.* **1965**, *47*, 2267-2278.

59. Mullis, K. B.; Faloona, F. A. *Meth. Enzym.* **1987**, *155*, 335-350.

60. Höss, M.; Dilling, A.; Currant, A.; Pääbo, S. *Proc. Natl. Acad. Sci.* **1996**, *93*, 181-185.

61. Handt, O.; Höss, M.; Krings, M.; Pääbo, S. *Experientia* **1994**, *50*, 524-529.

62. Schaefer, J.; Kramer, K. J.; Garbow, J. R.; Jacob, G. S.; Stejskal, E. O.; Hopkins, T. L.; Speirs, R. *Science* **1987**, *235*, 1200-1204.

63. Kramer, K. J.; Hopkins, T. L.; Schaefer, J. This volume.

64. Hatcher, P. G.; van der Hart, D. L.; Earl, W. L. *Org. Geochem.* **1980**, *2*, 87-92.

65. Wilson, M. A.; Pugmire, R. J.; Zilm, K. W.; Goh, K. M.; Heng, S.; Grant, D. *Nature* **1981**, *294*, 648-650.

66. Knicker, H.; Fründ, R.; Lüdemann, H.-D. *Naturwissenschaften* **1993**, *80*, 219-221.

67. Bortriatynski, J. M.; Hatcher, P. G.; Knicker, H. In *Humic and Fulvic Acids*; Gaffney, J. S.; Marley, N. A.; Clark, S. B., Eds.; ACS Symp. Ser. 651; American Chemical Society: Washington, D. C., 1996; pp 57-77.

68. Knicker, H.; Kögel-Knabner, I. This volume.

69. Siebert, S.; Knicker, H.; Hatcher, M. A.; Leifeld, J.; Kögel-Knabner, I. This volume.

70. Derenne, S.; Knicker, H.; Largeau, C.; Hatcher, P. G. This volume.

71. Surewicz, W. K.; Mantsch, H. H. *Biochim. Biophys. Acta* **1988**, *952*, 115-130.

72. Le Gal, J.-M.; Manfait, M.; Theophanides, T. *J. Mol. Struct.* **1991**, *242*, 397-407.

73. Harris, P. I.; Chapman, D. *TIBS* **1992**, *17*, 328-333.

74. Hadden, J. M.; Chapman, D.; Lee, D. C. *Biochim. Biophys. Acta* **1995**, *1248*, 11-122.

75. Edwards, H. G. M.; Russell, N. C.; Weinstein, R.; Wynn-Williams, D. D. *J. Raman Spect.* **1995**, *26*, 911-916.

76. Dauphin, Y.; Marin, F. *Experientia* **1995**, *51*, 278-283.

77. Budzikiewicz, H.; Djerassi, C.; Williams, D. H. *Structure Elucidation of Natural Products by Mass Spectrometry*; Holden-Day: San Francisco, CA, 1964.

78. Porter, Q. N. *Mass Spectrometry of Hetrocyclic Compounds*; General Heterocyclic Chemistry Series; Wiley: New York, NY, 1985; 2nd edition.

79. Jennings, W. *Analytical Gas Chromatography*; Academic Press: Orlando, FL, 1987.

80. *Mass Spectrometry of Biological Materials*; McEwen, C. N.; Larsen, B. S., eds.; Practical Spectroscopy A Series; Marcel Dekker, Inc.: New York, NY, 1990, Vol. 8.

81. McLafferty, F. W.; Turecek, F. *Interpretation of Mass Spectra*; University Science Books: Mill Valley, CA, 1993; 4th edition.

82. Jarman, M. *J. Anal. Appl. Pyrolysis* **1980**, *2*, 217-223.

83. Bracewell, J. M.; Robertson, G. W. *J. Anal. Appl. Pyrolysis* **1984**, *6*, 19-29.

84. van der Kaaden, A.; Boon, J. J.; de Leeuw, J. W.; de Lange, F.; Schuyl, P. J. W.; Schulten, H.-R.; Bahr, U. *Anal. Chem.* **1984**, *56*, 2160-2165.

85. de Leeuw, J. W.; van Bergen, P. F.; van Aarssen, B. G. K.; Gatellier, J.-P. L. A.; Sinninghe Damsté, J. S.; Collinson, M. E. *Phil. Trans. R. Soc. Lond. B* **1991**, *333*, 329-337.

86. Boon, J. J. *Int. J. Mass Spectrom. Ion Processes* **1992**, *118/119*, 755-787.

87. Larter, S. R.; Horsfield, B. In *Organic Geochemistry Principles and Application*; Engel, M. H.; Macko, S. A., Eds.; Topics in Geobiology 11; Plenum Press: New York and London, 1993; pp 271-287.

88. van Bergen, P. F.; Collinson, M. E.; Briggs, D. E. G.; de Leeuw, J. W.; Scott, A. C.; Evershed, R. P.; Finch P. *Acta Bot. Neerl.* **1995**, *44*, 319-342.

89. Stankiewicz, B. A.; van Bergen, P. F.; Duncan, I. J.; Carter, J. F.; Briggs, D. E. G.; Evershed, R. P. *Rapid Comm. Mass Spectrom.* **1996**, *10*, 1747-1757.

90. Stankiewicz, B. A.; Hutchins, J. C.; Thomson, R.; Briggs, D. E. G.; Evershed, R. P. *Rapid Comm. Mass Spectrom.* **1998**, *11*, in press.

91. Reeves III, J. B.; Francis, B. A. This volume.

92. Macko, S. A.; Engel, M. H.; Qian, Y. *Chem. Geol.* **1994**, *114*, 365-379.
93. Tuross, N.; Fogel, M. L.; Hare, P. E. *Geochim. Cosmochim Acta* **1988**, *52*, 929-935.
94. Schimmelmann, A.; DeNiro, M. J. *Contributions Marine Sci.* **1986**, *29*, 113-130.
95. Schimmelmann, A.; DeNiro, M. J.; Poulicek, M.; Voss-Foucart, M.-F.; Goffinet, G.; Jeuniaux, Ch. *J. Archaeol. Sci.* **1986**, *13*, 553-566.
96. Velinsky, D. J.; Pennock, J. R.; Sharp, J. H.; Cifuentes, L. A.; Fogel, M. L. *Mar. Chem.* **1989**, *26*, 351-361.
97. Bebout, G. E.; Fogel, M. L. *Geochim. Cosmochim. Acta* **1992**, *56*, 2839-2849.
98. Fogel, M. L.; Cifuentes, L. A. In *Organic Geochemistry Principles and Application*; Engel, M. H.; Macko, S. A., Eds.; Topics in Geobiology 11; Plenum Press: New York and London, 1993; pp 73-98.
99. Macko, S. A.; Fogel, M. L.; Hare, P. E.; Hoering, T. C. *Chem. Geol.* **1987**, *65*, 79-92.
100. Pennock, J. R.; Velinsky, D. J.; Ludlam, J. M.; Sharp, J. H.; Fogel, M. L. *Limnol. Ocean.* **1996**, *41*, 451-459.

ANALYTICAL TECHNIQUES IN STUDIES OF N-CONTAINING MACROMOLECULES

Chapter 2

Analysis of Intractable Biological Samples by Solids NMR

Karl J. Kramer[1], Theodore L. Hopkins[2], and Jacob Schaefer[3]

[1]Grain Marketing and Production Research Center, Agricultural Research Service, U.S. Department of Agriculture, Manhattan, KS 66502
[2]Department of Entomology, Kansas State University, Manhattan, KS 66506
[3]Department of Chemistry, Washington University, St. Louis, MO 63130

Solids NMR is a noninvasive analytical method that can be used to investigate the chemical compositions and covalent interactions that occur in biological samples intractable to conventional analytical approaches. Solids NMR techniques such as cross polarization, dipolar decoupling, magic angle spinning, magnetization dephasing, and isotopic enrichment have been used to obtain high resolution spectra that provided information about the relative concentrations and internuclear distances between atoms in complex biological solids. Levels of proteins, chitin, catechols, lipids, pigments, and other organic constituents in composite materials were estimated. Covalent interactions between specific carbons and nitrogens have been detected by isotopic enrichment with labeled precursor molecules. We have used solids NMR in studies of polymeric and analytically intractable samples such as insect cuticular exoskeletons, egg cases, egg shells, silk cocoons, and marine coral skeletons. Evidence was obtained for stabilization mechanisms occurring primarily when quinones derived from catechol-containing compounds including catecholamines and o-diphenols with acid, aldehyde and alcohol side chains form adducts with functional groups of structural proteins and perhaps chitin. Solids NMR can be utilized for probing the compositions and covalent interactions of many other types of biological samples found in the biosphere and geosphere and, when combined with other analytical techniques, provides a powerful approach for elucidating the complex structures of biopolymeric materials.

Many invertebrates stabilize and strengthen skeletal structures by cross-linking of structural proteins, dehydration, and impregnation with chitin, minerals, and phenolic compounds (1-6). Some of the skeletal components are relatively stable, and they can be preserved during decay in the biosphere and geosphere. For example, Bass et al. (7) reported the selective preservation of

[2]Address correspondence to K. J. Kramer, GMPRC-ARS-USDA, 1515 College Avenue, Manhattan, KS 66502, U. S. A. Phone: 785-776-2711. Fax: 785-537-5584. Email: kjkramer@ksu.edu.

chitin during the anoxic decay of shrimp in a marine environment. Analysis using solid-state ^{13}C nuclear magnetic resonance (NMR) revealed that, whereas other organic components were highly degraded after only 8 weeks, chitin remained the major constituent of the biomass. Longer-term decay of chitinous tissues, however, did occur, with replacement of chitin by more resistant organic matter such as alkanes and alkenes derived from other sources. Therefore, depending on the type of sample and environmental conditions, solids NMR can monitor the levels of some N- and C-containing macromolecules and provide information about their physical states and decay processes.

This chapter reviews the results of solids NMR research conducted in our laboratories on biological sclerotized structures, such as insect cuticles, egg cases, egg shells, cocoons and marine coral skeletons, describes some of the progress made on problems in insect and coral biochemistry, and previews a few of the experiments that we hope to conduct in the future. Solids NMR also has been used to investigate the compositions of other types of complex organic materials found in plants, animals, and microbes and offers great potential for future development of more powerful analytical techniques. We have also used solids NMR to probe the types of covalent interactions that arise when some of these materials are assembled and stabilized. Because of the intractable nature of sclerotized structures, there is little quantitative data available about the structural composition or cross-linking chemistry. This review focuses on compositional and developmental information sought by those interested in the chemistry of organic materials found in nature, rather than on details of the analytical methods.

NMR techniques were developed initially (8) to investigate nuclear properties of several different elements (e.g. 9). Today, NMR is a powerful tool for solving problems in structural chemistry and, most recently, medical imaging. With the development of line-narrowing and gradient-imaging methods in the last 20 years, NMR spectroscopy has blossomed into a major technique in solid-state materials science and now provides a noninvasive approach to investigate detailed structures of heterogeneous materials such as C- and N-containing macromolecules and other polymeric composites.

Sclerotization is a complex process used by insects and other invertebrate animals to confer stability and mechanical versatility to their cuticular exoskeletons and certain other proteinaceous structures. Cross polarization (CP), dipolar decoupling, magic angle spinning (MAS), magnetization dephasing, and isotopic enrichment provide high resolution ^{13}C and ^{15}N NMR spectra that yield information about the types and relative concentrations of carbon and nitrogen atoms as well as internuclear distances and covalent bonding between specific carbons and nitrogens. Relative amounts of protein, chitin, catechols, lipids, pigment, and oxalate have been estimated, and covalent interactions between protein nitrogens and catechol carbons detected. The results of these solids NMR studies together with those of traditional chemical analyses support the hypothesis that sclerotization of protective structures in insects, corals, and other invertebrates occurs primarily when quinones derived from various compounds containing the catechol moiety (including catecholamines and o-diphenols with acid, aldehyde, or alcohol side chains) form cross-links and adducts with functional groups of proteins incorporated into these structures.

Solids NMR methodology

NMR measures the radio frequencies emitted and the rates of realignment of nuclear spins in a magnetic field after atomic nuclei absorb energy from radio frequency (rf) pulses. The single-resonance, natural abundance, ^{13}C NMR spectrum of a solid usually consists of a single, broad, featureless line, with a width of approximately 20 kHz (Fig. 1, top). The major source of line broadening in this situation is the static dipolar interaction between carbons and nearby protons (*10*). These protons include covalently bonded methine, methylene, or methyl protons, together with more distant, indirectly bonded protons. Dipolar interactions depend upon the orientations of internuclear vectors with the applied magnetic field. In a crystal powder or an amorphous material, all orientations occur, resulting in a broad distribution of dipolar splittings.

Dipolar decoupling removes this broadening in a straightforward way. If the ^{13}C resonance is observed in the presence of a strong rf field at the Larmor resonance frequency of the protons, the protons undergo rapid transitions or spin flips, which cause the time-averaged dipolar field generated by the protons at the carbon nucleus to disappear. This process is analogous to the more familiar scalar decoupling used to remove spin-spin splitting from high resolution ^{13}C NMR of liquids (*11*). The only difference is the strength of the rf fields used in the two experiments. Dipolar decoupling requires rf fields greater than the local fields experienced by the protons arising from ^1H-^1H and ^1H-^{13}C interactions. For a typical solid, this might require an rf field of 40 kHz, or about 10 times as large as that necessary to perform ordinary scalar decoupling (*10*).

With dipolar decoupling, the ^{13}C NMR spectrum of a solid begins to show signs of improved resolution (Fig.1, middle). A typical spectrum now has a width of approximately 15 kHz at a ^{13}C Larmor frequency of 50 MHz with a few spectral features clearly evident. The spectrum is not, however, of liquidlike high-resolution quality. The remaining broadening is due to chemical shift anisotropy (CSA). The magnetic field at a carbon nucleus depends upon the shielding or screening afforded by the surrounding electron density. In general, the surrounding electron density is not symmetric. Thus, the chemical shift of, for example, a carbonyl carbon in an ester group, depends upon whether the C-O carbonyl is lined along the magnetic field, is perpendicular to it, or is in some other orientation. In a liquid with rapid molecular motion, only an average or isotropic chemical shift is observed. In a single crystal, a single chemical shift may be observed, but its value depends upon the orientation of the crystal relative to the magnetic field (*12*). In an amorphous solid or crystal powder, on the other hand, a complicated CSA line shape is observed that arises from the sum of all possible chemical shifts.

For a typical solid with a variety of chemically different carbons, the ^{13}C NMR spectrum is a sum of different CSA patterns, having somewhat different shapes and, most importantly, having different isotropic centers. Overlapping CSA patterns for carbonyl carbons, aromatic carbons, and aliphatic carbons destroy the resolution that was expected from dipolar decoupling (Fig. 1. middle).

Fortunately, a method is available to regain the lost resolution. The broadening arises from restrictions placed on molecular motion in the solid. In the laboratory, we can supply a kind of molecular motion by mechanically

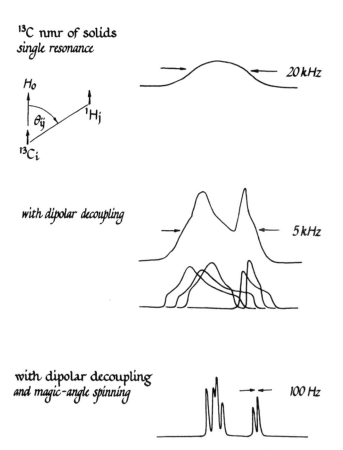

Figure 1. Schematic ^{13}C NMR lineshapes for a disordered sample in the solid
state. Single-resonance spectra are broadened by strong, orientation-
dependent ^{1}H-^{13}C dipolar interactions (top). These are removed by
resonant irradiation of the protons, which reveals overlapping asymmetric
lineshapes (middle) arising from the orientation dependence of the ^{13}C
chemical shift. High-speed mechanical rotation of the solid at the magic
angle produces liquid-like lineshapes (bottom). Adapted from references
10 and 23.

rotating the sample about the diagonal of a cube whose edges are provided by the rectilinear coordinate system defined by the applied static magnetic field (*13*). The rotation axis is at half the tetrahedral angle relative to the applied field, the so-called "magic angle" of 54.7%. This mechanical rotation interchanges axes and internuclear directions relative to the magnetic field and, therefore, has many of the same averaging properties as isotropic motion (Fig. 1, bottom). Fast MAS is achieved using the same kind of gas-supported bearings that are used in high speed centrifuges. The combination of dipolar decoupling and fast MAS affords resolution that usually is limited only by variations of bulk susceptibility within the sample, typically about 0.5 ppm.

Solids NMR experiments combine cross polarization for sensitivity enhancement of the signal to noise ratio (*10*) with dipolar decoupling to remove dipolar interactions from protons and MAS for high resolution of chemical shifts (*14-17*). Assignments of carbon chemical shifts are made by comparison with literature values or chemical shifts of standard compounds. Table 1 lists the chemical shift assignments in the ^{13}C CP-MAS NMR spectra of organic materials. For example, carbon chemical shifts for chitin were assigned by comparison to carbon solution and solid ^{13}C NMR spectra of 2-acetamido-2-deoxyglucopyranoside and crab chitin (Fukamizo et al., 1986). In general, signals from chitin, protein, catechol, and lipid carbon dominate the natural abundance ^{13}C solid-state NMR spectra. Concentrations of carbons can be estimated by extrapolation of signal intensity to zero contact time and comparison with the signal from an external standard compound. Difference spectra from single and double cross-polarization (DCP) experiments allow measurement of heteronuclear coupling between two stable isotopes that are within approximately 2 Å of each other (*18*). A relatively recent technique for MAS solid-state NMR, rotational echo double resonance (REDOR; *19, 20*), is more sensitive than DCP in measuring long-range heteronuclear interactions up to about 5 Å for ^{13}C-^{15}N, 8 Å for ^{13}C-^{31}P and 12 Å for ^{13}C-^{19}F. For REDOR analysis, magnetization on one rare spin rf channel is dephased by rotor-synchronized pulses on a second isotope channel, while interactions with protons are suppressed by dipolar decoupling. The extent of dephasing has a simple relation to the strength of the dipolar coupling and the internuclear distance. REDOR provides a direct measure of heteronuclear dipolar coupling between isolated pairs of labeled nuclei. In a solid with a ^{13}C-^{15}N labeled pair, for example, the ^{13}C rotational echoes that form each rotor period following a ^1H-^{13}C cross polarization transfer can be prevented from reaching full intensity by insertion of a ^{15}N π pulse each half rotor period. The REDOR difference (the difference between a ^{13}C NMR spectrum obtained under these conditions and one obtained with no ^{15}N π pulses) has a strong dependence on the ^{13}C-^{15}N dipolar coupling and, hence, the ^{13}C-^{15}N internuclear distance.

REDOR is described as double resonance even though three radio frequencies (for example, ^1H, ^{13}C and ^{15}N) are used, because the protons are removed from the important evolution part of the experiment by resonant decoupling. The dephasing of magnetization in REDOR arises from a local dipolar ^{13}C-^{15}N field gradient and involves no polarization transfer. REDOR has no dependence on ^{13}C or ^{15}N chemical shift tensors and does not require resolution of a ^{13}C-^{15}N coupling in the chemical shift dimension.

Another technique that enables high-resolution NMR analysis of specific

Table 1. Chemical shift assignments of resonances in the CP-MAS ^{13}C NMR spectra of insect support structures.

Resonance no.	Values (ppm)*	Assignment
1	172	Carbonyl carbons of chitin, protein, lipid, and catechol acyl groups
2	170	Carbonyl carbons of oxalate
3	155	Phenoxy carbon of tyrosine, guanidino carbons in arginine
4	144	Phenoxy carbons of catechols
5	129	Aromatic carbons
6	116	Tyrosine carbons 3 and 5, imidazole carbon 4, catechol carbons 2 and 5
7	104	G1cNAc carbon 1
8	82	GlcNAc carbon 4
9	75	G1cNAc carbon 5
10	72	G1cNAc carbon 3
11	60	GlcNAc carbon 6, amino acid α-carbons
12	55	GlcNAc carbon 2, amino acid α-carbons
13	44	Aliphatic carbons of amino acids, catechols, and lipids
14	33	Aliphatic carbons of amino acids, catechols, and lipids
15	23	Methyl carbons of chitin, protein, lipid, and catechol acyl groups; amino acid methyne carbons
16	19	Methyl carbons of amino acids and lipids

*Values relative to external tetramethylsilane reference.

atoms in compounds is selective isotopic enrichment, which helps to minimize the background of both isotopic natural abundance and similar chemical shifts that occur in rather complex molecular assemblies or mixtures. Labeling has been used to examine the chemical compositions and structures of a variety of sclerotized proteinaceous materials from insects (21) and marine mussels (22), all of them not easily amenable to analysis by conventional solution-based techniques. We have employed labeling with detection by MAS solids NMR for measuring distances between pairs of spin-1/2 heteronuclei. In particular, our method of choice determines specific internuclear separations between dilute $^{13}C,^{15}N$ spin pairs that can be incorporated into samples by isotopic enrichment. The low natural abundance of ^{13}C and ^{15}N (1.11 and 0.37 atom %, respectively) makes these spins ideal candidates for site-specific isotopic labeling. Internuclear distances between spin pairs can be determined from the $1/r^3$ distance dependence of the dipolar interaction. A sample is needed that can efficiently incorporate labels into specific atom pairs, while at the same time not scrambling the labels into other types of bonding.

Solids NMR experiments can not only be difficult to conduct but also expensive. Major costs are incurred from construction of spectrometers and probes, and, to a lesser degree, the synthesis or purchase of compounds with specific atoms enriched with NMR-active nuclei. Although a number of commercial instruments are available, many of them are primarily solution-based instruments adapted for solid samples rather than designed solely for the analysis of solids, which limits their performance. Also, rather large samples (~1 micromole of label) are needed for solids NMR analysis, and in the case of isotopic enrichment, a biological system that incorporates micromolar amounts of a label into appropriate chemical structures usually is required.

Solids NMR applications

Schaefer and Stejskal (23) first obtained natural abundance ^{13}C NMR spectra of good resolution from completely solid amorphous samples. Comparable natural abundance ^{15}N NMR spectra are generally not of usuable quality because signal levels are reduced by an order of magnitude. In addition, ^{15}N is four times less abundant than ^{13}C and three times less sensitive as an NMR nucleus. Peter et al. (24) first used solids NMR to study the compositions of sclerotized insect cuticles. Similar natural abundance ^{13}C NMR spectra were collected for pupal cases and exuviae of several insect species, which generally indicated comparable compositions of protein, chitin, and catechols (Fig. 2). Those results were interpreted to be consistent with the hypothesis that the sclerotization of cuticle occurs by the denaturation of structural proteins by polyphenolic compounds, but the actual mechanism of sclerotization was not determined.

Subsequently, Schaefer et al. (18) and Merritt et al. (25) reported the results of a study on the composition and several heteronuclear interactions of pupal and adult moth cuticles from the tobacco hornworm, Manduca sexta. Selective ^{13}C and ^{15}N isotopic enrichment of cuticle by injection of appropriately labeled precursor amino acids and the catecholamine dopamine (Fig. 3) enabled the detection of covalent linkages between ring nitrogens of protein histidyl residues and ring carbons of the catecholamine, which is a precursor of quinonoid tanning

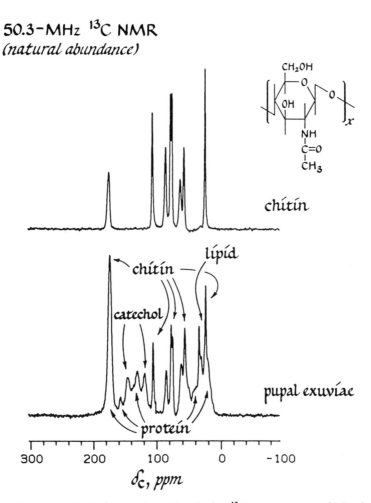

Figure 2. Cross-polarization magic-angle spinning ^{13}C NMR spectra of *Manduca sexta* chitin (top) and pupal exuviae (bottom). Adapted from ref. 18.

22

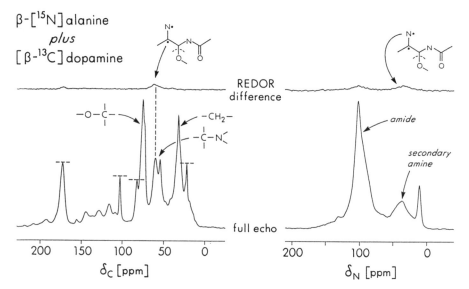

Figure 3. ^{13}C NMR spectra (left) and ^{15}N NMR spectra (right) of *Manduca sexta* pupal cuticle labeled by ß-[^{15}N]alanine and [ß-^{13}C]dopamine after eight (left) and four (right) rotor cycles of dephasing with magic-angle spinning at 3205 Hz. The REDOR difference spectra arise exclusively from ^{13}C-^{15}N directly bonded pairs. The line assignments in the REDOR difference spectra are for the carbon (top left) and nitrogen (top right) marked by the solid circles. The dotted arcs in the structures indicate that the assignment is independent of whether the attached groups are present. Line assignments for the major high-field peaks in the full-echo ^{13}C NMR spectrum are based on chemical shifts. The heights of peaks in the full-echo ^{13}C NMR spectrum (bottom, left) arising exclusively from natural-abundance ^{13}C are indicated by horizontal dashed lines. Adapted from ref. 25.

agents. These results supported the hypothesis that the stiffening of insect cuticle during sclerotization is correlated with the deposition of protein and chitin polymers and their cross-linking by quinonoid derivatives of catecholamines. Since then, many other types of insect support structures have been examined using solids NMR to gain insight into the mechanisms whereby insects assemble and stabilize these materials.

Williams et al. (*26*) compared ^{13}C NMR spectra of *Heliothis virescens* (tobacco budworm) cuticles dissected from ß-^{13}C-labeled, tyrosine-injected and noninjected insects. Incorporation of the ^{13}C-labeled tyrosine into the cuticle had little effect on the chemical shift of the labeled atom, indicating that the side chain ß-carbon was not modified heavily during cuticle morphogenesis. The authors hypothesized that the ß-carbon of tyrosine or one of its metabolites is not involved in protein cross-linking reactions during *Heliothis* cuticle sclerotization.

Chemical compositions. Samples analyzed for carbon composition include insect chitin, egg shells, cocoons, egg cases, cuticles, and marine coral skeletons. These materials often contain a few percent of water and ash, both of which are not quantifiable by NMR analysis, but instead are determined by gravimetric analysis. ^{13}C NMR was used for carbon compositional analysis, and natural abundance ^{13}C-CP-MAS spectra were obtained on samples that were cleaned of adhering tissues, washed with either deionized water or a dilute detergent solution, frozen, lyophilized, and finally ground into a powder of a particle size approximately 40 mesh or smaller. The samples analyzed ranged from the relatively homogeneous chitin preparation to the highly heterogeneous insect cuticles and coral skeletons.

Chitin. Chitin that was extracted from either crab or insect cuticle by exhaustive boiling in 1 M sodium hydroxide was one of the first insect materials examined by solids NMR (*24*; *27*). It is the major polysaccharide in cuticle and is composed of ß(1→4) linked 2-deoxy-2-acetamido-D-glucopyranosyl residues (GlcNAc). For example, α-chitin preparation from larval cuticle of the tobacco hornworm is >99% homogeneous, with the remainder being residual amino acids. The CP-MAS ^{13}C-NMR spectrum of α-chitin (Fig. 2) consists of eight well-defined resonances indicating a high structural homogeneity, which parallels a high degree of order as suggested by the line width of the signals (27, *28*). Little or no glucosamine/chitosan is present, because the degrees of N-acylation in acetic anhydride-treated and untreated samples are comparable, as inferred from zero contact time-extrapolated intensities of the methyl and other carbon signals. The relative carbon intensities are near theoretical values for GlcNAc.

In the spectrum of intact hornworm larval cuticle, eight of the major resonances observed are primarily due to carbon atoms from chitin (Table 1 and Fig. 1). The rest of the resonances are primarily from protein carbons. Thus, solids NMR demonstrated that the major organic component of hornworm larval cuticle is chitin with a high degree of chemical homogeneity and also that the conditions of hot alkali used to extract chitin from the cuticle and digest away the protein do not cause significant deacylation.

Solids NMR can be used to predict and examine conformation and to monitor the enzymatic hydrolysis of chitin (*27-32*). Subtle differences between chitin polymorphs can be detected. The spectrum of ß-chitin has seven instead of the eight resonances seen in the case of α-chitin (*28*). The signals of the C-3 and C-5 carbons of ß-chitin merge into a single resonance centered at 77 ppm, which causes a chemical shift variation of C-3, suggesting involvement of this carbon in unique structural features depending on the polymorph. The carbonyl carbon signal of α-chitin at 175 ppm exhibits an asymmetric profile, indicating the occurrence of two populations of acetamido groups. This does not occur with ß-chitin, whose carbonyl carbon signal is sharper and more symmetrical than that observed for the α-form. Most of the other ß-chitin signals, however, are broader than the corresponding ones of α-chitin. Regarding chitin hydrolysis, solids NMR can detect higher oligomers formed during the initial stages of hydrolysis, which are not otherwise observable by conventional methods (*32*). Moreover, the method is sensitive and selective enough to indicate which specific type of enzyme (endo- or exo-cleaving) of the chitinolytic enzyme complex initiates the hydrolysis.

Insect egg shells, cocoons, and egg cases. Other structural materials studied using solids NMR include insect egg shells, cocoons, and egg cases. These are relatively homogeneous in organic components and lack chitin, as evidenced by resonances at 44, 116, 129, and 155 ppm (Table 1, *21,33*). Most of these materials are composed of >95% protein, which may be stabilized by interpolypeptide cross-links. For example, moth cocoon silk consists primarily of two proteins: the thread-like protein, fibroin, and the glue-like protein, sericin (*34*). Fujiwara and Kobayashi (*35*) analyzed fibroin composition and conformation using solids NMR. Other types of cocoons from wild silk moths are lightly tanned and contain low levels (0.4-1%) of catecholic compounds, in addition to protein (*33*). These compounds may add strength to the structures by cementing or cross-linking the silk proteins together.

Egg cases of praying mantids contain approximately 93% protein and 7% phenolics (*33*). Cockroach egg cases are similar in organic composition, except that some also contain the two-carbon dicarboxylic acid, oxalic acid, which probably is associated with calcium and may further render the egg case hard and brittle or act as an antibiotic factor (36, *37*). Some cockroaches (American and Oriental) deposit egg cases that contain about 7-8% oxalate (170 ppm, Table 1), whereas egg cases of others (German) contain little or no oxalate.

Insect cuticles. Insect cuticle is a composite material, primarily a polymeric structure of protein and chitin chains with lesser amounts of catechols, lipids, minerals, pigments, and water. It has been studied extensively by solids NMR. The compositions of moth, beetle, and cockroach cuticles and some of the changes in composition that occurred when the exoskeleton is stabilized have been determined. The peptide backbone α-carbon signals at 55 and 60 ppm (Table 1) are used to estimate protein levels, whereas signals at 104 ppm (chitin carbon 1) and 82 ppm (chitin carbon 4) are representative of chitin content, because they are due to single carbon types, are well resolved from other resonances, and appear as prominent, well-defined signals in the spectral background of whole cuticular material. The aromatic catecholic ring carbon signal at 144 ppm and the methylene carbon signal at 33 ppm are used to estimate catechol and lipid levels, respectively.

Larval cuticles are generally very flexible and largely unpigmented; many pupal cuticles are stiffer and darker in coloration; and adult cuticles, excluding scales, are thin, lightly colored, stiff, and adapted for flight behavior (*6, 18*). One of the major organic differences detected by solids NMR analysis between stiff and flexible cuticular structures is the presence of higher levels of catechols in the former type.

Schaefer et al. (18) determined changes in chemical composition during sclerotization of insect cuticles using solids ^{13}C NMR. An unsclerotized cuticle spectrum is composed primarily of protein and lipid carbon signals. A sharp signal at 33 ppm indicates that a significant amount of lipid is present in untanned cuticle, perhaps as the newly secreted waxy layer of the epicuticle. After sclerotization has occurred, increased levels of chitin and catechols are apparent in contrast to a decreased percentage of lipid. The ^{13}C spectrum of pupal exuviae, which is the outermost and most heavily sclerotized part of the pupal cuticle, is generally similar to that of a fully tanned cuticle, except that the level of catechols is increased fourfold. This suggests that insect cuticles are sclerotized in part by incorporation of catechols into the protein-chitin-lipid composite of newly ecdysed soft cuticles.

The chemical composition of adult cuticles from six coleopteran species was estimated by solids ^{13}C NMR (38). The adult cuticles are dark brown in color. Of all cuticles examined to date, adult beetle cuticles contain some of the highest levels of catechols, ranging from 16 to 33%. In addition to protein, chitin, catechols, and lipid, the presence of melanic pigment in the exoskeleton was confirmed by solids NMR difference spectral analysis of elytra (wing covers) from wild type and *black* mutant strains of the red flour beetle, *Tribolium castaneum*. The ^{13}C NMR difference spectrum obtained by subtracting the spectrum of a wild type from that of a *black* mutant strain showed that both types of elytra have comparable levels of protein, chitin, and lipid, but that the *black* elytra have more carbons characteristic of melanin in two regions, 0-30 ppm and 80-160 ppm. In addition, wild type elytra have more carbons characteristic of ß-alanine (negative carbon resonances at 35 and 175 ppm). These results agree with those obtained by wet chemical analysis, which revealed that *black* mutants have higher levels of dopamine, a melanin precursor, and lower levels of ß-alanine and N-ß-alanyldopamine relative to wild type strains (39, 40). These chemical differences lead to an overproduction of melanin when the excess dopamine in the *black* strain is oxidized.

Another species that offers a comparison of mutant cuticular phenotypes for NMR analysis is the Mediterranean fruit fly, *Ceratitis capitata*. In the mutant *white pupa*, the puparium fails to tan, but normally tanned larval and adult cuticular structures develop. The mutant puparium is fivefold less resistant to compression than the wild type strain (41). Solid state ^{13}C NMR analysis reveals that more chitin but less ß-alanine and fewer catechols are present in the mutant's cuticle relative to the wild type cuticle. Apparently, the *white pupa* mutant is defective in a transport mechanism that shuttles catecholamines into the puparial cuticle, thus preventing normal sclerotization and pigmentation.

Solids NMR analysis was used to compare sclerotized and mineralized types of cuticles of two dipteran species. The puparial cuticle of the house fly, *Musca domestica*, is stabilized primarily by sclerotization, in which catecholic metabolites accumulate in and presumably cross-link the protein/chitin matrix of the larval cuticle (42). The face fly, *M. autumnalis*, on the other hand, hardens its puparial cuticle by depositing calcium and magnesium phosphates into it. Whereas mineral salts constitute more than 60% of the face fly puparial exuviae, they make up only 3% of the house fly puparial exuviae (43). Protein, chitin, and catechols make up approximately 90% of the house fly cuticle, whereas they account for less than 30% of the face fly cuticle. The results demonstrate that dipterans use both catechols and minerals for stabilization of puparial cuticle, with the house fly relying primarily on sclerotization and the face fly on mineralization.

Marine corals and mussels. CP-MAS ^{13}C NMR spectra of skeletal components of individual colonies of the New Zealand black coral, *Antipathes fiordensis*, have a marked similarity to spectra of insect cuticles (*44*). The organic content is approximately 70% protein, 10% chitin, 3-15% diphenol, and 5% lipid by weight. The chitin content of the Caribbean black coral, *A. salix*, is about double that of *A. fiordensis* (*45*). Small compositional differences exist between the older load-bearing skeletal base portions and the younger pinnules or tips. Apparently, these marine corals become stabilized by cross-linking of structural proteins to other proteins and possibly to chitin.

Holl et al. (*22*) examined the covalent cross-linking of proteins in a glue secreted by the marine mussel, *Geukensis demissa*, by ^{13}C and ^{15}N solids NMR using specific labeling with ^{13}C and ^{15}N-lysine. They found no evidence for the involvement of lysyl amino groups in covalent cross-links or ionic complexes in the glue. Proteins containing the catechol amino acid, 3,4-dihydroxyphenylalanine (DOPA), appear to become cross-linked as the glue hardens (*5*). REDOR dephasing experiments, however, using plaques and threads from the byssus of the mussel *Mytilus edulis* labeled by sea-water exposure to ^{13}C- and deuterium-labeled tyrosine revealed that only about one- tenth of the tyrosine rings are within 4 Å of each other or rings of DOPA (*46*). No evidence was seen for the formation of covalent C-C linkages between or among tyrosine or DOPA rings in either plaques or threads.

Heteronuclear interactions. Intermolecular cross-linking of proteins is considered to be one of the primary mechanisms for sclerotization of insect cuticles, silk structures, egg shells, egg cases, and marine coral skeletons. Because of the intractable nature of these structures, however, little direct evidence for a cross-linked structure has been reported until recently (6, 47). Solids NMR has been extremely useful for probing the cross-linked structures in the pupal cuticle of the tobacco hornworm, and the data support the cuticle model originally proposed by Pryor (*48, 49*), which depicts protein chains cross-linked by quinonoid derivatives of catechols. NMR analyses of *M. sexta* pupal cuticle containing protein labeled with 1,3-[^{15}N$_2$]-histidine demonstrates that a side chain histidyl nitrogen becomes covalently linked to ring or side chain carbons of the dopamine moiety of N-ß-alanyldopamine during sclerotization (18, 50). After the pupal cuticle is labeled with both 1,3-[^{15}N$_2$]-histidine and ring- [^{13}C$_6$] dopamine, the double cross-polarization spectrum reveals that one of the aromatic catecholamine carbon atoms is bonded covalently to a ring nitrogen of histidine (*18*). The CP-MAS ^{15}N NMR spectral data of pupal cuticle labeled by injection of L-1,3-[^{15}N$_2$]-histidine also support the formation of carbon-nitrogen cross-links. The ^{15}N NMR spectrum of unsclerotized cuticle shows protonated and nonprotonated histidyl ring nitrogen peaks at 140 and 225 ppm, respectively, as well as a natural abundance amide nitrogen peak at 100 ppm. A new signal is observed at 155 ppm that builds up during the time course of sclerotization and becomes a major nitrogen resonance in the exuviae. Dipolar dephasing produced by delayed ^{1}H decoupling has little effect on the new signal, which shows the nitrogen to be nonprotonated. Its chemical shift indicates that a histidyl nitrogen is attached to either an aromatic or aliphatic carbon. Taken together, the ^{13}C and ^{15}N NMR data are consistent with the formation of an aromatic carbon-nitrogen bond via a mechanism whereby an imidazole nitrogen attacks a phenyl carbon of an o-quinone derivative of the catecholic compound.

Cross-links between a histidyl ring nitrogen and the ß-carbon (C7) of dopamine were postulated to occur in insect cuticle as a result of ß-sclerotization or quinone methide sclerotization (4, 51). Double cross-polarization NMR spectroscopy was too insensitive for detecting this type of linkage because of inherently slow C-N polarization transfer and fast spin-lock relaxation for the labeled dopamine carbon (8, 52); therefore, REDOR experiments were conducted to circumvent these technical problems. REDOR provided an order of magnitude improvement in sensitivity and was used to identify covalent bond formation between the 7-carbon of dopamine and the ring nitrogen of histidine (53). In cuticle labeled by both 7-^{13}C dopamine and ring $^{15}N_2$ histidine, only the resonance at $\delta_C = 60$ had a ^{13}C REDOR difference signal above the ^{15}N natural abundance level. This signal arose from the formation of a protein-catecholamine 7-carbon covalent bond. The weak difference signal at 120 ppm was due to natural abundance carbons in ^{15}N-labeled histidine. About two-thirds of the bound histidyl residues in the exocuticular protein were to be linked covalently to catecholamine ring carbons, and about one-third to the 7-carbon in pupal cuticle.

Recently, Xu et al. (47) isolated from acid hydrolysates of insect cuticle four adducts of dopamine and histidine, which apparently are formed from Michael 1,4- and 1,6-addition reactions of two quinonoid sclerotizing agents, N-ß-alanyldopamine quinone and N-ß-alanyldopamine quinone methide, with both of the imidazole nitrogens of histidyl residues of cuticular proteins. The adducts were identified as 7-N^τ-, 6-N^τ-, 7-N^π-, and 6-N^π-histidyldopamines. Thus, both side chain nitrogens of histidine react with either ring or side chain carbons of N-ß-alanyldopamine, and the 7-adducts are approximately twice as abundant as the 6-adducts. These structural identifications extended earlier results from solids NMR spectroscopic analyses, which first detected both ring and side chain catecholamine-histidine bonding in insect cuticle (18; 53).

Schaefer et al. (18) also proposed that covalent bonds exist between cuticular protein and chitin. Solid-state ^{15}N NMR analysis of chitin prepared by alkali extraction of 1,3-[$^{15}N_2$]-histidine-labeled M. sexta pupal exuviae revealed an ^{15}N chemical shift expected for a substituted imidazole nitrogen cross-linked structure. Apparently, the chitin is not coupled directly to protein, but instead to a catecholic carbon, which serves as a bridge between protein and chitin. Because the evidence for a protein-chitin linkage in cuticle is indirect, additional work is needed to obtain more substantive direct evidence for such a linkage.

NMR experiments with double-labeled ^{13}C-^{15}N-ß-alanine indicated that the terminal amino group of N-ß-alanyldopamine also was involved in covalent bonding (53). Recently, solids NMR detected a novel type of intercatechol linkage in insect cuticle (25). REDOR ^{13}C and ^{15}N NMR, in combination with ^1H-^{13}C dipolar modulation and ^{15}N-^{15}N dipolar restoration at the magic angle, revealed covalent bonding between the terminal nitrogen of one N-ß-alanyldopamine molecule to the ß-carbon of another, forming a protein-catechol-catechol-protein cross-link. This type of dimeric C-N catechol linkage is different from the C-O bridges between 3,4-dihydroxyl groups of catechols and α and ß carbons of catechols present in dimers isolated several years ago from insect cuticles (54).

Use of dopamine labeled with ^{13}C in the α-position of the side chain provided evidence of covalent bonding to the α-carbon, but the nature of the respective adducts was not established (53). The formation of oligomeric forms of catecholamines may occur through the α-carbon, ß-carbon, and/or ring carbons via addition reactions with quinonoid compounds, resulting in larger cross-linking agents and space fillers in the cuticular matrix (54, 55). REDOR spectra of insect cuticle showed oxygen substitution at both α and ß carbons of N-ß-alanyldopamine, consistent with the formation of intercatechol oxygen brides in dimeric or oligomeric forms of catecholamines (25).

In summary, solids NMR analysis provides direct evidence for covalent modifications to the ring carbons, a carbon, ß carbon, and the terminal nitrogen of N-ß-alanyldopamine in the pupal cuticle of M. sexta. The central assumption of this chemical functionality is that the oxidation products of catecholamines serve as cross-linking agents for proteins or chitin in a variety of ways, making sclerotized cuticle structurally heterogeneous. Although solids NMR has provided good evidence for portions of the cross-linked structure of the insect cuticle, the supramolecular structure is still incomplete, and little information has been obtained about the nature of the transient intermediates that serve as cross-linking agents. Solids NMR data are consistent with the hypothesis that quinones and quinone methides serve as cross-linking agents, but other reactive intermediates are also possible.

Future work

Progress in our understanding of the molecular organization of sclerotized structures has been aided greatly by the results of studies using solids NMR spectroscopy. The NMR data generally support the model first proposed by Pryor (1, 48, 49), which states that sclerotization is a process in which certain proteinaceous structures, such as insect cuticles, are stabilized by cross-linking, dehydration, and impregnation with phenolic metabolites. Natural abundance ^{13}C NMR has yielded valuable and unique information regarding the composition of sclerotized structures and their compositional variations at different developmental stages or for different mutants. Incorporation of specific ^{13}C and ^{15}N labels with the use of advanced techniques such as REDOR has allowed us to test hypotheses regarding the covalent linkages formed during sclerotization (22, 50, 53). In the future, we hope to explore further the central role played by catecholamines in tissue sclerotization using recently developed NMR techniques. Solids NMR analysis after ^{13}C, ^{15}N, and ^{17}O labeling of samples with precursors will be used to identify additional adducts and cross-links, particularly those involving the α-carbon and amino group of dopamine. Several relatively new magnetic resonance methods, including transferred echo double resonance (TEDOR) ^{13}C and ^{15}N NMR (56, 57); extended dipolar modulation (XDM) ^{13}C NMR (58); dipolar rotational spin-echo (DRSE, 20); XY-8 dipolar recovery at the magic angle (DRAMA) ^{15}N NMR (59); and simple excitation for the dephasing of rotational-echo amplitudes (SEDRA, 60), will be utilized to detect single and multiple heteronuclear connectivities in skeletal structures. TEDOR is a rotor synchronized solid-state technique that selects dipolar coupled spins from among the background of uncoupled spins by a coherence transfer from one spin of the heteronuclear pair to the other (56). Triplets like $-^{13}C-^{15}N-^{13}C-$, $-^{15}N-^{13}C-^{15}N-$, and $-^{13}C-^{17}O-^{13}C-$ can be detected specifically by double coherence transfers.

Additional solids NMR experiments can be designed to detect carbon-oxygen and carbon-carbon cross-links in samples. Little direct evidence for C-O or C-C linkages between proteins, chitin, and catecholamines has been reported to date. However, residual protein and catecholamines were detected in chitin extracted from insect cuticles (18), suggesting the presence of C-O cross-links to chitin. Because the evidence for a protein-chitin linkage in cuticle is weak (18), additional research should be conducted to obtain more substantive support for such a linkage. Cuticular samples labeled with either 3,4-$^{17}O_2$-dopamine or $^{17}O_5$-N-acetylglucosamine and either ^{13}C-labeled amino acids, catecholamines, or N-acetylglucosamine can be examined using solids NMR for the presence of $^{13}C-^{17}O$ linkages between catecholamines, proteins, and chitin. A REDOR-based method to monitor indirectly the chemical connectivity of quadrupolar nuclei such as deuterium and ^{17}O can be utilized for ^{17}O-labeled cuticle samples (61). For investigating C-C bonding, another NMR technique that was developed for measuring $^{13}C-^{13}C$ distances and is similar to the REDOR method but is called combined-rotation with nutation (CROWN) NMR can be used (62). The power and specificity of these and other novel solid-state NMR techniques will allow direct testing of hypotheses about the biochemistry of insect and marine invertebrate structures as well as other types of organic materials found in the geosphere and biosphere, which has not been possible in the past.

Acknowledgements

This research was supported in part by National Science Foundation grants MCB-9418129 to KJK and TLH, and MCB-9604860 to JS. Cooperative investigation between the Agricultural Research Service, Kansas Agricultural Experiment Station (Contribution no. 98-69-J) and Washington University. Mention of a proprietary product does not constitute a recommendation by the USDA. The Agricultural Research Service, USDA is an equal opportunity/affirmative action employer and all agency services are available without discrimination.

References

1. Pryor, M. G. M. Sclerotization. In: *Comparative Biochemistry*. Florkin, M.; Mason, H. S., Eds. Academic Press, New York, N.Y. **1962**, *4B*, pp 371-396.
2. Brown, C. H. *Structural Materials in Animals*. Wiley, New York, N.Y. **1975**, p. 448.
3. Brunet , P. C. J. The metabolism of the aromatic amino acids concerned in the cross-linking of insect cuticle. *Insect Biochem.* **1980**, *10*, 467-500.
4. Andersen, S. O. Sclerotization and tanning of the cuticle. In: *Comprehensive Insect Physiology, Biochemistry, and Pharmacology;* Kerkut, G. A. and Gilbert L. I., Eds.; Pergamon Press, Oxford. **1985**, *3*, pp 59-74.
5. Waite, J. H. The phylogeny and chemical diversity of quinone-tanned glues and varnishes. *Comp. Biochem. Physiol.* **1990**, *97B*, 19-29.
6. Hopkins, T. L.; Kramer, K. J. Insect cuticle sclerotization. *Ann. Rev. Entomol.* **1992**, *37*, 273-302.

7. Baas, M.; Briggs, D. E. G.; van Heemst, J. D. H.; Kear, A. J.; de Leeuw, J. W. Selective perservation of chitin during the decay of shrimp. *Geochim. Cosmochim. Acta.* **1995,**. *59*, 945-951.

8. Rabi, I. I.; Millman, S.; Kusch, P.; Zacharias, J. R. The molecular beam resonance method for measuring nuclear magnetic moments. *Phys.Rev.* **1939**, *59*, 526-539.

9. Becker, E. D. A brief history of nuclear magnetic resonance. *Analyt. Chem.* **1993**, *65*, 295A-302A.

10. Stejskal, E. O.; Memory, J. D. *High Resolution NMR in the Solid State. Fundamentals of CP/MAS.* Academic Press, New York, N.Y. **1994**.

11. Wüthrich, K. *NMR of Proteins and Nucleic Acids.* Wiley & Sons, New York, N.Y. **1986**.

12. Haeberlen, U. *High Resolution NMR in Solids. Selective Averaging.* New York, NY. **1976**, pp 16-31.

13. Andrew, E. R. The narrowing of NMR spectra of solids by high-speed specimen rotation and the resolution of chemical shift and spin multiplet structures for solids. *Proc. Nucl. Magn. Reson. Spectrosc.* **1971**, *8*, 1-89.

14. Hartmann, S. R.; Hahn, E. L. Nuclear double resonance in the rotating frame. *Phys. Rev.* **1962**, *128*, 2024-2042.

15. Lurie, F. M.; Slichter, C. P. Spin temperature in nuclear double resonance. *Phys. Rev.* **1964**, *133*, A1108-A1122.

16. McArthur, D. A.; Hahn, E. L.; Walstedt, R. E. Rotating-frame nuclear double-resonance dynamics. *Phys. Rev.* **1969**, *188*, 609-638.

17. Pines, A.; Gibby, M. G.; Waugh, J. S. Proton-enhanced NMR of dilute spins in solids. *J. Chem. Phys.* **1973**, *59*, 569-590.

18. Schaefer, J.; Kramer, K. J.; Garbow, J. R.; Jacob, G. S.; Stejskal, E. O.; Hopkins, T. L.; Speirs, R. D. (1987) Aromatic cross-links in insect cuticle: detection by solid state ^{13}C and ^{15}N NMR. *Science.* **1987**, *235*, 1200-1204.

19. Gullion, T.; Schaefer, J. Detection of weak heteronuclear dipolar coupling by rotational-echo double-resonance NMR. *Adv. Magn. Reson.* **1989**,*13*, 57-83.

20. Bork V.; Gullion, T.; Hing, A. Measurement of [13]C-^{15}N coupling by dipolar-rotational spin-echo NMR. J. Magn. Reson. **1990**, *88*, 523-528.

21. Kramer, K. J.; Hopkins, T. L.; Schaefer, J. Applications of solids NMR to the analysis of insect sclerotized structures. *Insect Biochem. Molec. Biol.* **1995**, *25*, 1067-1080.

22. Holl, S. M.; Hansen, D.; Waite, J. H.; Schaefer, J. Solid-state NMR analysis of cross-linking in mussel protein glue. *Arch. Biochem. Biophys.* **1993**, *302*, 255-258.

23. Schaefer, J.; Stejskal, E. O. Carbon-13 nuclear magnetic resonance of polymers spinning at the magic angle. J. *Am. Chem. Soc.* **1976**, *98*, 1031-1032.

24. Peter, M. G.; Grun, L.; Forster, H. CP/MAS-^{13}C-NMR spectra of sclerotized insect cuticle and of chitin. *Angew. Chem. Int. Ed. Engl.* **1994**, *23*, 638-639.

25. Merritt, M. E.; Christensen, A. M.; Kramer, K. J.; Hopkins, T. L.; Schaefer, J. Detection of inter-catechol-protein cross-links in insect cuticle by carbon-13 and nitrogen-15 NMR. *J. Amer. Chem. Soc.* **1996**, *118*, 11278-11282.

26. Williams, H. J.; Scott, A. I.; Woolfenden, W. R.; Grant, D. M.; Vinson, S. B.; Elzen, G. W.; Baehrecke, E. H. (1988) *In vivo* and solid-state ^{13}C nuclear magnetic resonance studies of tyrosine metabolism during insect cuticle formation. *Comp. Biochem. Physiol.* **1988**, *89B*, 317-321.

27. Fukamizo, T.; Kramer, K. J.; Mueller, D. D.; Schaefer, J.; Garbow, J.; Jacob, G. S. Analysis of chitin structure by nuclear magnetic resonance spectroscopy and chitinolytic enzyme digestion. *Arch. Biochem. Biophys.* **1986**, *249*, 15-26.

28. Focher, B.; Naggi, A.; Torri, G.; Cosani, A.; Terjevich, M. Structural differences between chitin polymorphs and their precipitates from solution-evidence from CP-MAS carbon-13 NMR, FT-IR and FT-Raman spectroscopy. *Carbohydr. Polym.* **1992**, *17*, 97-102.

29. Saito, H.; Tabeta, R.; Hirano, S. Conformation of chitin and N-acyl chitosans in solid-state as revealed by ^{13}C cross polarization/magic angle spinning (CP/MAS) NMR spectroscopy. *Chem. Lett.* **1981**, 1479-1482.

30. Saito, H.; Ando, I. High-resolution solid-state NMR studies of synthetic and biological macromolecules. *Annu. Rep. NMR Spectroscopy*. **1990**, *21*, 209-290.

31. Tanner, S. F.; Chanzy, H.; Vincendon, M; Roux, J. C.; Gaill, F. High-resolution solid-state carbon-13 nuclear magnetic resonance study of chitin. *Macromolecules*. **1990, 23**, 3576-3583.

32. Rajamohanan, P. R.; Ganapathy, S.; Vyas, P. R.; Ravikumar, A.; Desphande, M. V. Solid-state CP/MASS ^{13}C-NMR spectroscopy: a sensitive method to monitor enzymatic hydrolysis of chitin. *J. Biochem. Biophys. Methods.* **1996**, *32*, 151-163.

33. Kramer, K. J.; Bork, V.; Schaefer, J.; Morgan, T. D.; Hopkins, T. L. Solid state ^{13}C nuclear magnetic resonance and chemical analyses of insect noncuticular sclerotized support structures: mantid oothecae and cocoon silks. *Insect Biochem.* **1989a**, *19*, 69-77.

34. Seifter, S.; Gallop, P. M. The structural proteins. In *The Proteins* (2nd ed.), Neurath, H., Ed.; Academic Press, New York, N.Y. **1966**, *4*, pp 153-458.

35. Fujiwara, T.; Kobayashi, Y. Conformational study of ^{13}C-enriched fibroin in the solid state using the cross polarization nuclear magnetic resonance method. *J. Mol. Biol.* **1986**, *187*, 137-140.

36. Kramer, K. J.; Christensen, A. M.; Morgan, T. D.; Schaefer, J.; Czapla, T. H.; Hopkins, T. L. Analysis of cockroach oothecae and exuviae by solid-state ^{13}C NMR spectroscopy. *Insect Biochem.* **1991**, *21*, 149-156.

37. Yoshida, M.; Cowgill, S. E.; Wightman, J. A. Roles of oxalic and malic acids in chickpea trichome exudate in host-plant resistance to *Helicoverpa armigera. J. Chem. Ecol.* **1997**, *23*, 1195-1210.

38. Kramer, K. J.; Morgan, T. D.; Hopkins, T. L.; Christensen, A. M.; Schaefer, J. Solid state ^{13}C NMR and diphenol analyses of sclerotized cuticles from stored product Coleoptera. *Insect Biochem.* **1989b**, *19*, 753-757.

39. Kramer, K. J.; Morgan, T. D.; Hopkins, T. L.; Roseland, C. R.; Aso, Y.; Beeman, R. W.; Lookhart, G. L. Catecholamines and ß-alanine in the red flour beetle, *Tribolium castaneum*. Roles in cuticular sclerotization and melanization. *Insect Biochem.* **1984**, *14*, 293-298.

40. Roseland, C. R.; Kramer, K. J.; Hopkins, T. L. Cuticular strength and pigmentation of rust-red and black strains of *Tribolium castaneum*.

Correlation with catecholamine and ß-alanine content. *Insect Biochem.* **1987** *17*, 21-28.

41. Wappner, P.; Kramer, K. J.; Hopkins, T. L.; Merritt, M.; Schaefer, J.; Quesada-Allué, L. A. *White pupa*: a *Ceratitis capitata* mutant lacking catecholamines for tanning the puparium. *Insect Biochem. Molec. Biol.* **1995**, *25*, 365-373.

42. Roseland, C. R.; Grodowitz, M. J.; Kramer, K. J.; Hopkins, T. L.; Broce, A. B. Stabilization of mineralized and sclerotized puparial cuticles of muscid flies. *Insect Biochem.* **1985**, *15*, 521-528.

43. Kramer, K. J.; Hopkins, T. L.; Schaefer, J. Insect cuticle structure and metabolism. In *Biotechnology for Crop Protection*; Hedin, P.; Menn, J. J.; Hollingworth, R. M. ,Eds. ACS Symposium Series 379, American Chemical Society, Washington, D. C. **1988**, pp 160-185.

44. Holl, S. M.; Schaefer, J.; Goldberg, W. M.; Kramer, K. J.; Morgan, T. D.; Hopkins, T. L. Comparison of black coral skeleton and insect cuticle by a combination of carbon-13 NMR and chemical analyses. *Arch. Biochem. Biophys.* **1992**, *292*, 107-111.

45. Goldberg, W. M.; Hopkins, T. L.; Holl, S. M.; Schaefer, J.; Kramer, K. J.; Morgan, T. D.; Kim, K. Chemical composition of sclerotized black coral skeletons (Coelenterata: Antipatharia): a comparison of two species. *Comp. Biochem. Physiol.* **1994**, *107B*, 633-643.

46. Klug, C. A; Burzio, L. A.; Waite, J. H.; Schaefer, J. In situ analysis of peptidyl DOPA in mussel byssus using rotational-echo double-resonance NMR. *Arch. Biochem. Biophys.* **1996**, *333*, 221-224.

47. Xu, R.; Huang, X.; Hopkins, T. L.; Kramer, K. J. Catecholamine and histidyl protein cross-linked structures in sclerotized insect cuticle. *Insect Biochem. Molec. Biol.* **1997**, *27*, 101-108.

48. Pryor, M. G. M. On the hardening of the ootheca of *Blatta orientalis*. *Proc. Roy. Soc. London Ser.* **1940a**, *B128*, 378-393.

49. Pryor, M. G. M. On the hardening of the cuticle of insects. *Proc. Roy. Soc. London Ser.* **1940b**, *B 128*, 393-407.

50. Christensen, A. M.; Schaefer, J.; Kramer, K. J. Comparison of rotational echo double resonance and double cross polarization NMR for detection of weak heteronuclear dipolar coupling in solids. *Magnetic Reson. Chem.* **1991a**, *29*, 418-421.

51. Sugumaran, M. Molecular mechanisms for cuticular sclerotization. *Adv. Insect Physiol.* **1988**, *21*, 179-231.

52. Kramer, K. J.; Hopkins T. L. Tyrosine metabolism for insect cuticle tanning. *Arch. Insect Biochem. Physiol.* **1987**, *6*, 279-301.

53. Christensen, A. M.; Schaefer, J.; Kramer, K. J.; Morgan, T. D.; Hopkins, T. L. Detection of cross-links in insect cuticle by REDOR NMR spectroscopy. *J. Amer. Chem. Soc.* **1991b**, *113*, 6799-6802.

54. Andersen, S. O.; Jacobson, J. P.; Repstorff, P. Coupling reactions between amino compounds and N-acetyldopamine catalyzed by cuticular enzymes. *Insect Biochem. Molec. Biol.* **1992a**, *22*, 517-527.

55. Andersen, S. O.; Peter, M. G.; Roepstorff, P. Cuticle catalyzed coupling between N-acetylhistidine and N-acetyldopamine. *Insect Biochem. Molec. Biol.* **1992b**, *22*, 459-469.

56. Hing, A. W.; Vega, S.; Schaefer, J. Transferred echo double resonance NMR. *J. Magn. Reson.* **1992**, *96*, 205-209.

57. Hing, A. W.; Vega, S.; Schaefer, J. Measurement of heteronuclear dipolar coupling by transferred echo double resonance NMR. *J. Magn. Reson.*

58. Hing, A. W.; Schaefer, J. Two dimensional rotational-echo double resonance of val_1-[1-^{13}C]gly_2-[^{15}N]ala_3-gramicidin A in multilamellar dimyristoylphosphatidylcholine dispersions. Biochemistry. **1993**, *32*, 7593-7604.

59. Klug, C. A.; Zhu, W.; Merritt, M. E.; Schaefer, J. Compensated XY8-DRAMA pulse sequence for homonuclear dephasing. *J. Magn. Reson.* **1994**, *A109*, 134-136.

60. Gullion, T.; Vega, S. A simple magic angle spinning NMR experiment for the dephasing of rotational echoes of dipolar coupled homonuclear spin pairs. *Chem. Physics Lett.* **1992**, *194*, 423-428.

61. Gullion, T. Detecting 13C-17O dipolar interactions by rotational-echo adiabatic-passage, double-resonance NMR. *J. Magn. Reson.* **1995**, *117*, 326- 329.

62. Joers, J. M.; Rosanske, R.; Gullion, T.; Garbow, J. R. Detection of dipolar interactions by CROWN NMR. *J. Magn. Reson.* **1994**, *A106*, 123-130.

Chapter 3

On-Line Determination of Group-Specific Isotope Ratios in Model Compounds and Aquatic Humic Substances by Coupling Pyrolysis to GC–C–IRMS

G. Gleixner and H.-L. Schmidt

Lehrstuhl für Allgemeine Chemie and Biochemie, TU München, D-85350 Freising-Weihenstephan, Germany

Bulk isotope ratios of organic matter provide relevant information about the sources and processes involved in their synthesis. More information is obtained from the intermolecular and intramolecular distributions of isotopes in individual compounds. Valuable corresponding information can be obtained by their partial degradation. We have used a system for the controlled pyrolysis of small amounts of organic matter by a Curie point pyrolyzer connected to a GC-C-IRMS device. The system was applied to refractory organic acids from aquatic ecosystems yielding peaks with characteristic δ-values differences. The assignment of some pyrolysis products and their δ^{13}C-values relating to cellulose (average mean -23.5 ‰ ± 1.0 ‰) and lignin (average mean -29.4 ‰ ± 5.8 ‰) revealed plants as source material. The high enrichment of 4-methyl-phenol (δ^{13}C = -23.2 ‰) suggests secondary (bacterial or fungal) sources of this compound. The δ^{15}N-values of pyrolysis products from chitin are similar to their bulk values.

Humic substances (HS) are the predominant form of organic matter in peat and soil, and can also be isolated from clear surface and ground waters (1). The primary origin of these materials is "biomass" from plants, which is itself a complex conglomerate of different polymers, additionally containing many monomeric primary and secondary natural compounds (2). It is known that this mixture, after plant death, undergoes oxidative and hydrolytic degradation by microorganisms (mainly fungi and bacteria), accompanied by secondary structural changes, e.g. aromatizations, Maillard product formation (3, 4) or photochemical degradation (5). Biodegradation forms new structural units in bacterial or fungal biomass, which may be accompanied with selective preservation of biomolecules (6).

To isolate and characterize humic substances, standardized extraction procedures have been developed (7), yielding the soluble humic and fulvic acid fractions, and additionally, in soils and peat, an insoluble residue, humin. The molecular weight of fulvic acids is in general smaller than that of humic acids, covering the range from 100 to 700 Da with a maximum at 350 Da (8). Furthermore the oxygen content (30 and 40 %) of fulvic acids generally exceeds that of humic acids, their carbon content varies between 45 and 50 %, their hydrogen content between 3 and 5 % and their nitrogen content between 0.5 and 4 % (9). However, because of the complex nature and the high molecular weight of humic substances it is so far not possible to describe a defined structure. Nevertheless structural elements have been identified in pyrolysis products, e.g. furan, benzene, phenol, aliphatic chains, pyrrole and pyridine (10, 11). O-alkyl-, aryl- and alkyl-carbon groups were found by ^{13}C-NMR experiments (12, 13) and amino-N was detected by ^{15}N-CPMAS-NMR experiments (14). On the base of these results, models for the structure of humic substances have been proposed (15), but they are not generally accepted (16). On the other hand procedures of their formation cannot yet be deduced, because various structural elements may originate from different precursors and different synthetic phases mentioned before.

To obtain a better understanding about the ecological function of humic substances, it is important to elucidate their origin and the mechanisms of their formation. A powerful tool to reconstruct chemical or biochemical reactions involved in the formation or decomposition of organic compounds is the detection of isotope discriminations resulting from these reactions. Isotopic patterns of natural products reflect these events and processes (17, 18) and in addition can provide information about the primary source material (19, 20). For example the isotopic content of individual alkanes isolated from sediments was decisive to identify their source (21). First attempts to use isotopic abundances of humic substances in order to elucidate their composition showed that their bulk ^{13}C content is determined by the carbohydrate / lignin ratio as obtained by NMR. This is in line with the relative depletion in ^{13}C of phenylpropanes as compared with carbohydrates (17). More detailed information about HS structure and humification processes is to be expected from individual isotopic abundances of defined fragments, because these may coincide with corresponding precursor structures. Furthermore, a comparative analysis of the isotopic content of fragments from possible source materials and corresponding HS will help to identify the preservation of defined structural elements. Finally isotopic shifts in individual fragments can give hints on the mechanisms involved in the synthesis of polymer material.

In order to obtain such fragments Curie-point pyrolysis under controlled conditions for the degradation of HS can be used with high reproducibility (22, 23) and, in combination with on-line GC/MS (gas chromatography/mass spectrometry) and on-line GC-C-IRMS (gas chromatography-combustion-isotope ratio mass spectrometry), the fragments can be identified and analyzed for their isotopic content. The reliability and reproducibility of this method has been shown by determinations of the isotopic content of tetramethylsilyl derivatized off-line pyrolysates (24) and off-line tetramethylammonium hydroxide thermolysis products (25). Furthermore on-line pyrolysis of GC-eluates (*26*) in GC-C-IRMS devices have already demonstrated that pyrolysis of these compounds takes place without molecular rearrangements and / or isotopic fractionation.

This paper deals with the development of a simple procedure to study the isotopic content of molecular fragments of HS using on-line Curie-point pyrolysis GC/MS to identify pyrolysis products, and parallel on-line Curie-point pyrolysis GC-C-IRMS to obtain isotopic data of these products. The first results are discussed with reference to primary material and metabolic reactions involved in the biosynthesis of humic substances.

Materials and Methods

Origin and purification of humic substances. Fulvic acids (FA) were isolated according to the XAD-8 method (*7*) from a number of sites in Germany: I, a brown water lake "Hohlohsee" (HO 10 FA) in the Black Forest, II, a groundwater source (FG 1 FA) near Hannover, III, a soil percolate (BS 1 FA) near Bayreuth, IV, a lake in the brown coal mining area (SV 1 FA) near Leipzig and V, a biological sewage plant (ABV 2 FA) near Karlsruhe. Additionally HS from the "Hohlohsee" (HO 12 K) were concentrated by ultrafiltration (0.45 μm > HS > 4 kD). These six samples are reference materials from a special research program of the DFG "Refractory organic acids in water", and sample codes are originating from there. Chitin and N-acetyl glucosamine were purchased from Sigma Chemical Co., St. Louis, USA and Aldrich-Chemie, Steinheim, Germany, respectively.

Multielement isotope determination. The bulk isotopic content of H, C, N and O from aquatic fulvic acids was determined by standard procedures (*27*).

Carbon and nitrogen isotopic analyses were performed with an elemental analyzer RoboPrep-CN (*28*), Europe Scientific, Crewe, Great Britain, connected to a VG MM 903 isotope ratio mass spectrometer, Vacuum Generators, Middlewich, Great Britain. 3 mg FA for C analysis and 10 mg FA for N analysis were converted in a Dumas-combustion unit at 1273 K with chromiumoxide and copperoxide to CO_2 and N_2, these were gas chromatographically separated and measured "on-line" by IRMS.

Oxygen isotopic analysis was performed under a controlled pyrolysis conditions (*29*). About 4 mg FA were placed in Ag capsules, and the bound oxygen was quantitatively converted at 1373 K to CO, which was, after gas chromatographic purification, on-line transferred to a VG MM 903 IRMS.

Hydrogen isotopic analysis. 25 mg FA were off-line combusted in a stream of synthetic air. The oxidation water was cryogenically isolated and reduced at Uranium (1073 K, 8 min) to H_2. The gas was transferred to a SIRA 24 IRMS, (VG, Middlewich, Great Britain) for isotope ratio determination.

Calculation of δ-values and isotopic corrections. All isotopic ratios (R = [heavy isotope] / [light isotope]) were expressed in the δ-scale relative to an international standard using equation 1.

$$\delta \text{ element } [‰] = (R_{sample} - R_{standard}) / R_{standard} * 1000 \tag{1}$$

International standards used for H and O were V-SMOW, for carbon V-PDB and for nitrogen AIR. All $\delta^{13}C$-values were corrected for their ^{17}O content after Craig (30).

Curie-point pyrolysis - gas chromatography/mass spectrometry (Py-GC/MS). The freeze dried and homogenized humic substances were pyrolyzed in a Type 0316 Curie-point pyrolyzer (Fischer, 53340 Meckenheim, Germany). The total heating time was 9.9 s and the final pyrolysis temperatures employed was 773 K. The principle, potential and limitations of Py-GC/MS of humic fractions and dissolved organic matter (DOM) have been described previously (31). The column and GC conditions were identical to those in the GC-C-IRMS system (see below).

Curie-point pyrolysis - gas chromatography - conversion - isotope ratio mass spectrometry (Py-GC-C-IRMS). Between 0.1 and 2 mg of HS were placed into a ferromagnetic sample holder within a high frequency coil, which was directly connected to the GC-inlet system (Figure 1). After flushing the system with He, the sample was inductively heated in a few milliseconds to the Curie-temperature of the ferromagnetic probe (773 K) and the pyrolysis products formed were immediately transferred to the GC-column for separation (injector and interface temperature 523 K). The pyrolysis products were separated on a BPX 5 column (50m x 0.32 mm, film thickness 1.0 μm, Scientific Glass Engineering, 64331 Weiterstadt, Germany) using a temperature program of 305 K for 5 min, in 5 K/min to 573 K. The column outlet was coupled to a combustion furnace converting the pyrolysis products to CO_2, N_2 and H_2O (CuO, NiO and PtO at 1 233 K). The water was trapped, and CO_2 and N_2 were on-line transferred to the IRMS. The $\delta^{13}C$- and $\delta^{15}N$-values were determined using an isotope ratio mass spectrometer (Delta S, Finnigan MAT, 28127 Bremen, Germany).

Results and Discussion

Global isotopic content of FA isolated from aquatic ecosystems. The δ^2H-values of the FA varied between -100 and -145 ‰ (Table I). These values are negative in comparison with those of recent terrestrial biomass, indicating a relatively high content of aliphatic or aromatic hydrocarbons (32). The $\delta^{13}C$-values of the FA were between -20 and -26 ‰, indicating C3-plant material as original source (33). The relative high $\delta^{13}C$-values of the FA from the sewage plant (ABV 2 FA) are consistent with a fast turnover rate or high trophic level of this material, as within food chains every trophic level is isotopically enriched relative to the previous one. The $\delta^{18}O$-values of the FA (between 13 and 16 ‰) indicate a relative enrichment as compared with recent plant material and with corresponding HA (values not shown). This is probably due to the secondary introduction of oxygen from O_2 ($\delta^{18}O$ = +22,4 ‰ not from H_2O $\delta^{18}O$ ~ -5 ‰) in the course of the formation of FA in the aquatic system.

38

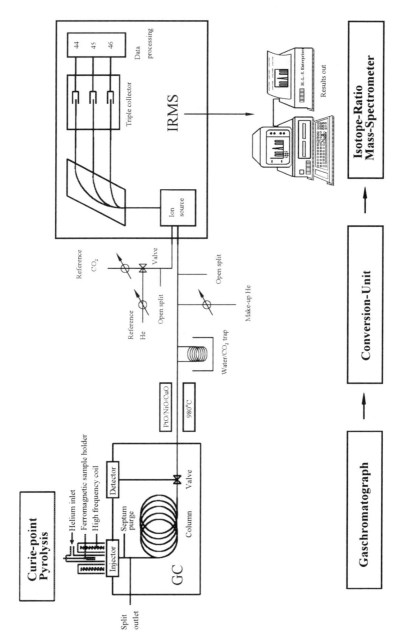

Figure 1: Schematic of the Py-GC-C-IRMS system used

The δ^{15}N-values of the FA between -7 and +27 ‰ cover the whole natural abundance range. As yet we have no satisfactory explanation for these results.

The open question in these results is, whether the bulk δ-values observed are solely due to the primary source(s) indicating that the δ-value differences of the structural units are identical or similar in HS and its possible source materials, or whether they arise partially from secondary transformations implying isotope discriminations and non-statistical isotope distributions. In the latter case differences would preferably be observed in some of the pyrolysis products. Corresponding investigations are planned by partial pattern analysis of individual compounds, e.g. carbohydrates and lignins, and their mixtures.

Pyrolysis and isotopic abundance of fragments from aquatic humic substances. In order to produce defined fragments of the humic substances for isotopic analysis it was necessary to use a method which allowed controlled degradation. Curie-point pyrolysis is such a method already used in the analysis of defined organic compounds (*22, 26*). Because of the small amounts of material available (concentration of dissolved organic carbon (DOC) varies only between 1 mg/l and 35 mg/l (brown water), only 50 % of which can be isolated as FA), it was obligatory that the method used should provide reproducibly molecular fragments using only a mg of FA. Furthermore no derivatisation method introducing additional atoms to the pyrolysis products should be applied. Under these aspects an on-line Py-GC-C-IRMS (Figure 1) was adapted and optimized.

Identification of pyrolysis products. The column used for identification in the Py-GC/MS was identical to that used in the Py-GC-C-IRMS device. Pyrolysis and separation were performed under identical conditions. The MS was scanned from 50 to 300 m/e and pyrolysis products were identified from the analysis of their fragmentation pattern using the Wiley and NBS library (*31*).

Carbon isotope ratio in pyrolysis products. After pyrolysis, GC separation and on-line combustion the system yields a trace for m/e 44 (CO_2) and for the ratio 45/44 ($^{13}CO_2/^{12}CO_2$, Figure 2), from which the δ^{13}C-values were calculated. The δ-values of the identified pyrolysis products are shown in Table 2; mainly derivatives of furan, benzene and phenol were identified. A good peak separation is the main problem in these complex chromatograms, but as can be seen from the ratio trace, most peaks are separated. Using individual background calculation of ISODAT software, the standard deviations of the identified peaks were between 0.4 and 1.4 ‰, with exception of phenol. This is probably due to the existence of multiple sources of this pyrolysis product.

The δ^{13}C-values of the identified peaks were in the normal C3-plant range (between -22 and -35 ‰). Among them the average mean value for all furans and acetic acid, which are supposed to be polysaccharide pyrolysis products (*10*), is -23.5 ‰ ± 1.0 ‰. This is in agreement with δ-values for such precursors (*34*). The average mean δ^{13}C-value of all phenols and methyl-benzene, which are assumed to be pyrolysis products of lignin (*10*), is -29.4 ‰ ± 5.8 ‰. This value is identical to lignin oxidation products from C3-plants (-30.4 ‰ ± 3.9 ‰, *35*), and the depletion

Table I. Stable hydrogen, carbon, nitrogen and oxygen content of fulvic acids (FA) isolated from aquatic ecosystems. [‰] versus V-SMOW for hydrogen and oxygen, V-PDB for carbon and AIR for nitrogen. Standard deviations for C, N and O below 0.5 ‰, for H below 3 ‰. n.d. = not determined

Origin	Code	δ^2H	$\delta^{13}C$	$\delta^{15}N$	$\delta^{18}O$
Brown water lake	HO 10 FA 3.95	-102.4	-24.9	-8.4	14.9
Soil percolate	BS 1 FA	-129.8	-26.1	7.3	14.8
Ground water	FG 1 FA	-144.2	n.d.	n.d.	16.5
Sewage plant	ABV 2 FA	-110.4	-19.6	0.3	13.4
Brown coal waste water	SV 1 FA	-123.4	-24.3	27.2	15.3

Table II. $\delta^{13}C$-values and standard deviations from identified pyrolysis products of a brown water lake concentrated by ultrafiltration (HO 12 K). Peak numbers correspond to Fig. 2

Peak No.	Name	$\delta^{13}C$-value	s.d.
4	2-methyl-furan	-24.09	1.31
5	acetic acid	-23.35	0.35
9	toluene	-30.25	0.49
10	furanon	-22.90	0.74
11	2-furaldehyde	-22.23	0.92
11	2-cyclopentene-1-on		
17	5-methyl-2-furaldehyde	-24.87	0.36
19	phenol	-29.20	3.70
22	4-methyl-phenol	-23.22	0.98
24	2-methoxy-phenol	-34.75	0.54

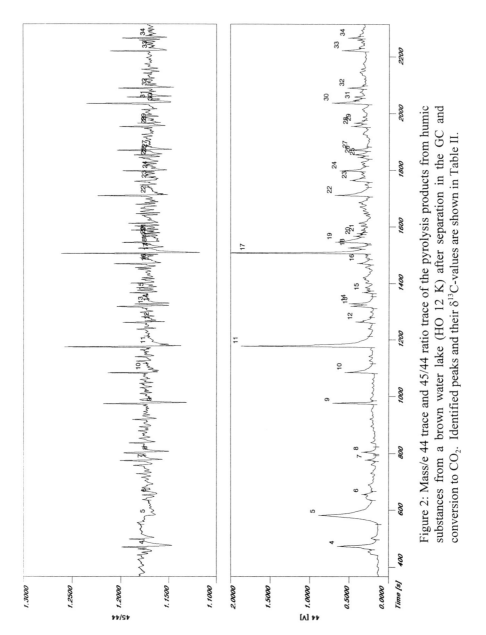

Figure 2: Mass/e 44 trace and 45/44 ratio trace of the pyrolysis products from humic substances from a brown water lake (HO 12 K) after separation in the GC and conversion to CO_2. Identified peaks and their $\delta^{13}C$-values are shown in Table II.

relative to carbohydrates (5.9 ‰) is identical to that found in recent plant material (*17*). Hence these fragments could originate from unmodified carbohydrates and lignin from primary sources (vascular plants).

Some phenol derivatives, e.g. 4-methyl-phenol, are relatively ^{13}C-enriched in comparison with the bulk of the phenols. This may indicate the origin of these pyrolysis products of the HS from secondary material, e.g. from newly built biomass of fungi or bacteria. This is consistent with independent observations, namely that carbohydrates from wood degrading fungi are enriched in ^{13}C relative to source wood (*18*).

Nitrogen isotopes in pyrolysis products. Nitrogen was the second element to be tested in the course of this study. Generally the nitrogen content in organic substances is small in comparison with carbon, and most of the nitrogen is located in amino groups of proteins or in structural polymers, e.g. chitin or keratin. While the pyrograms of plant biomass are dominated by polar N-containing pyrolysis products appearing in the early parts of the chromatogram (among others ammonia and acetonitrile, *11*), it is reported that pyrograms of humic material also contain pyrolysis products in the later part of the chromatogram (among others methyl-pyrroles, pyridines, pyrazine and nitriles, *11*, *36*). These nitrogen containing fragments may originate from amino sugars, e.g. chitin, from chlorophyll and other tetrapyrrols, amino acids or DNA bases. In principle they could have been synthesized by decomposers in the humification process and therefore are expected to be enriched in ^{15}N per trophic level by 3 ‰ (*37*). On the other hand they can selectively remain from source materials having unchanged δ^{15}N-values. Comparative analysis of the isotopic content of pyrolysis products from source materials and corresponding humic materials would support these informations.

At first we tested the system for fragmentation of nitrogen containing fragments by the pyrolysis conditions applied. For these experiments we used a natural derivative of a purine ring, caffeine, of known nitrogen isotopic content as model substance (*38*). 40 µg of the compound were pyrolysed. The δ^{15}N-value and retention time (rt) of the main peak obtained from the Py-GC-C-IRMS system (-1.2 ‰ ± 0.4 ‰) were identical to those obtained by the classical reference method (δ^{15}N = -1,45 ‰ in EA-IRMS). This indicates that caffeine was not fragmented but just evaporated. Therefore we pyrolyzed coffee powder and analyzed the caffeine peak. The obtained δ^{15}N-value +1.2 ‰ was in the range known for natural caffeine (*38*). These results prove that isotopic ratios from the system are reliable for components which can directly be evaporated from complex mixtures.

Secondly we pyrolysed standard chitin with a δ^{15}N-value of -1.25 ‰ ± 0.3 ‰. Figure 3 shows the m/e 28 and the 29/28 ratio trace of the Py-GC-C-IRMS device from the pyrolysis of standard chitin. So far, non of the pyrolysis products has been directly identified, but corresponding investigations performed by others (*39*), suggest acetamide to be the main peak (peak 5, Figure 3). The δ^{15}N-values determined so far are mostly in the range from -0.2 to -4.2 ‰ with standard deviations between 0.2 and 1.4 ‰ and a mean average for these peaks of -2.2 ‰. This value is in comparison to the high standard deviation identical to the standard

Table III. δ^{15}N-values and standard deviations from unidentified pyrolysis products of standard chitin. Peak numbers correspond to Fig. 3

Peak No.	δ^{15}N-value	s.d.
5	-8,9	3,0
6	-0,2	1,0
7	-2,0	1,4
8	-4,2	0,3
9	-1,2	0,8
10	-2,0	1,4
11	-2,6	0,6
13	-1,8	0,4
14	-3,7	0,2

44

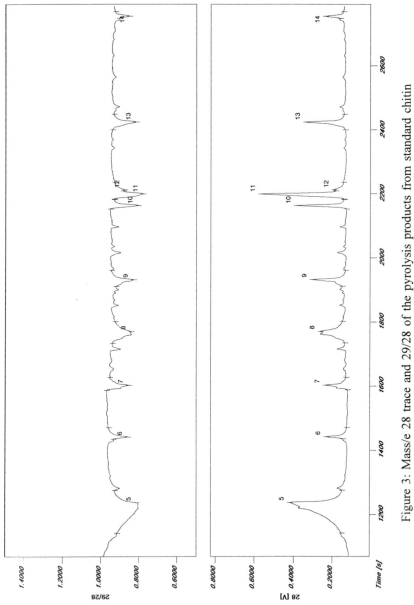

Figure 3: Mass/e 28 trace and 29/28 of the pyrolysis products from standard chitin after separation in the GC and conversion to N_2. $\delta^{15}N$-values of unidentified pyrolysis products see Table III.

chitin, but also a small isotope effect of the pyrolysis step could be implied. This will be investigated in future.

Conclusion

The system developed for on-line Py-GC-C-IRMS provided to be a valuable tool for the investigation of structural elements and genesis of complex macromolecules like aquatic humic substances. Comparative analysis of recent plant material and humic substances will help to elucidate relationships and information on HS formation. Furthermore the method will be useful to investigate intramolecular isotope distributions of defined compounds, and, applied to known precursors and products, it will help to elucidate reaction mechanisms and isotope effects.

Acknowledgments

This work is supported by the special research program "Refraktäre Organische Säuren in Gewässern" (ROSIG) from the DFG (Gl 626/1-4). Reference fulvic acids of the program were kindly provided by F. H. Frimmel and G. Abbt-Braun, Karlsruhe, Germany. The Py-GC/MS measurements and identification of pyrolysis products were kindly performed by H.-R. Schulten, Taunusstein, Germany. Excellent technical assistance from S. Danzer is gratefully acknowledged.

Literature Cited

1. Humic Substance II - Search of the Structure; Hayes, M. H. B.; MacCarthy, P.; Malcom, R. L.; Swift, R. S. Eds.; Wiley, Chichester, 1989.
2. *The Chemistry of Natural Products*; Thomson, R. H. Ed.; Blackie Academic & Professional: Glasgow, Great Britain, 1993.
3. Freeman K. H.; Boreham, C. J.; Summons, R. E.; Hayes, J. M. *Org. Geochem.* **1994,** *21*, 1037.
4. Ikan, R. et al. *Sci.Total. Environ.* **1992,** *117/118*, 1.
5. Opshal, S.; Benner, R. *Nature* **1997,** *386*, 480.
6. van Bergen, P. F.; Collinson, M. E.; Briggs, D. G. E.; de Leeuw, J. W.; Scott, A. C.; Evershed, R. P.; Finch, P. *Acta Bot. Neerl.* **1995,** *44*, 319.
7. International Humic Substance Society (IHSS) 1981, Standards and Reference Committee, Denver, Colorado.
8. McIntyre C.; Batts B. D.; Jardine, D. R. *J. Mass. Spectrom.* **1997,** *32*, 328.
9. Abbt-Braun, G.; Schmiedel, U.; Frimmel, F. H. *Vom Wasser* **1990,** *75*, 59.
10. Schulten, H.-R.; Leinweber, P. *J. Anal. Appl. Pyrolysis* **1996,** *38*, 97.
11. Schulten, H.-R,; Sorge-Lewin, C.; Schnitzer M. *Biol. Fertil. Soils* **1997,** *24*, 249.
12. Fründ, R.; Guggenberger, G.; Haider, K.; Knicker, H.; Kögel-Knaber, I.; Lüdemann, HD.; Luster, J.; Zech, W.; Spiteller, M. *J. Plant Nut. Soil Sci.* **1994,** *157*, 175.
13. *Techniques and Applications of Nuclear Magnetic Resonance Spectroscopy in Geochemistry and Soil Science;* Wilson, M. A., Ed.; Pergamon Press, Oxford, 1987.

14. Knicker H.; Lüdemann, H.-D.; Haider, K. *Europ. J. Soil Sci.* **1997**, *48*, 431.
15. Schulten, H.-R.; Plage, B.; Schnitzer, M. *Naturwiss.* **1991**, *78*, 311.
16. de Leeuw, J. W.; Hatcher, P. G. *Naturwiss.* **1992**, *79*, 331.
17. Schmidt, H.-L.; Gleixner G. In *Stable isotopes: Integration in Biological, Ecological and Geochemical Processes,* Griffiths, H., Ed., BIOS: Oxford, England, 1997.
18. Gleixner, G.; Danier, H.-J.; Werner, R. A.; Schmidt H.-L. *Plant Physiol.* **1993**, *102*, 1287-1290.
19. Eglinton, T. I.; Benitez-Nelson, B. C.; Pearson, A.; McNicol, A. P.; Bauer, J. E.; Druffel, E. R. M. *Science* **1997**, *277*, 796.
20. Freeman, K. H. *Science* **1997**, *277*, 777.
21. Spooner, N.; Rieley, G.; Collister, J. W.; Lander, M.; Cranwe, P. A. *Org. Geochem.* **1994**, *21*, 823.
22. Meuzelaar, H. L. C.; Haverkamp J.; Hileman, F. D. *Pyrolysis Mass Spectrometry of recent and fossil biomaterials*; Elsevier, Amsterdam, 1982.
23. Saiz-Jiminez, C. *Org. Geochem.* **1995**, *10*, 955.
24. Goñi, M. A.; Eglinton, T. I. *J. High Resol. Chromat.* **1994**, *17*, 476.
25. Pulchan, J.; Abrajano, T. A.; Helleur, R. *J. Anal. Appl. Pyrolysis* **1997**, *42*, 135.
26. Corso, T. N.; Brenna, J. T. *Proc. Nat. Acad. Science* **1997**, *4*, 1049.
27. Schmidt, H.-L. In *Messung von radioaktiven und stabilen Isotopen;* Simon, H., Ed.; Springer-Verlag, Berlin 1974, 291.
28. Barrie, A.; Davies, J. E.; Park, A. S.; Workman, T. C. *Spect. Int.* **1989**, *1*, 34.
29. Kornexl, B.; Werner, R. A.; Roßmann, A.; Schmidt, H.-L. *Z. Lebensm. Unters. Forsch.* **1997**, *205*, 19.
30. Craig, H. *Geochim. Cosmochim. Acta* **1957**, *12*, 133.
31. Schulten, H.-R.; Schnitzer M. *Soil Science,* **1992**, *153*, 205.
32. Schmidt, H.-L.; Kexel, H.; Butzenlechner, M.; Schwarz, S.; Gleixner, G.; Thimet, S.; Werner, R. A.; Gensler, M. In *Stable isotopes in the biosphere*; Wada, E.; Yoneyame, T. ; Minagawa, M.; Ando, T.; Fry, B. D., (Eds.); Kyoto University Press, Japan, 1995, 17.
33. O'Leary, M. H. In *Stable isotopes in the biosphere*; Wada, E.; Yoneyame, T. ; Minagawa, M.; Ando, T.; Fry, B. D., (Eds.); Kyoto University Press, Japan, 1995, 78.
34. Macko, S. A.; Helleur, R.; Hartley, G.; Jackman, P. *Org. Geochem.* **1990**, *16*, 1129.
35. Goñi, M. A.; Eglinton, T. I. *Org. Geochem.* **1996**, *24*, 601.
36. Marbot, R. *J. Anal. Appl. Pyrolysis* **1997**, *39*, 97.
37. Balzer, A.; Gleixner, G.; Grupe, G.; Schmidt, H.-L.; Schramm, S.; Turban-Just, S. *Archeometry* **1997**, *33*, 415.
38. Weilacher, T.; Gleixner, G.; Schmidt H.-L. *Phytochem.* **1996**, *41*, 1073.
39. Stankiewicz, B. A.; van Bergen, P. F.; Duncan, I. J.; Carter, J. F.; Briggs, D. E. G.; Evershed, R. P. *Rapid Commun. Mass Spectrom.* **1996**, *10*, 1747.

Chapter 4

Pyrolysis–Gas Chromatography for the Analysis of Proteins: With Emphasis on Forages

James B. Reeves, III, and Barry A. Francis

Nutrient conservation and Metabolism Laboratory, Agricultural Research Service, U.S. Department of Agriculture, Building 200, Room 124, BARC East, Beltsville, MD, 20705

Pyrolysis-gas chromatography has been shown to be very useful in the analysis of forages, particularly for the lignin and carbohydrate fractions. However, the same efforts have not shown a similar abundance of information on the proteins present, even for forages containing almost 30% protein.   Despite the fact that pyrolysis of isolated proteins produces an abundance of information, only small amounts of phenylacetonitrile, indole, methylindole, a methylphenol, dimethyl- or ethylpyrrole, and methanethiol are generally reported for forages. Reexamination of the subject indicates that many more products are produced, but have been missed due to the low levels found, or because many do not contain nitrogen and may have been assumed to originate from other sources, such as lignin. In conclusion, results indicate that more information on proteins is present in pyrograms of forages than has been recognized.

Ruminants, such as sheep and dairy cows, are often fed diets containing large amounts of lignocellulosic materials in the form of forages such as legumes (alfalfa, clovers, etc..) or grasses (orchardgrass, tall fescue, etc.). While ruminants, unlike humans and many other animals, are able to digest much of the lignocellulosic material present in these forages, they still, like humans and other animals, require a source of nitrogen for building proteins, nucleic acids, etc.. Thus, the composition of animal feedstuffs is extremely important both with respect to the nutritional well being of the animal and to the farmer trying to make a profit. Because of the importance of forage composition to animal nutrition, forages have been extensively studied in efforts to improve animal performance by improving the nutritional value of their feedstuffs. The same can be said for the assay procedures used to determine forage composition.

47

In addition, the role of nitrogen in animal diets, of which protein is the principle source, has taken on an importance beyond the nutritional well-being of the animal, due to the role of animal wastes in pollution of water sources, ground water, streams, etc. (*1*). Considerable research is being carried out on various components of the "feed to animal to waste" cycle shown in Fig. 1. As can be seen, organic forms of nitrogen, of which protein is the primary fraction, occupy a prominent position in this cycle. Considerable efforts are now being expended on understanding the various pathways and final disposition of the organic nitrogen fractions in the feed and manure. For example, at present, the rate and degree to which organic nitrogen in manure is converted to the inorganic forms useable by plants is largely unknown, with estimates ranging from 0 to 50% being converted on a time scale of a few months (Personal communications). With forages comprising a large fraction of both the feed used for growing animals and the subsequent manure produced, it is very important to be able to determine their composition and contribution to the various pathways in the cycle shown.

Due to the need for rapid and practical tests, most of the analyses carried out on the feeds and wastes involved in animal nutrition are based on empirical methods using extractive techniques, such as the extraction of fiber with neutral or acidic solutions of detergent (*2*). These measures have then been related to the nutritional status of ruminants (*3*). Similarly, the protein in forages is determined as total nitrogen, either by combustion (*4*) or by digestion (*5*) and the total nitrogen is converted to protein by the use of conversion factors (*5*). While some efforts have been made to study forage components in a more detailed fashion, such as the use of nitrobenzene oxidation to determine lignin composition (*6*), such efforts are hampered by the complex nature of the materials in question and the need for different methodologies for each of the individual components present (proteins, lignins, soluble carbohydrate monomers and polymers, insoluble carbohydrates, waxes, etc., *7*). It is easy to see how useful a single procedure capable of simultaneously determining all or many of the various components would be. While near-infrared reflectance spectroscopy has been extensively investigated for determining forage composition and is capable of determining many different components simultaneously, it functions by relating spectral information to other procedures, such as fiber determinations, in a correlative manner (*8*). The resulting relationships are then used to determine the composition of other similar samples for which only spectra are available. While this procedure is extremely useful where large numbers of similar samples are to be assayed, the need for developing calibrations for each component of interest and the required number of samples for such development make it less than ideal for research studies, where sample sets are often small and vary greatly in composition from one set to the next.

Pyrolysis-Gas Chromatography-Mass Spectrometry (PY-GCMS) would appear to be potentially an extremely useful tool for providing more detailed qualitative information on the composition of forages and wastes, without the need to conduct a multitude of different assays to obtain information on all the different components present. Indeed, it would allow the use of a single method for the study of the fate of the nitrogen fraction of forages as they pass from the feed, through the animal, and

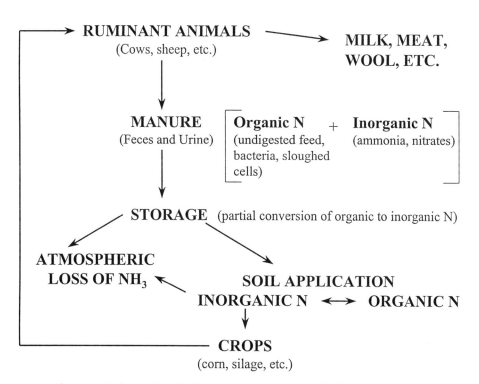

Figure 1. Pathways for distribution of nitrogen in production of ruminants.

subsequently to manure, etc.. (Fig. 1), while also providing information on the fate of the non-nitrogen components (carbohydrates, lignin, etc..). However, while PY-GCMS has been applied to the study of forages by many researchers (9-13), it would appear to have a serious weakness when it comes to the nitrogen fraction. The primary form of nitrogen in forages is protein, and very few products are reported in the literature with respect to nitrogen and forages. The objective of this report is to reexamine this problem in more detail.

Methods

All spectra presented were produced using a Chemical Data System Pyroprobe model 2000 equipped with an AS-2500 Pyrolysis Autosampler (Chemical Data System Inc, Oxford PA, USA). Pyrolysis was carried out at 600°C for 10 sec using sample aliquots of approximately 0.5 mg in quartz tubes. The pyrolyzate was swept directly into a Finnigan MAT GCQ gas chromatograph coupled to a Finnigan MAT GCQ ion trap mass spectrometer (Finnigan MAT, San Jose CA, USA). The gas chromatographic column was a DB-5ms (J&W Scientific, Inc, Folsom, CA, USA) (30 m x 0.25 mm i.d., 0.25 μm film thickness) operated from 50 to 300° C at 5 °C/min holding the initial temperature for 10 min. The injector temperature was 300° C. The PY/GC interface temperature was 200°C. Carrier gas (He) was held at a constant velocity of 40 cm/sec. with the split vent flow set to 100 ml/min. The GC/MS transfer line temperature was 300° C and the ion source temperature was 200° C. Mass spectra were obtained by electron impact at 70 eV from 10 or 40 to 650 *m/z* (1 scan/sec). Peak identification (Table 1) was based on mass spectral interpretation and published libraries of mass spectra of lignocellulose pyrolyzates (9-13). When using tetramethylammonium hydroxide (TMAH) the sample to be pyrolyzed was placed in the pyrolysis tube and approximately 5 μl of TMAH solution (25% by wt. in water) added. Finally all samples were finely ground using a WIG-L-BUG single ball mill (Crescent Dental MFG, Co. Lyons, IL, USA).

Pyrolysis Products Reported in the Literature for Forage Proteins

Previous efforts by the authors and results in the literature have reported only a few products as originating from the pyrolysis of the protein fraction of forages (9-13). Those being methanethiol from the sulfur containing amino acids, indole and methylindole from tryptophan, 4-methylphenol from tyrosine, phenylacetonitrile from phenylalanine, and derivatives of pyrrole from proline. However, even for samples containing a reasonable amount of protein (i.e, red clover hay with a protein content of 14.5%) several of these products were, if found at all, present in relatively small amounts, compared with other products produced from lignin or carbohydrates. In the case of methanethiol, resolution from the other low molecular weight products produced is also a problem, although the use of a different GC column or temperature program could probably be used to counteract this problem.

Considering that there are 20 amino acids commonly found in proteins, and for other components, such as lignin, it has been observed that even after chemical extraction of the majority of the lignin fraction by sodium chlorite (*14*) one can still find markers for lignin (*9*), the question remains as to what happens to all the products one might expect (quantitatively and qualitatively) from the pyrolysis of proteins in forages. The lack of protein pyrolysis products, and the relatively low amounts found for those reported, is especially bewildering when considered in light of the excellent correlations found between pyrolysis products and protein concentrations in a study of 67 different forage type materials using a packed column (*15*). While identification of the products was not carried out, correlations (R^2) of pyrolysis products and protein concentration for the 67 samples were found at the 0.94 level for single product correlations and at the 0.96 level using 3 products. Interestingly, on an individual feed basis (5 different feeds studied), the best results were achieved for the samples containing the least amount of protein (vegetative corn and wheat plants). While it is possible that the protein concentration was determined by difference (that which was not something else must be protein), it is extremely unlikely that such high correlations would have been found if that were the case. Another possibility is that a low molecular weight and highly volatile product or products is/are produced during the pyrolysis of proteins, such as ammonia or a low molecular weight amine. Such products may have been detected using the packed column and flame detector, but have been missed by mass spectrometry because of the instrument settings used. Generally, the first minute or two of the pyrogram is ignored, as well as fragments with m/z of less than 40, to eliminate interferences from all the impurities in the carrier gas, water in the samples, and low molecular weight material produced by pyrolysis of almost anything.

The question then remains: Why pyrolysis of forages does not produce more evidence of proteins? Possibilities included: 1) Proteins in general only produce low yields of relatively few products, 2) the products are being produced, but are not being recognized as originating from proteins or are being missed entirely, and 3) forages behave differently during pyrolysis than other protein containing materials due to the way the protein is bound within the plant cell wall (*16*).

Pyrolysis Products of Amino Acids

Since proteins are polymers composed of amino acids joined by amide bonds, one might expect that analysis of the products produced by pyrolysis of amino acids (*17*) might be useful in searching for unidentified protein-derived pyrolysis products. However, our data from the pyrolysis of amino acids have shown that the major products produced are often not the same as those produced during the pyrolysis of proteins. This can be seen by comparing the pyrograms in Fig. 2 for the amino acids L-asparagine (A) and L-cysteine (B) and the results found for gelatin in Fig. 3. While the major product produced by pyrolysis of L-asparagine (1H-pyrrole-2,5-dione) was found at 885 scans and for L-cysteine (2-methylthiazolidine) at 555 scans, there were no major products in the pyrogram of gelatin in these regions. Examination of the mass-spectral data confirmed the absence these two products in the gelatin pyrogram (Data not shown). Similar results are found for most amino acids when comparing results obtained with

Table1. Identification of pyrolysis products shown in Figures 2-8.

Peak #	Compound name	Origin[1]	Ions (m/z)[2]
1	unknown	P	44,34,18,45,19,17
2	unknown	A[3]	58,60,59,45,47,61
3	unknown	?	43,45,61,42
4	unknown	?	43,75,45,31,29,42
5	unknown	P	41,43,39,58,27
6	pyrazole[4]	P	68,41,39,28,40
7	pentanenitrile[4]	P	41,39,54,43,27,84
8	pyrrole[5]	P	67,39,41,40,38,28
9	toluene[5]	P	91,92,65,39,63,51
10	an ethylpyrrole[4]	P	80,95,78,53
11	2-furaldehyde[6]	C	95,39,96
12	a methylpyrrole[5]	P	80,81,41,53,39
13	ethylbenzene[5]	P?	91,106,65,51,77,92
14	2-furanmethanol[6]	C	98,41,42,53,81,39
15	styrene[5]	P	104,78,103,77,51,50
16	5-methyl-2(3H)-furanone[4]	C	70,98,55,41
17	phenol[6]	P,C,L	94,66,39,65
18	4-methoxytoluene[5]	P[3]	122,77,121,91,107
19	trans 1-propenylbenzene[4]	P	117,115,91,94,69,65
20	3-methyl-1,2-cyclopentanedione[6]	C	112,84,41,55,69
21	unknown	P	80,91,53,92,123,81
22	unknown	?[3]	126,55,83,67,39,97
23	guaiacol[6]	L	109,124,81
24	a methylphenol[5]	P,L	107,108,77,79,80,51
25	unknown	P[3]	84,113,28,26,85,114
26	unknown	?	43,44,57,39,29,117
27	unknown	P[3]	41,127,42,39,112,58
28	unknown	P[3]	139,39,54,53,110,82
29	phenylacetonitrile[5]	P	90,117,89,116
30	a dimethoxybenzene[4]	?[3]	77,138,95,123,65
31	4-methoxystyrene[4]	P?[3]	91,134,119,65
32	unknown	?	109,107,55,122,82
33	4-methylguaiacol[6]	L	138,123,95,67
34	unknown	P[3]	42,127,142,56
35	unknown	?[3]	142,127,39,41,83,113
36	unknown	P[3]	130,42,98,88
37	4-vinylphenol[5]	L	120,91,65,51
38	3-phenylpropanenitrile[4]	P	91,65,131,92,39,51
39	unknown	?	91,120,133,132,117,118
40	unknown	?	109,82,84,55,129
41	1-methylindole[4]	P[3]	131,130,89,103,77
42	4-ethylguaiacol[5]	L	137,152,122,91,77

Table 1. *Continued*

43	unknown	P[3]	39,54,82,139,53,108
44	indole[5]	P	117,90,89,63
45	4-vinylguaiacol[5]	L?	150,135,77,107
46	L-proline, 1-methyl-5-oxo-, methyl ester[4]	P[3]	98,41,42,70
47	2,6-dimethoxyphenol[6]	L	154,65,93,139,96,111
48	a dimethylindole[4]	P[3]	144,145
49	eugenol[5]	L	164,77,103,149,133
50	1,2,4-trimethoxybenzene[4]	L[3]	125,153,168,93,110,65
51	3-methylindole[5]	P	130,131,77,103,51
52	L-proline,5-oxo-, methyl ester[4]	P[3]	84,41,39,28,56
53	a methoxybenzene acetic acid, methyl ester[4]	?[3]	121,77,91,180
54	unknown	?	125,42,98,69,139,166
55	2,6-dimethoxy-4-methylphenol[5]	L	168,107,153,125
56	trans isoeugenol[5]	L	164,149,77,103,131,121
57	unknown	?[3]	101,144,45,75,88,129
58	unknown	?	91,65,104,182
59	2,6-dimethoxy-4-vinylphenol[5]	L	180,165,137,91,77,122
60	unknown	?	107,108,136,80,53,81
61	a dimethoxybenzoic acid, methyl ester[4]	?[3]	165,196,79,77
62	unknown	?	170,185,153,143,128,98
63	unknown	?	98,97,140,41,42,167
64	unknown	?	94,164,136,66
65	unknown	P?	156,129,199
66	trans 2,6-dimethoxy-4-propenylphenol[5]	L	194,119,179,77,131,151
67	unknown	P?	186,93,130,103,65,38
68	unknown	?	138,123,70,151
69	acetosyringone[5]	L	191,196,153,138
70	unknown	?	70,97,125,44,168
71	trans coniferyl alcohol[5]	L	137,91,124,180
72	syringylacetone[5]	L	167,210,123
73	unknown	?	83,111,41,68,55,154
74	unknown	?	67,95,81,41,55,123
75	unknown	?	67,81,95,41,123
76	unknown	?	94,190,134,162,106
77	unknown	A[3]	67,55,81,41,96
78	unknown	A[3]	43,55,87,143,74,101
79	unknown	P?	70,154
80	unknown	P?	70,194
81	unknown	?	87,41,55,129,73

[1]P=Protein, C=Carbohydrate, L=Lignin, A=Artifact.
[2]Masses listed in decreasing abundance.
[3]Formed in presence of tetramethylammonium hydroxide.
[4]Unconfirmed library search result.
[5]Matched with published results.
[6]Verified with authentic compound.

54

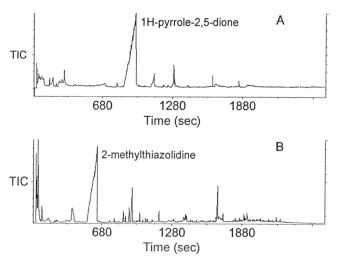

Figure 2. Pyrogram of L-Asparagine (A) and L-Cysteine (B) (TIC = Total ion current).

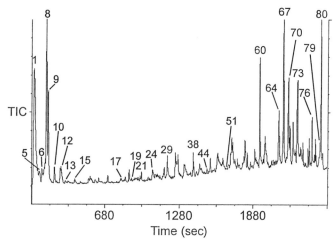

Figure 3. Pyrogram of gelatin with major peaks labeled as listed in Table 1 (TIC = Total ion current).

free amino acids to those obtain with proteins, and are due to the differences between free amino acids and the chemically bound forms found in proteins. When free, amino acids have the form:

$$NH_2-CH(R^1)-COOH$$

but in proteins they are bound as amide linked peptides in the form:

$$NH_2-CH(R^1)-CONH-CH(R^2)-CONH-CH(R^3)-CO .... NH-CH(R^X)\text{·}COOH$$

where $R^{(1, 2, 3..X)}$ are various aliphatic and aromatic moieties. It is easy to see why the products from proteins and amino acids could be different, with deamination and decarboxylation being more likely for amino acids, although some of the products such as indole and methanethiol are found for both materials (18). While the products resulting from amino acids might also be produced from free amino acids found in forages, the small amounts of free amino acids present means that such products would, at best, be minor contributors to forage pyrograms.

Pyrolysis of Isolated Proteins

Pyrolysis of isolated proteins of both plant (commercial soy powder) and non-plant origin (commercial gelatin) demonstrated that pyrolysis of proteins is not a problem in and of itself. As shown in Fig. 3., the pyrolysis of purified gelatin produces a multitude of products. These products are not unique to animal derived proteins as shown by the many similar products found (Fig. 4) for the pyrolysis of soybean extract (83% protein, 13% carbohydrate, 3% fat). Previously (18), we reported that while some of these products were also found in an alfalfa sample which contained 29.8% protein (Fig. 5), examination of the red clover hay pyrograms did not find any additional protein products.

As part of this effort, the results for both the alfalfa (Fig. 5) and the red clover hays (Fig. 6) were carefully reexamined. It was found that more of the protein-derived products were present in both pyrograms that previously reported by either the authors (18) or others (10-12), with 16 protein-derived products found for the alfalfa hay and 13 for the red clover hay. These results indicate that at least part of the explanation for the question of why reports on pyrolysis of forages do not show more protein-derived products, is simply a failure to recognize the products for what they are.

As shown in Fig. 6, forage pyrograms can be roughly divided into three regions: 1) The early scans (labeled matrix) dominated by low molecular products, such as acetic anhydride and carbon dioxide, which originate from a variety of sources, 2) a region dominated by products from carbohydrates, and 3) a region dominated by phenolic products from lignin. Note, however, that no single area contains products originating from proteins (13).

A few of the products (those eluting in the first two minutes) have not been reported because, due to the multitude of low molecular weight products produced during pyrolysis of almost any biological material, this early part of the pyrogram is generally ignored. One of these early products (Peak #1) appears to be a volatile, low molecular weight compound containing an amine group. This product was found by examining mass spectra for fragments with masses down to 10, something generally not

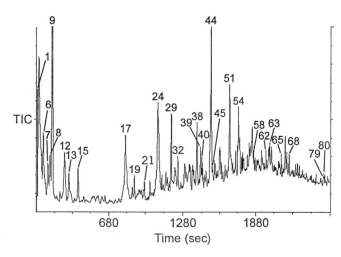

Figure 4. Pyrogram of soybean powder (protein content = 83%) with major peaks labeled as listed in Table 1 (TIC = Total ion current)

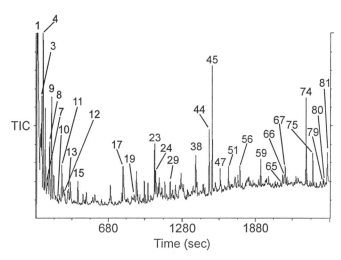

Figure 5. Pyrogram of an alfalfa hay, protein content = 29.8%, with major peaks labeled as listed in Table 1 (TIC = Total ion current).

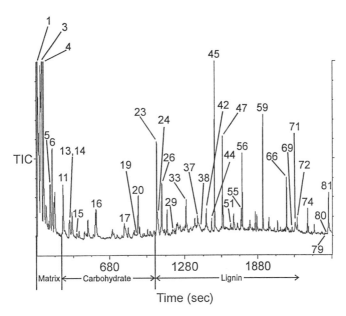

Figure 6. Pyrogram of a typical red clover hay, protein content = 14.5%, with major peaks labeled as to Table 1 (TIC = Total ion current).

done for pyrolysis of forage materials. It appears that this product is an amine or amide, rather than ammonia, but exact identification has not been possible due to the presence of many other low molecular weight products in the same region of the pyrogram. Other products, such as styrene and toluene, do not contain nitrogen and thus have either been ignored by the forage community or assumed, perhaps like other products, to originate from lignin, although other researchers have reported them as originating from protein (19).

However, these findings do not completely account for difficulties found when examining pyrolysates of forages for evidence of proteins. The alfalfa sample contained an unusually high protein content (29.8%) due to collection at a very early stage of growth, and yet only 16 out of 21 possible products (Table 1) were found. Normally alfalfa hay is harvested when the plants are more mature and thus has a protein content closer to 15% than 30%. Even for this high protein alfalfa, several of the products were present at low levels. For the red clover hay (14.5% protein), still fewer products were found, 14, and at even lower levels. Without a mass spectrometer, detection of many of these products would be impossible due to the small amounts present relative to the products generated from the non-protein components, indicating that the relative yield of products from proteins, as compared to the lignin and carbohydrate fractions of forages, may be a problem, and it is not that they have just been missed or ignored. The fact that the products originating from proteins are not found in one region on the pyrogram (Fig. 6) also complicates their determination.

Pyrolysis of Proteins in Other Materials and Possible Matrix Effects in Forages

While the detection of nitrogen compounds from the pyrolysis of forages may be a problem, this does not appear to be a problem for the pyrolysis of other materials. For example, it has been reported that up to 90% of the nitrogen fraction in soils is volatilized by pyrolysis resulting in many GC separable compounds (20, 21). Work by Stankiewicz and co-workers (19) showed similar results for proteins in invertebrate cuticles, with 44 compounds designated as originating from proteins. While some of these, such as vinylphenol, may also originate from lignin in forages and thus would not be distinctive markers for proteins, most were nitrogen-containing compounds for which this would not apply.

While van de Meent and co-workers (22), using synthetic constructed mixtures, reported the quantitative yield of total products from protein to be less than for carbohydrates or lignin, with relative responses of 1.00, 0.31 and 0.65, for carbohydrates, protein, and lignin respectively, even trace amounts of lignin in fractions extracted for their carbohydrates are easily detected (11). Thus, the products from a sample with a 14.5% protein content, such as the red clover hay shown in Fig. 6, should be easily detected, even with a relative yield of half that of lignin. Therefore, it appears that, at least for forages, the relative yield can be considerably less in natural samples.

One possible explanation for the lower than expected yield of protein pyrolysis products in the pyrograms of forages is that the pyrolysis products produced in the presence of the other forage components are either not the same, due to secondary reactions, as those produced when a pure protein is pyrolyzed (matrix effects), or are produced in lower amounts. Such reactions were reported by van de Meent and co-workers (*22*) for pyrolysis of mixtures of albumin, dextran and lignosulphonic acid. Van der Kaaden and co-workers (*23*) also reported that the presence of various salts during pyrolysis significantly changed the pyrogram of amylose. Forages and byproducts can contain significant amounts of ash (1.3% of the red clover hay) which might play a similar role.

As forage plants grow, the relative composition and morphological structure of the plant changes. With increasing maturity the relative amount of lignin in forages tends to increase, in order to give structural support to the plant, and the relative amount of protein decreases (*24*). Also, there appears to be an intimate structural relationship between lignin and protein in the various layers of plant cell walls, with protein being necessary for lignin disposition (*16*). Thus, one can imagine that with increasing forage maturity, the protein present becomes more and more intimately bound with lignin and other components of the cell wall, thus increasing the likelihood of matrix effects on product yield and possibly composition. Further efforts using a variety of samples will be needed to determine if this is indeed the case.

Preliminary Results Using Tetramethylammonium Hydroxide

Tetramethylammonium hydroxide is often used in conjunction with pyrolysis (often referred to as "thermally assisted hydrolysis and methylation gas chromatography" to obtain products not produced by pyrolysis without TMAH. For example, pyrolysis generally results in decarboxylation of carboxylic acids. With TMAH, the carboxylic acid is methylated, and thus the acid group can be retained (*12, 25*). In addition, as implied in the name "thermally assisted hydrolysis", the nature of the cleaving process is different (hydrolysis versus thermal cracking for pyrolysis without TMAH). The results for the soybean protein extract (pyrogram without TMAH shown in Fig. 4) and the red clover hay (pyrogram without TMAH shown in Fig. 6) using TMAH are shown in Figs. 7 and 8, respectively. As can be easily seen by comparing the results with TMAH (Figs. 7 and 8) with those without TMAH (Figs. 4 and 6), the products produced by the two different procedures are quite different in nature, as indicated by the differing retention times. Initial efforts indicate that the use of TMAH may be beneficial in obtaining more information on the proteins present in forages. Results indicate that both pyrolysis and thermally assisted methylation occurred, with both new products and also many of those produced during pyrolysis without TMAH found. At a minimum, seven additional protein derived products were found in the red clover pyrogram when TMAH was used, which were also found in similar pyrograms of gelatin and soybean protein extract. More may be present, but confirmation will require further work.

Summary

Pyrolysis-gas chromatography, with and without mass spectrometry, has been shown to be very useful in the analysis of forages, particularly for analysis of the lignin and

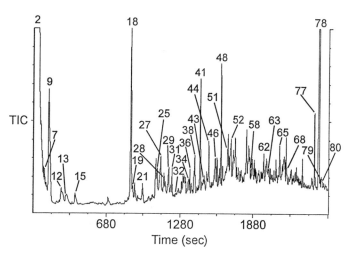

Figure 7. Pyrogram, using tetramethylammonium hydroxide, of soybean powder with major peaks labeled as listed in Table 1 (protein content = 83%, TIC = Total ion current).

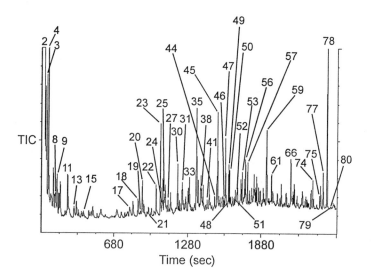

Figure 8. Pyrogram, using tetramethylammonium hydroxide, of a typical red clover hay, protein content = 14.5%, with major peaks labeled as listed in Table 1 (TIC = Total ion current).

carbohydrate fractions. However, the literature on the pyrolysis of forages does not report an abundance of information (qualitatively or quantitatively) for the proteins fraction in forages, despite the fact that pyrolysis of isolated proteins produces numerous products. Only small amounts of phenylacetonitrile, indole, methylindole, a methylphenol, methanethiol and dimethyl- or ethylpyrrole have generally been reported for forages. However, reports from the pyrolysis of isolated proteins and non-forage materials show an abundance of protein-derived products to be produced. A reexamination of this subject indicates the problem to be two-fold in nature: First, pyrolysis of forages often produces relatively small quantities of protein-derived products when compared with other components of forages (lignin, carbohydrates, etc..), isolated proteins or other protein containing material, such as invertebrate cuticles. And second, it is now apparent that many of the protein-derived products in forage pyrograms have simply not been reported, most likely due to the small amounts produced or because they do not contain nitrogen and were assumed to originate from other fractions, such as lignin. Efforts have also shown that the mass and scan cutoffs often used to avoid the compilation of the low molecular weight products produced during the pyrolysis of all biological materials (equivalent to eliminating the solvent front in other GC operations) results in some protein-derived products being missed. For example, examination of the first two minutes of pyrograms for masses down to 10 m/z found a low molecular weight compound containing an NH_2 group. It was also found that the use of TMAH results in a mixture of products containing both some of the products produced in the absence of TMAH and also new products from the use of the TMAH. Initial studies indicate that using TMAH may be of benefit to efforts to obtain more information on proteins in forage type materials. Finally, considering the importance of proteins and other nitrogen compounds in the evaluation of animal feeds (26), and the increasing interest in nitrogen compounds in the area of animal waste disposal and water pollution (1), further research and clarification is needed on the subject of the pyrolysis of forages and resulting protein-derived products.

REFERENCES

1.   Ross, C. C. (Ed.), *Proc. of the Seventh Inter. Sym. on Agric. and Food Processing Wastes*, Amer. Soc. Agric. Engin., 1995.

2.   Goering H. K.; Van Soest, P. J. *Forage fiber analysis (Apparatus, reagents, procedures, and some applications)*. Agric. Handbook No. 379, Agric. Res. Serv., USDA, Washington, DC, 1970.

3.   Van Soest, P. J *Nutritional Ecology of the Ruminant,* O & B Books, Inc., Corvallis, OR, 1982; pp. 230-248.

4.   AOAC Method 990.03, *Official Methods of Analysis*, AOAC Int., Gaithersburg, MD.

5.   AOAC Method 2.051, *Official Methods of Analysis*, AOAC Int., Gaithersburg, MD.

6.   Reeves, III, J. B. *J. Dairy Sci.* **1986**, 69, 71-76.

7.   Van Soest, P. J. *Nutritional Ecology of the Ruminant*, O & B Books, Inc., Corvallis, OR, 1982; pp. 75-138.

8.   Williams, P.; Norris, K. (Eds.), *Near-Infrared Technology in the Agricultural and Food Industries,* Amer. Assoc. of Cereal Chemists, Inc. St. Paul, MN, 1987.

9.   Reeves, III, J. B.; Galletti, G. C. *J. Anal. Appl. Pyrolysis*, **1993**, 24, 243-255.

10. Galletti. G. C.; Bocchini, P. *Rapid Commun. Mass Spectrom*, **1995**, 9, 815-826.

11. Ralph, J.; Hatfield, R. D. *J. Agric. Food Chem.*, **1991**, 39, 1426-1437.

12. Morrison, III, W. H.; Mulder, M. M. *Photochemistry*, **1994**, 35, 1143-1151.

13. Reeves, III, J. B.; Galletti, G. C. *Org. Mass Spectrom.*, **1993**, 28, 647-655.

14. Collings, G. F.; Yokoyama, M. T.; Bergen, W. G. *J. Dairy Sci.*, **1978**, 61, 1156-1160.

15. Reeves, III, J. B. *J. Dairy Sci.*, **1990**, 73, 2394-2403.

16. Iiyama, K.; Lam, T. B. T.; Meikle, P. J. ; Ng, K; Rhodes, D. I.; Stone, B. A., in H. G. Jung, D. R. Buxton, R. D. Hatfield and J. Ralph (Eds.), *Forage Cell Wall Structure and Digestibility*, American Society of Agronomy, Inc., Madison, WI, 1993, pp. 647-649.

17. Chiavari, G.; Galletti, G. C. *J. Anal. Appl. Pyrolysis,* **1992**, 24, 123-137.

18. Reeves, III, J. B.; Francis, B. A. *J. Anal. Appl. Pyrolysis*, **1997**, 40-41, 243-265.

19   Stankiewicz, B.A.; van Bergen, P. F.; Duncan, I. J.; Carter, J. F.; Briggs, D. E. G.; Evershed, R. P. J *Rapid Comm. Mass Spectrom.*, **1996**, 10, 1747-1757.

20. Saiz-Jimenez, C.; de Leeuw, J. W. *J. Anal. Appl. Pyrolysis*, **1986**, 9, 99-119.

21. Schulten, H.-R.; Sorge, C.; Schnitzer, M. *Biol. Fertil. Soils*, **1995**, 20, 174-184.

22. van de Meent, D.; de Leeuw, J. W.; Schenck, P. A. *J. Anal. Appl. Pyrolysis,* **1982**, 4, 133-142.

23. van der Kaaden, A.; Haverkamp, J; Boon, J. J.; de Leeuw, J. W. *J. Anal. Appl. Pyrolysis*, **1983**, 5, 199-220.

24. Reeves, III, J. B. *J. Dairy Sci.*, **1987**, 70, 1583-1594.

25. Challinor, J. M. *J. Anal. Appl. Pyrolysis*, **1996**, 37, 1-13.

26. Van Soest, P. J. *Nutritional Ecology of the Ruminant,* O & B Books, Inc., Corvallis, OR, 1982; pp. 230-248.

PROTIENS AND DNA: THEIR OCCURRENCE AND FATE IN THE BIO- AND GEOSPHERE

Chapter 5

Biogeochemistry of Organic Nitrogen Compounds

Jeffrey L. Bada

Scripps Institution of Oceanography, University of California at San Diego,
La Jolla, CA 92093–0212

Nitrogen containing organic compounds represent the second most abundant reservoir of nitrogen on the surface of the Earth. However, the organic compounds that make up this global nitrogen pool are not well characterized. Although amino acids and the nitrogenous bases of nucleic acids make up only a few percent of the total organic nitrogen reservoir, the geochemical reactions of these compounds have been extensively studied. Because hydrolysis reactions are rapid on the geologic time scale, both proteins and nucleic acids (DNA and RNA) are not preserved for more than 10^3 to 10^5 years in most environments. The racemization reaction of amino acids converts the L-amino acids present in the biosphere into a racemic mixture (D/L amino acid ratio = 1.0) in the geosphere in less than 10^6 years. Anhydrous conditions, such as those that may be associated with amber entombed insects, may retard both biopolymer hydrolysis and racemization. Condensation reactions between amino acids and sugars, including sugars at apurinic sites in nucleic acid fragments, likely result in the incorporation of these compounds into geopolymers such as humic acids. Although rearrangement reactions in geopolymers may scramble the original molecular structures, part of the global organic nitrogen inventory was originally derived from amino acids and nucleic acid bases.

The largest surface reservoir of nitrogen on the Earth after atmospheric N_2 is nitrogen-containing organic compounds (1). Although this organically bound nitrogen was originally produced by biological processes, in the geochemical environment a complex series of reactions have altered the biogenic organic nitrogen molecules and scrambled their original molecular architecture. As a result, the organic nitrogen in surficial reservoirs contains molecules with complex geochemical histories, ages and sources.

In living organisms the bulk of the organic nitrogen is in the form of protein bound amino acids (2). For example, over 50 % of the dry weight of an average *E. Coli* bacterial cell consists of protein. The ratio of protein to RNA is about 5, while the protein/DNA ratio is around 20 (2). The ratio of protein to other forms of organic nitrogen such as porphyrins is generally less than that of nucleic acids, with the exception of photosynthetic organisms in which the porpyhrin chlorophyl may be a significant organic nitrogen component.

An inventory of organic nitrogen containing molecules on the Earth's surface is given in Table I. As expected, amino acids do indeed constitute the bulk of the organic nitrogen in the biosphere, yet they make up only a few percent of the total organic nitrogen in the various geologic reservoirs. What molecules then account for

Table I. The inventory of organic nitrogen and amino acids on the surface of the Earth [based on (1, 3, 4)].

Reservoir	organic N $(10^{15}$ g N)	amino acids $(10^{15}$ g of AA)	amino acid N (AAN)*	% of AAN of org N
biosphere	10-14	<100	<15	~100
sediments	$4\text{-}6 \times 10^5$	$0.1\text{-}3 \times 10^5$	$0.2\text{-}5 \times 10^4$	0.3-12.5
soils	60-300	<10	<1.5	<0.5-2.5
ocean	200-500	20-30	3-5	0.6-2.5
polar ice	$(0.5\text{-}1)^+$	0.03	0.005	

* An average amino acid % N of 15 % was used in this estimate
+ Estimated based on a % AAN of organic N = 1-2 %

the majority of the organic-bound nitrogen in the various reservoirs outside the biosphere? The nitrogenous bases of nucleic acids (i. e., adenine, guanine, cytosine, thymine and uracil) can not account for the non-amino acid organic nitrogen because they are present in insignificant quantities in the largest reservoir, sediments [see (5), and references therein], and because several of the bases have limited stabilities on the geologic time scale (6). Other possible candidate compounds include porphyrins, amino sugars, amines and the organic nitrogen components of geopolymers such as humic substances and kerogen. Porphyrins derived from chlorophyl represent only roughly 10^{-2} % of the total organic carbon in sediments (7), so they can not account for the bulk of the organic surficial nitrogen inventory. Amino sugars, such as N-acetyl-glucosamine, are major components of chitin which appears to be exceptionally well preserved in some geologic deposits (for example, see the chitin chapters in This Volume), but amino sugars are about 10 times less abundant than amino acids in many marine sediments (8). This leaves organic nitrogen bound in geopolymers as the major component of the Earth's surficial organic nitrogen pool. Although most of the organic nitrogen components in geopolymers were originally derived from biogenic amino acids, amino sugars, etc., geochemical reactions have extensively altered the biologic molecular structures producing macromolecules which are unique to the geologic environment. In addition, ammonia released during the long-term diagenesis of organic material can become geochemically incorporated into kerogen (9), and this thus represents another way nitrogen can be sequestered in the global surficial organic nitrogen reservoirs.

It is not the intent of this brief review to consider the biogeochemistry of all the various components of the Earth's surficial organic nitrogen inventory. Rather, the focus will be on proteins and nucleic acids and their amino acid and base components. These organic nitrogen compounds were selected because of the essential role they have in biochemistry and because their reactions under geochemical conditions are fairly well established.

The Geochemical Fate of Proteins and Amino Acids

All known organisms use proteins to construct cell walls and other structural components and for biochemical catalysts. Only twenty different amino acids are incorporated into proteins during biosynthesis even though a huge number of different amino acids are theoretically possible, most of which have been synthesized in the laboratory. In addition, only L-amino acids are found in most proteins. The fate of this unique mix of biogenic amino acids can be traced as it is released from the biosphere and becomes incorporated into the other amino acid reservoirs on the surface of the Earth.

Protein Hydrolysis. Proteins are prone to natural hydrolysis reactions which limit their survival both *in vivo* and in geologic environments. Hydrolysis under geochemical conditions can take place by three competing reactions, with the simplest being the cleavage of an internal peptide bond [see (3) and (10) and references therein]. Peptide bonds containing certain amino acids such as aspartic acid and serine are more rapidly cleaved than ones containing valine and leucine. An internal aminolysis reaction, which yields a diketopiperazine (cyclic dipeptide), is a significant hydrolysis pathway for amino acids at the N-terminal position of proteins. This pathway becomes more significant as an overall hydrolysis pathway as internal hydrolysis produces an increasingly larger number of peptide fragments. The final reaction is the hydrolysis of an amino acid at the C-terminal position, and this pathway may be favored when the adjacent amino acid residue is strongly electron withdrawing. The importance of hydrolysis at the C-terminal position also increases in significance as internal hydrolysis fragments the original protein.

Because of hydrolysis, proteins are not preserved in most geologic environments for periods in excess of a few million years. In the carbonate matrix, proteins are nearly completely hydrolyzed to smaller peptides and free amino acids in about a million years on the ocean floor (11), and in around 10^5 years in surface environments (12). In bones, hydrolysis of the main protein component, collagen, is even more rapid and little intact collagen remains in bones after only $1-3 \times 10^4$ years, except for those in cool or dry depositional environments (13). Proteins present initially as minor components of bone are more somewhat resistant to hydrolysis than collagen and these may be preserved longer than collagen (14). However, one of these, osteocalcin, is still apparently degraded on time scales of 10^5 years in temperate environments (15).

These protein survival times imply that no detectable protein residues should be preserved in very old fossils such as dinosaur remains, although there have been reports suggesting otherwise (16, 17). Unless there are hydrolysis protection mechanisms, however, these claims should be viewed with skepticism. One apparent protection mechanism is dehydration, and proteins in amber entombed insects may be exceptionally well preserved even over time scales of 10^7 to 10^8 years (18, 19). In general, however, the prospects of retrieving from most types of ancient biogenic specimens protein fragments which have retained meaningful sequence information appears remote.

Amino Acid Degradation. Unless encapsulated in biominerals such as carbonate shells or in other protective substances like amber, free amino acids liberated during protein hydrolysis are rapidly consumed by organisms and thus do not persist for extended times in geologic environments. A good example of the efficiency of this biological uptake is the presence of low concentration of free amino acids in the oceans, especially in older deep waters (20). In biominerals, especially shells, free amino acids produced from protein hydrolysis are retained for extended geologic periods although they too eventually diffuse out of the mineral matrix and are lost. Because the mineral matrix is more porous, the amino acid content of bone decreases rapidly on a time scale of only 10^3 to 10^4 years.

Although catalysis such as metal ions, carbonyl compounds and oxygen greatly accelerate the non-biological decomposition of amino acids, in general chemical decomposition reactions, such as decarboxlyation, are extremely slow and would never take place on the surface of the Earth even over very long periods of geologic time (3). There are some exceptions, however. For example, asparagine and glutamine undergo rapid irreversible deamidation yielding aspartic and glutamic acids and ammonia. This reaction has a half-life at $37°C$ and neutral pH of a few days to years in peptides (21) and its occurrence in living mammals may play an important role in determining the *in vivo* life-time of proteins [see (22) for example]. Thus, asparagine and glutamine would not persist under geochemical conditions on the Earth's surface for any appreciable length of time. Other rapid chemical decomposition reactions involve the β-substituted amino acids, serine, threonine and cysteine, which decompose by β-elimination, reversible aldol cleavage and decarboxylation (11, 23). Of these three reactions, β-elimination dominates. Dehydration of serine yields racemic alanine as does the β-elimination of HS⁻ from cysteine. Racemic α-aminobutyric acid, an amino acid not found in proteins, is produced from threonine. These reactions are rapid, and occur in decades in living mammals (24), and over geologic periods of 10^3 to 10^6 years in the oceans and carbonate sediments (11, 23, 25). Another rapid geochemical reaction is the decomposition of arginine to urea and the non-protein amino acid ornithine (26). This reaction has also been observed in fossils 10^4 to 10^5 years old (27). Because their decomposition reactions are rapid on the geologic time scale, the presence of serine, threonine, cysteine or arginine in fossil specimens more than 10^6 years old (unless they are preserved in protective materials such as amber) is usually an indication that recent secondary amino acid contamination is present.

The non-protein amino acids β-alanine and γ–aminobutyric acid occur in sediments (28), but they are not present in biogenic carbonates such as foraminifrial tests (29), which suggests that they are not directly produced from the diagenesis of protein amino acids. Also, the β-decarboxylation of aspartic acid, which would yield β-alanine, has not been observed in heated aqueous solutions (30). The bacterial mediated degradation of aspartic and glutamic acids is apparently the source of β-alanine and γ-aminobutyric acid in sediments. Both of these amino acids appear to be very stable with respect to both chemical and biological destruction because they are the most abundant amino acids present in abyssal marine clay sediments (28).

Racemization. Amino acids both in the free and bound state undergo racemization, a reaction in which a pure amino acid enantiomer is converted into a racemic mixture [see (3) and references therein]. Thus, even though only the L-enantiomers of the amino acids are used exclusively in biochemistry, once isolated from active biological processes, racemization begins and increasing amounts of D-amino acids are produced with increasing isolation time. Amino acid racemization takes place rapidly under natural conditions, and as a result D-amino acids have been detected in the metabolically inert tissues of long-lived mammals, in fossils and in sediments. Theoretically, given enough time and without the introduction of more recent amino acid contaminants, amino acids isolated from active biological processes should eventually completely racemize (i. e., equal amounts of L- and D-amino acids would be present). Even in cold environments on the surface of the Earth where racemization rates are the slowest, amino acids should be totally racemized on time scales of a few million years. Thus, even though one of the most distinctive features of the biosphere is the presence of only L-amino acids, the amino acids in the other reservoirs on the Earth's surface consist of a mixture of both the D- and L-enantiomers. Given that the residence times of amino acids in the various reservoirs outside the biosphere are likely substantial (probably in the range of 10^3 to $>10^6$ years), the average D/L ratio on the surface of the Earth may be closer to the racemic value of 1 rather that the value of 0 which characterizes the biosphere.

There are exceptions with respect to the susceptibility of amino acids to undergo racemization. For example, the racemization rate of L-isoleucine at the β-carbon, which yields the non-protein amino acids L-alloisoleucine and D-isoleucine, is considerably slower than any other amino acid racemization reactions (31). The racemization of isoleucine at the β-carbon has a half-life of roughly 10 million years in surface fossils, and 100 million years in deep ocean sediments. Thus, the presence of L-alloisoleucine and D-isoleucine in very old fossil specimens such as dinosaur remains would be an excellent indicator that some ancient biomolecules are still preserved. Another exception to the rule is in amber entombed insects where only L-amino acids have been detected, even in specimens as old as 10^8 years (19, 32). This unusual preservation is apparently the result of the anhydrous nature of the amber matrix (amino acid racemization rates are greatly retarded in the absence of water).

The geochronological applications of the amino acid racemization reaction have been extensively investigated [see (3) and (13) and references therein]. The main limitation of this dating method is that the extent of racemization in a fossil is a function not only of time, but it also depends on variables such as temperature, humidity, extent of protein hydrolysis, the type of fossil matrix and the presence of secondary amino acid contaminants. However, when the amino acid racemization rate can be calibrated at a location using specimens whose ages have been determined by independent dating techniques, absolute racemization-based ages can be calculated. The racemization dating of carbonate fossils has an upper dating limit of about 10^5 to 10^6 years, and is particularly useful in dating carbonate fossils which are difficult or impossible to date by other methods. The dating of fossil bone material using racemization has proven to be more problematic than biogenic carbonates because the main protein initially present in the bone matrix, collagen, is prone to hydrolysis and the liberated peptides and the amino acids are not retained by the bone matrix. Because bones are an open system with regard to amino acid migration, secondary amino acids can easily become incorporated into the bone matrix. However, teeth, especially the enamel component, have been found to be an excellent material for the amino acid racemization-based dating of mammalian fossils specimens. The aspartic acid racemization has also been found to take place in living mammals, and the extent of *in vivo* racemization in teeth and the eye lens nucleous can be used to assess the biological age of many long-lived mammal species.

Condensation Reactions. Amino acids, and the free amino group of small peptides and proteins, undergo condensation with sugars (the free aldehyde form), a reaction generically known as the browning or Maillard reaction [see (33) and references therein]. Amino sugars would also be components in this reaction because they contain both aldehyde and amine components. The Maillard reaction has been extensively investigated because it occurs during the preparation of some foods and in the tissues of living mammals, and the reaction products may alter the properties of the affected proteins. The products include high molecular weight components (several thousand daltons), collectively known as melanoidins, which have some characteristics similar to humic acids, geopolymers which make-up a large fraction of sedimentary organic carbon. Maillard (34) first suggested early in this century that this sugar-amino acid condensation reaction was the source humic acids on the Earth, and more recent studies have tended to support this view (35, 36). However, there are concerns as to whether Maillard-type reactions could take place in the oceans, in other natural waters, in fossils and in sediments due to the low concentrations of amino acids and sugars which are present.

An important aspect of sugar-amino acid condensation is that this reaction provides a possible geochemical pathway for making geomacromolecules which contain organic nitrogen components originally derived from amino acids and amino sugars. The major difference between this geopolymer and proteins is that the carbon-nitrogen bond may be more stable than the peptide bond, especially when an aminoketose produced by the Amadori rearrangement is the major product. Thus, whereas proteins are rapidly hydrolyzed to free amino acids and small peptides, the amino acids (and amino sugars as well) bound in Maillard reaction products are apparently much more resistant to hydrolysis (37) and could thus be preserved over long geologic time periods. This resistance to hydrolysis thus could explain why the bulk of the organic nitrogen in surficial reservoirs outside the biosphere is bound in geopolymers possibly generated by Maillard type reactions.

The Geochemical Fate of Nucleic Acids and Their Component Bases

All organisms use nucleic acids for the storage of genetic information (generally, this is DNA although some viruses use RNA) and for the translation of this genetic information for protein synthesis (RNA). Both DNA and RNA thus could be very important biomolecules if they were preserved in the fossil record for extended periods of time. For example, diseases associated with RNA viruses could be traced historically from human remains and the genetic information in preserved DNA sequences could be used to evaluate the species affiliations of fossil specimens.

RNA Survival. Of the two nucleic acids used in biology, RNA is the most unstable because of the presence of the phosphodiester bond involving the 2'-hydroxyl group of ribose. The presence of active hydrolytic enzymes after cell death greatly limits the postmortem survival of RNA. In geologic environments, based on heating experiments at pH 7 using single-stranded viral RNA, the half-life for cleavage of the phosphodiester bond has been estimated to be <1000 years at 0°C (38). Thus, even if RNA molecules survive enzymatic degradation, intact RNA sequences would not be expected to survive in most natural environments for more than a few centuries. Some recent attempts to characterize viral RNA from human tissues preserved from a 1918 global flu epidemic have been successful (39), but the prospect of extending this type of study further back in time appears to be remote.

DNA Survival. DNA is also a fragile molecule and it has been suggested that DNA should not be preserved in geologic environments for more than 10^4 years (40-42). Oxidation of the DNA bases is one important reaction (41), especially with respect to the ability of the PCR (polymerase chain reaction) technique to correctly amplify DNA sequences (43). Oxidation reactions, however are not apparently directly involved in the hydrolysis of the DNA backbone itself.

Depurination, involving the hydrolysis of the deoxyribose/adenine or guanine bond, followed by rapid chain breakage, is thought to be the main reaction in the fragmentation of DNA in the geologic environment (40-42). The cleavage of the N-glycosyl bond is around 20 times faster for purines in comparison to pyrimidines at neutral pH. The release of the purines generates apurinic sites in the DNA chain. At the apurinic sites, the free aldehyde form of the 2' deoxyribose group is involved in a β-elimination reaction which rapidly breaks the DNA backbone at the 3'-phosphodiester bond of the apurinic sugar. Studies of the rate of depurination in aqueous solution (44) have provided the basis for the conclusion that under geochemical conditions DNA should be broken into fragments containing only a few hundred base-pairs in around 10^4 years in temperate environments.

Since the advent of the PCR technique for the routine amplification of trace amounts of DNA, there have been many reports on the isolation of DNA sequences from fossil specimens [for example, see (42)]. Some of these, for example the recovery of DNA from a ~$4x10^4$ year old Siberian mammoth (45) and from the $35x10^4$ year old "Ice Man" found in an Alpine glacier (46), are consistent with the expected geologic survival time of DNA, while others, such as the claim of preservation of meaningful DNA fragments in Miocene plant material (47) and in dinosaurs bones (48), clearly are not. There have also been conflicting reports on the successful amplification of DNA from insects preserved in amber as old as 10^8 years [for example, see (49-52)], which greatly exceeds the geologic time period estimated for DNA survival. The survival time of genetically meaningful DNA sequnces under geochemical conditions remains an area of considerable debate.

Predicting DNA Survival Using Amino Acid Racemization. Direct analyses for the presence of DNA sequences in fossil specimens could provide a means of evaluating DNA survival. However, because of the minute amounts of DNA that are predicted to be present (on the order of only a few thousand molecules or less), this is not technically feasible at the present time. An alternative method for judging DNA survival is required, and amino acid racemization seems to provide the bases for such a technique. The rates and activation energies of depurination and aspartic acid racemization at neutral pH are nearly identical over the temperature range of $45°$ to $120°C$ (53). This similarity in the kinetic parameters of the two reactions lead to the suggestion that amino acid racemization can be used to monitor DNA survival indirectly (53). Recent studies of the extent of racemization and the recovery of DNA fragments indicate that when the D/L aspartic acid ratios are less than 0.1 to 0.15, meaningful DNA sequence information is preserved (32, 54). This level of aspartic acid racemization predicts that DNA should only be preserved in geologic environments for periods of a few thousand years in temperate regions and no more than around 10^5 years in colder, high latitude areas. These survival periods are remarkably similar to those estimated from the laboratory based depurination studies of DNA (43, 44).

It is now recommended that before DNA amplification is attempted, especially on valuable human remains, the extent of aspartic acid racemization be evaluated in order to determine the prospect that the sample contains endogenous DNA. This approach was recently used in the successful recovery of DNA from a Neanderthal skeleton (55). On the other hand, the racemizaiton test has not helped settle the issue of DNA survival in amber entombed insects. The D/L aspartic acid ratios of insects in amber are essentially unchanged from those in modern insects (19, 32, 53), which appears to support the claims that meaningful DNA sequence information are preserved in these ancient specimens (49, 50). Recent efforts of verify that DNA is indeed preserved for extraordinary periods of geologic time in amber entombed organisms have failed, however (52).

Degradation of Nucleic Acid Bases. Compared to amino acids, the bases of nucleic acids liberated by reactions such as depurination and depyrimidation have limited geochemical stabilities. For example, cytosine, adenine and guanine all undergo loss of their exocyclic amino groups (deamination) at rates which are rapid on the geologic time scale (6). The products are uracil from cytosine, hypoxanthine from adenine and xanthine from guanine. Hypoxanthine and xanthine are also geologically unstable and decompose at rates which are comparable to those of adenine and guanine. The reaction involving cytosine is the fastest (6), and it is unlikely that this base would persist in geologic environments for more than a few thousand years. In nucleic acids, cytosine deamination also takes place, although the rate is estimated to be much slower than those for depurination and depyrimidation (56). Thus, in DNA sequences preserved in fossil specimens deamination may have converted some cytosine residues to uracil.

Hypoxanthine and xanthine are not found in biospheric nucleic acids, but they have been found to increase in concentration in recent lake sediments over time scales of 10^3 to 10^4 years, the result of adenine and guanine deamination (57). Because of the instability of cytosine, adenine and guanine under geochemical conditions, their reported presence (and the absence of hypoxanthine and xanthine) in Pliocene and Miocene sediments (5) could indicate the input of recent biogenic contamination, although there may be preservation mechanisms for these bases that are presently largely unknown (see below).

Condensation Reactions. Because three (cytosine, adenine and guanine) of the five bases in DNA and RNA have exocyclic primary amino groups, they can undergo Maillard-type condensation reactions with sugars (58), and thus become incorporated into melanoidin or humic acid-like compounds. However, compared to amino acids, this purine-pyrimidine/sugar condensation reaction has not been extensively investigated and it is not known whether the three bases can be incorporated into geopolymers and thus perhaps be stabilized with respect to decomposition. It is also possible that DNA fragments in which depurination has taken place may be incorporated into geopolymers by Maillard-type reactions. The free aldehyde form of the deoxyribose residues at the apurinic sites in DNA can react with primary amines (59), and this "nicking" of DNA at apurinic sites provides a potential geochemical reaction pathway for attaching amino acids and peptides to DNA sequences. The products of this amino acid-apurinic site reaction, especially ones which contain Amadori rearrangement components and nucleic acid sequences rich in pyrimidines, could have enhanced geologic stabilities because the β-elimination reaction at the apurinic site could be substantially retarded. Whether these kind of reaction products would yield amplifable DNA sequences when used in the PCR technique is not known, but it is conceivable that nucleic acid sequences bound in geopolymers could generate false DNA positives in samples where no meaningful DNA fragments are actually preserved.

Conclusions

Our understanding of the geochemical reactions involving proteins and nucleic acids, and their component amino acids and bases, is extensive. This information can in turn be used to predict the fate and survival of these molecules once they exit the biosphere and become part of the other surifical organic nitrogen reservoirs on the Earth. A comparable level of understanding of the other organic nitrogen compounds would enhance our knowledge of the cycling, fate and storage of nitrogen containing organic molecules on the surface of the Earth.

References

1. Jaffe, D. A. In *Global Biogeochemical Cycles*; Butcher, S. S., Charlson, R. J., Orians, G. H., Wolfe, G. V., Eds; Academic Press: New York, NY, **1992**; pp. 263-284.
2. Brock, T. D.; Smith, D. W.; Madigan, M. T. *Biology of Microorganisms*; Prentice-Hall, Englewood Cliffs, NJ, **1984**; p. 133.
3. Bada, J. L. *Phil. Trans. R. Soc. Lond. B* **1991**, *333*, 349.
4. Bada, J. L.; Brinton, K. L. F.; McDonald, G. D.; Wang, X. In *Circumstellar Habitable Zones*; Doye, L., Ed.; Travis House Publications, Sunnyvale, CA, **1996**; pp. 299-304.
5. Shimoyama, A.; Hagishta, S.; Harada, K. *Geochem. J.* **1988**, *22*, 143.
6. Levy, M.; Miller, S, L. *Abstracts 213th National Meeting Am. Chem. Soc.* **1997**, *GEOC*, XXX.
7. Eckardt, C. B.; Keely, B. J.; Waring, J. R.; Chicarelli, M. I.; Maxwell, J. R. *Phil. Trans. R. Soc. Lond. B* **1991**, *333*, 339.

8. Belluomini, G.; Branca, M.; Calderoni, G.; Schnitzer, M. *Org. Geochem.* **1986**, *9*, 127.
9. Schimmelmann, A.; Wintsch, R. P.; Lewan, M. D.; DeNiro, M. J. This volume.
10. Collins, M. J.; Walton, D.; King, A. This volume.
11. Bada, J. L. ; Man, E. H. *Earth Sci. Revs.* **1980**, *16*, 21.
12. Serban, A.; Engel, M. H.; Macko, S. A. *Org. Geochem.* **1987**, *13*, 1123.
13. Bada, J. L. *Ann. Rev. Earth Planet. Sci.* **1985**, *13*, 241.
14. Masters, P. M. *Geochim. Cosmochim. Acta* **1987**, *51*, 3209.
15. King, K.; Bada, J. L. *Nature* **1987**, *281*, 135.
16. Gurley, L. R.; Valdez, J. G.; Spall, W. D.; Smith, B. F.; Gillette, D. D. *J. Prot. Chem.* **1991**, *10*, 75.
17. Muyzer, G.; Sandberg, P.; Knapen, M. H. J.; Vermeer, C.; Collins, M.; Westbroek, P. *Geology* **1992**, *20*, 871.
18. Cano, R. J.; Boruck, M. K. *Science* **1995**, *268*, 1060.
19. Wang, X. S.; Poinar, H. N.; Poinar; G. O.; Bada, J. L. In *Amber, Resinite and Fossil Resin*, Anderson, K. and Crelling, J. C., Eds.; ACS Symposium Series 617, American Chemical Society: Washington, D. C., **1995**; pp. 255-262.
20. Lee, C.; Bada, J. L. *Earth Planet. Sci. Letts.* **1975**, *26*, 61.
21. Robinson, A. B.; Rudd, C. J. *Curr. Top. Cell. Regul.* **1974**, *8*, 247.
22. Harding, J. *Adv. in Protein Chem.* **1985**, *37*, 247.
23. Bada, J. L.; Ho, M.-S.; Man, E. H.; Schroeder, R. A. *Earth Planet. Sci. Lett.* **1978**, *41*, 67.
24. Masters, P. M. *Calcif. Tissue Int.* **1985**, *37*, 236.
25. Bada, J. L.,; Hoopes, E. A. *Nature* **1979**, *282*, 822.
26. Murray, K.; Rasmussen, P.; Neustaedter, x. x.; Luck, J. M. *J. Biol. Chem.* **1965**, *240*, 705.
27. Galatik, J.; Galatik, A.; Blazej, A. *Chemicke Listy* **1988**, *82*, 623.
28. Schroeder, R. A.; Bada, J. L. *Earth Sci. Revs.* **1976**, *12*, 347-391.
29. Schroeder, R. A. *Earth Planet. Sci. Letters* **1975**, *25*, 274.
30. Bada, J. L.; Miller, S. L. *J. Am. Chem. Soc.* **1970**, *92*, 2774.
31. Bada, J. L.; Zhao, M.; Steinberg, S.; Ruth, E. *Nature* **1986**, *319*, 314.
32. Poinar, H. N.; Hoss, M.; Bada, J. L.; Pääbo, S. *Science* **1996**, *272*, 864.
33. Ledl, F.; Schleicher, E. *Angew. Chem. Int. Ed. Engl.* **1990**, *29*, 565.
34. Maillard, L. C. *C. R. Acad. Sci.* **1913**, *156*, 148.
35. Hoering, T. C. *Carnegie Inst. Washington Yearbook* **1973**, *72*, 682.
36. Hedges, J. I. *Geochim. Cosmochim. Acta* **1978**, *42*, 69.
37. Abelson, P. H.; Hare, P. E. *Carnegie Inst. Washington Yearbook* **1969**, *68*, 297
38. Miller, S. L.; Orgel, L. E. *The Origins of Life on Earth*; Prentice-Hall: Eaglewood Cliffs, NJ, **1974**.
39. Taubenberger, J. K.; Reid, A. H.; Krafft, A. E.; Bijwaard, K. E.; Fanning, T. G. *Science* **1997**, *275*, 1793.
40. Pääbo, S.; Wilson, A. C. *Curr. Biol.* **1991**, *1*, 45.
41. Lindahl, T. *Nature* **1993**, *362*, 709.
42. Poinar, H. N. This volume.
43. Pääbo, S.; Irwin, D. M.; Wilson, A. C. *J. Biol. Chem.*. **1990**, *265*, 4718.
44. Lindahl, T.; Nyberg, B. *Biochem.* **1972**, *11*, 3610.
45. Taylor, P. G. *Mol. Biol. Evol.* **1996**, *13*, 283.
46. Handt, O.; Richards, M.; Trommsdroff, M.; Kilger, C.; Simanalen, J.; Georgiev, O.; Bauen, K.; Stone, A.; Hedges, R.; Schaffner, W.; Utermann, G.; Sykes, B.; Pääbo, S. *Science* **1994**, *264*, 1775.
47. Golenberg, E. M.; Giannasi, D. E.; Clegg, M. T.; Smiley, C. J.; Durbin, M.; Henderson, D.; Zurawski, G. *Nature* **1990**, *344*, 656.
48. Woodward, S. R.; Weyand, N. J.; Bunnell, M. *Science* **1994**, *265*, 1229.
49. Desalle, R.; Gatesy, J.; Wheeler, W.; Grimaldi, D. *Science* **1992**, *257*, 1933.

50. Cano, R. J.; Poinar, H. N.; Roubik, D.; Poinar, G. O. Jr. *Med. Sci. Res.* **1992**, *20*, 619.
51. Lindahl, T. *Cell* **1997**, *90*, 1.
52. Austin, J. J.; Ross, A. J.; Smith, A. B.; Fortey, R. A.; Thomas, R. H. *Proc. Roy. Soc. Lond. B* **1997**, *264*, 467.
53. Bada, J. L.; Wang, X. S.; Poinar, H. N.; Pääbo, S.; Poinar, G. O. Jr. *Geochim. Cosmochim. Acta* **1994**, *58*, 3131.
54. Cooper, A.; Poniar, H. N.; Pääbo, S.; Radovcic, J. Debenath, A.; Caparros, M.; Barroso-Ruiz, C.; Bertranpetit, J.; Nielson-Marsh, C.; Hedges, R. E. M.; Sykes, B. *Science* **1997**, *277*, 1021.
55. Krings, M.; Stone, A.; Schmitz, R. W.; Krainitzki, H.; Stoneking, M.; Pääbo, S. *Cell* **1997**, *90*, 19.
56. Lindahl, T.; Nyber, B. *Biochem.* **1974**, *13*, 3405.
57. Dungworth, G.; Thijssen, M.; Zuurveld, J.; Van der Velden, W.; Schwartz, A. W. *Chem. Geol.* **1977**, *19*, 295.
58. Lee, A. T.; Cerami, A.; *Mutation Res.* **1990**, *238*, 185.
59. Pierre, J.; Laval, J. *J. Biol. Chem.* **1981**, *256*, 10217.

Chapter 6

The Geochemical Fate of Proteins

M. J. Collins[1], D. Walton[2], and A. King[1]

[1]Fossil Fuels and Environmental Geochemistry (Postgraduate Institute), NRG, Drummond Building, University of Newcastle upon Tyne, Newcastle upon Tyne NE1 7RU, United Kingdom
[2]Division of Earth Sciences, University of Derby, Kedleston Road, Derby DE22 1GB, United Kingdom

Proteins represent the major pool of nitrogen in the biosphere, but undergo rapid biodegradation and efficient microbial recycling so that few recognizable remnants survive into the geological record. If these remnants become incorporated into geopolymers, then their origin becomes increasingly difficult to identify. Despite the relative instability of proteins, sites which exclude enzymolysis reveal that proteins are sufficiently robust to survive in a recognizable form in many burial environments over millennia, and many of their constituent amino acids for millions of years. Decomposition occurs *via* a series of reactions, the most rapid of which are sulphydryl oxidation, deamidation and dehydration, but some of these reactions can be slowed significantly by reducing the conformational flexibility of the protein. Ultimately, amino acids will decompose completely to produce light hydrocarbons, the pattern of which, in some biominerals, is consistent with a protein origin.

There are an estimated 3,000 different proteins coded within a bacterial genome and up to 80,000 in the human (*1*). The importance of proteins stems from the modular nature of their composition. By reacting together a base (amino group) with a carboxylic acid it is possible to build up long chains of amino acids held together by stable (peptide) bonds. Each of the twenty or so amino acids possesses a different side chain, allowing a wide variety of structural and functional molecules to be constructed (see below). Nitrogen found within the peptide bone and some side chains, represents the major pool of organic nitrogen in the biosphere (*2*). Since the amounts of bioavailable inorganic nitrogen (principally as NH_4^+ and NO_3^-) are often limited, organic nitrogen and (hence protein) is efficiently recycled. Nevertheless, as Abelson (*3*) revealed over forty years ago some proteins lost to the geosphere are left as recognisable remnants in the form of trace amounts of amino acids found in sediments and fossils. The purpose of this chapter is to explore what is known of the rates and fates of the geochemical decomposition of proteins.

The Geochemistry of Organic Nitrogen

Most nitrogen available to the biosphere exists as highly un-reactive gaseous dinitrogen (N_2), which is difficult to fix biologically. The net effect is that nitrogen may be a limiting nutrient and bioavailable forms are heavily recycled (*4*). Although the amide bond within amino acids is very stable over biological time-scales (the estimated half-life of Gly-Gly bond at 15°C is 5,000 years, *5*) the modular structure of proteins and the requirement for biologically available nitrogen promotes efficient enzymatic recycling (proteolysis). The efficiency with which protein, and specifically protein nitrogen, is recycled is indicated by the low levels of free amino acids in most environments (*2*) and suggests that very little protein will survive the microbial mill in a recognisable form.

Attempts to construct mass balances of nitrogen or total organic matter in sediments usually observe a rapid decline in the proportion of biochemically recognisable nitrogen (e.g. *6, 7*) - Schnitzer's 'Group I nitrogen (*8*)' - coupled indirectly to an increase in biochemically unrecognisable (Group II) forms. In almost all published studies of sedimentary or soil nitrogen Group II nitrogen accounts for some 33-50% of the total N (e.g. *6, 9-15*). The proportion of recognisable (Group I) amino-N is also related to the age and oxidation history of the sample. In soil, amino-N constitutes 31 - 41% of total N (*8*), but the percentage is lower in shallow shelf sediments 20-55% (*8, 16-19*) and lowest of all in deep ocean sediments (*6, 11, 20*). It is interesting to note from studies of the C/N ratio of organic matter, a general increase in the proportion of nitrogen during the process of 'humification' of soils and a corresponding decrease during the same process of polycondensation in marine sediments, such that despite very different starting ratios, the C/N ratio in both soil and sediments is surprisingly similar (*21*). The process of 'humification' and the characterisation of nitrogen from the resulting high molecular weight geopolymers are of great significance, and the subject of a number of chapters (*22, 23, 24*).

Fate of Proteins

Most proteins are synthesised in small quantities, where they perform roles as catalysts, pumps and other attendant machinery of cellular operation. Structural proteins form extracellular networks or fibres and have a relatively simple, highly repetitive, structure (*25*); they are often very abundant, and may exist in an almost pure form (e.g. tendon is almost 100% type I collagen). Their properties and ease of isolation has made for their widespread exploitation by man. The use of collagen, (leather, parchment and glue), keratin, (wool), fibroin (silk), can be traced back over many millennia, from historical records and even preserved prehistoric artefacts.

Proteins are defined in terms of their function, and thus the variably degraded 'proteins' identified in ancient samples are not, strictly speaking, proteins at all. Type I collagen because it comprises over 90% of bone protein, is the most widely studied ancient 'protein', and can be used as an example. Following transcription, proto-collagen molecules are formed, which then undergo a variety of modifications, such as the excision of signal peptides, and changes to specific residues, to produce the functional collagen triple helix (*26*). Collagen or other 'proteins' extracted from archaeological bone or other sources are very highly modified and degraded; to describe them using a biochemical term which relates to a functional biomolecule is misleading. We would advocate the use of quotation marks to distinguish 'collagen' extracted from ancient samples from the functional molecule.

Analytical Approaches. As proteins degrade in the geosphere, biochemical techniques, such as immunoassays or stains specific for proteins (*27*), become increasingly controversial (e.g. *28*) as a means of detection and analysis. However, the value of these biochemical techniques in geochemical protein extracts has never been systematically explored.

An illustration of the problems encountered by the geochemist can be made using the most common approach for the detection and quantification of ancient 'proteins' and their breakdown products, namely the analysis of amino acids and by ignoring all but the initial step, acid hydrolysis. Some peptide bonds (such as those containing aspartic acid (Asp) are particularly acid labile (*29*), while others containing more hydrophobic residues required more aggressive chemical attack. Hydrolysis is a subtle balance between the need to release the maximum yield of amino acids and the defunctionalisation and polymerization caused by prolonged exposure to hot concentrated acid. Nitrogen-rich so called 'highly resistant biopolymers' isolated from bacteria which have been claimed as the major of organic nitrogen to the geosphere (*30*) are now thought to be, at least in part, artefacts of prolonged acid hydrolysis (*31*). This finding has ramifications for the whole of the selective preservation hypothesis (*32*) following the finding that similar patters of aliphatic hydrocarbon pyrolystes can be derived from melanoidins prepared in the presence of lipids (*33*).

Amino acid containing geopolymers are more resistant to hydrolysis than proteins (*2, 34*), and still yield amino acids after multiple hydrolyses. A further complication is that yields of amino acids also increase when sediment is digested with hydrofluoric acid (*11, 35*), suggesting that an association with sediments impedes the efficiency of either hydrolysis or extraction. Despite the complications associated with hydrolysis of geopolymers, however, it is rare in geochemical analyses to see more than one hydrolysis time reported, or steps taken to remove the mineral phase. The extent by which the observed loss of amino acids is related to inefficient extraction probably merits further investigation.

Exceptional Preservation. For proteins or amino acids to survive in the geological record they have to escape microbial recycling (*36*). Sites which retard the rate of microbial decay are those which retard biodegradation or somehow preclude the action of proteolytic enzymes or the enzymes themselves. Biodegradation rates are reduced at environmental extremes, although in such environments there are usually organisms adapted to exist here. Man has developed a number of food preservation techniques based upon these such as pickling and salting, however the two most efficient solutions are dehydration and cooling. Water is the essential medium for biochemical reactions and microbial decomposition will not occur in a totally anhydrous environment. It is hard to find environments totally lacking in chemically reactive water, but some of the most spectacular organic preservation occurs in environments with limited chemically available water, such as the tundra (37), arid deserts (38), amber (2) and tar pits (39).

The rates of chemical reactions and microbial metabolism are reduced at low temperatures, although some micro-organisms are still active below 0°C (40). Waterlogged (anoxic) sites seem to have improved organic preservation, suggesting low levels of microbial activity (41). It is not clear why anoxic conditions enhance preservation, as a wide variety of heterotrophic anaerobes exist, but it would seem that complete mineralisation rarely occurs in the absence of oxygen (see below).

The importance of exceptional preservation is as an information resource, albeit one which is poorly exploited, in particular in understanding the fate of proteins in environments from which microbial action is excluded. The most commonly investigated environment is proteins trapped within both biologically mediated (42), and presumably also, biologically induced minerals; the calcified skeletons of molluscs and foraminifera having formed the basis of most studies (43). Most proteins within biominerals are rapidly lost (in decades to tens of millennia, depending upon ultra-structure) but a small fraction appears to be trapped over much longer time scales (44). These so-called intra-crystalline proteins, operationally defined as those which survive prolonged chemical oxidation, may represent 0.97 Mt C yr^{-1} entering the geosphere or 0.7% of the annual C input (45). The dissolution of carbonate or leaching of amino acids which as been observed in deep sea biominerals (46) could provide potential substrate for subsurface micro-organisms (6). These biomineral entrapped proteins enable investigations of decomposition within semi-closed systems, although the kinetics of decomposition are complicated by the failure to distinguish between inter- and intra-crystalline fractions.

Mineralised collagen represents a wholly inter-crystalline system, which is therefore very prone to leaching and contamination. Non-mineralised collagen such as tendon and skin is rapidly biodegraded, its survival in most cases is as a mineralised tissue, where it is preserved as the mineral protects from enzymolysis (47). Thus the observation that significant 'collagen' is present over many millennia (Figure 1), indicates that the molecule is chemically robust over these time-scales in these burial environments.

Bone represents an insignificant input of protein into the geosphere, but pores within sediment grains offer a similar type of protection, on a globally significant scale (48). A number of studies have indicated a strong correlation between organic carbon and mineral surface area (48, 49), approximating to 0.75g organic carbon m$^{2-1}$. Sorption to mineral surfaces appears to slow down the rate of enzymolysis (50-53), but it is totally excluded within pores whose throat size is too small to permit their entry (estimated to be 8 nm diameter, 48). The protection mechanism is not absolute and much lower loadings are observed in abyssal sediments exposed to prolonged oxidation; see (54) for a synthesis of this work. Further work is required to characterise the sorbed organic matter, but intriguingly, archaeological pot sherds, which might be considered the ideal sites for sorption of organic matter, have yielded more lipids than proteins (55).

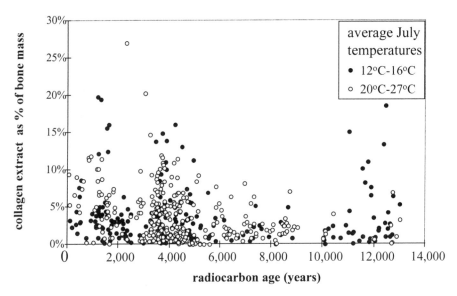

Figure 1. Comparative collagen yields, from low and high latitude sites. Yields are very variable, but high collagen yields are possible over long time-scales. Note the considerable scatter of data and the somewhat higher collagen yields in bones of >10,000 years at cooler latitudes. Data with kind permission from R. E. M. Hedges.

Protein Decomposition

Proteins can be viewed at three levels of organisation; structure, sequence and amino acid composition. The information available to a biochemist declines with decreasing levels of organisation, from complex folded quaternary and tertiary structure, *via* repeated secondary structural motifs to sequence and ultimately the composition of the amino acid building blocks themselves.

Loss of Three-dimensional Structure. Changes in the higher order structure can occur relatively easily in most proteins, leading to a loss of function. It would seem highly improbable that tertiary or higher level organisation would survive over geological time, in anything but structural proteins, such as collagen (see above). In the latter case, lack of solubility in water (and hence retention of the triple helix) is almost certainly a pre-requisite of long-term survival. It has been claimed that haemoglobin extracted from Neolithic stone tools were sufficiently well preserved to permit extraction and re-crystallization (*56*), but this claim is highly controversial (*57*).

Certain chemical transformations are so rapid as modify high-level structure and hence to seriously hamper the working life of proteins *in vivo*. One of the most unstable residues (and therefore one of the most widely studied) is asparagine (Asn), in particular when it is adjacent to a serine (Ser) or glycine (Gly) residue within a flexible region of a polypeptide chain. The deamidation of Asn promotes structural changes to the proteins, leading to loss of function (*58*). Isomerization of Asp *via* a cyclic succinimide (*59*) must be a significant physiological problem, as a repair enzyme has been isolated (*60*). Other alterations which alter tertiary structure include non-enzymatic glycation, which leads to alteration in long-lived proteins such as collagen (*61*) and hemoglobin (*62*) and dephosphorylation and hydrolysis both of which are observed in dentine phosphophoryns (*63*). Such structural modifications will hamper biochemical methods (such as protein sequencing) that have been proposed for the analysis of ancient 'proteins' (*64*).

We can learn something about the factors which will accelerate the loss of structure by considering their prevention. Modern chrome tanning of leather, for example, increases the stability of collagen a variety of different mechanisms, the most effective of which is the introduction of covalent cross-links between adjacent polypeptides, which reduces the flexibility of triple helices and fibres (*65*). A more subtle biotechnological approach has been adopted in an attempt to increase the robustness of bio-engineered proteins, namely the structural characterisation of proteins operating in extreme environments. Bacteria, known collectively as extremeophiles, survive in conditions which normally would be harmful to protein structure (thus life itself) include extremes of salinity, pH, pressure, irradiation and temperature (*66*). Of these conditions, it its those enzymes from temperature tolerant (thermophilic) organisms which have received the most attention. Comparison of thermophilic enzymes with related proteins in organisms which lack tolerance to high temperatures reveals surprisingly subtle differences which have the effect of reducing conformational flexibility (*67*) thereby reducing the rate of decomposition reactions such as deamidation of asparagine (*68*).

One extreme environment in which some proteins cannot only operate but which confers enhanced stability is non-aqueous solvents (*69*). Since hydrophobic residues of the protein core would become exposed if the protein denatured, organic solvent systems would seem intuitively to promote denaturation. The enhanced stability of proteins would appear to arise because the solvent system impedes the volume increase which occurs during protein unfolding. One may suspect that other configurations which hinder unfolding, such as mineralised collagen, may have similarly enhanced thermal stability (*70*), indeed it is interesting to note that hydrolysis of collagen does not occur within unstable regions of sequence (see below) but regions of greatest flexibility within the collagen triple helix itself (*71*).

Intriguingly the protein for which the greatest claims of long-term stability have been made (*72, 73*) and the only ancient 'protein' which has been sequenced (*74*), is osteocalcin, a small protein, so flexible that when in aqueous solution it does not give a single NMR structure (*75*). Claims for the exceptional preservation of this molecule rest on the assumption that once sorbed to apatite the protein is very stable, but this hypothesis has not been tested experimentally.

Peptide Bond Decomposition. The next step in decomposition, following the loss of three dimensional structure is loss of the sequence itself. The peptide bond linking the monomeric units in a protein is variably stable and degrades hydrolytically *via* three mechanisms: internal cleavage (*76*; preferentially at bonds containing Asp, Ser and Thr), aminolysis (likely to be prevalent at the N-terminal position, *77*) and amide bond hydrolysis at the C-terminus (*78*) poorly constrained. The rate of hydrolysis depends on the chemical characteristics of the amino acids on either side of the bond, available water and temperature.

Most studies of the influence of the nature of the residues on either side of the peptide bond on hydrolysis are conducted at extremes of pH. These investigations were particularly common in the 1950's when selective chemical hydrolysis was being investigated as an aid to protein sequencing (*79*). Aspartic acid containing bonds are prone to hydrolysis in mild acid, while at neutral pH Asn-Ser (*80*) and Asn-Pro (*81*) bonds are prone to spontaneous decomposition.

At neutral pH, the formation of a diketopiperazine and its subsequent hydrolysis at the up stream bond leads to a foreshortening of the polypeptide by two residues (*2*). There has been relatively little investigation of this topic despite its importance not only in terms of peptide bond hydrolysis, but also in racemization as the planar diketopiperazine can undergo much more rapid interconversion from a D to an L enantiomer than a highly constrained peptide (*82*).

Water promotes the rapid degradation of proteins by hydrolysis. Preservation of protein nitrogen is better if the protein is discrete from any large included source of water (*83*). Even in protected environments such as shell and bone, there may be water present up to several percent by mass (*84, 85*). Decomposition of some amino acids also yields water (see below), which would be available to continue the degradative process.

The temperature of the site of preservation controls the overall rate of any of the hydrolysis reactions. The activation energy of peptide bond hydrolysis is estimated to be 26-28 kcal mole^{-1} (*86-88*) which equates to the rate of hydrolysis increasing by an order of magnitude with a 20°C rise in temperature. Laboratory estimates of the rate of hydrolysis from eggshell proteins (*88*) predict rates in other biominerals such as deep sea foraminifera (*89*) and Upper Cenozoic brachiopods (*90*) with surprising accuracy, but the rate of release of Asp from coral proteins rich in Asp and Ser are an order of magnitude faster (*91*). Dentine proteins rich in these two residues are unstable (*92*), presumably in part due to the high number of Asx-Ser bonds prone to succinimide formation and cleavage (*80*). Unstable sequences in acidic proteins from biominerals may have a functional role in mineral growth, the 'suicide motifs' promoting chain scission which in turn alters the mineral binding characteristics of the polymer.

The total amino acid concentration declines in fossil biominerals, either due to leaching from the skeleton or further diagenetic alteration. There is commonly a residual fraction of bound amino acids (e.g. *81, 89, 90*), although it is not yet clear what these bound residues represent - residues within diagenetic polymers, or unusually stable peptide bond pairs or oligopeptides

Amino Acid Decomposition

The most common method of assessing the stability of amino acids has been by heating aqueous solutions of the pure amino acids (pyrolysis experiments) (*92, 93*). The kinetics of decomposition of are complicated by multiple pathways of reaction (Figure 2), the presence of other amino acids, as well as other inorganic and inorganic consitituents. Mixing amino acids has been shown to increase the rate of their decomposition indicating some inter-reaction of the samples (*92*), as has the presence of calcium carbonate (*94, 95*). The products of pyrolysis reactions have generally been identified by their reaction to ninhydrin (e.g. *92, 93*) or by derivatization to *n*-butyl N-trifluoracetic acids (*95*). Through this methodology, only a limited number of the potential degradative products may be recognised.

Decomposition Pathways. The major pathways of initial degradation of amino acids are discussed elsewhere (*2*), the importance of such studies being to understand the long term stability of the compounds, (*98-100*), the origin of decomposition products such as α-aminobutyric acid (*96*) and the exploitation of reaction kinetics for dating and other purposes, (e.g. racemization, *2*; oxidation, *101*). The use of amio acid oxidation for dating is a novel application which uses the sulfur amino acid cystine, which is abundant in the structural protein keratin. Oxidation of cystine to cysteic acid in wool carpets and other textiles over the historical past offers a potential means of authentication, although the free radical mechanism of oxidation is catalysed by both *uv* irradiation and the presence of metal oxides (*102*) which may mean that the rate is highly variable.

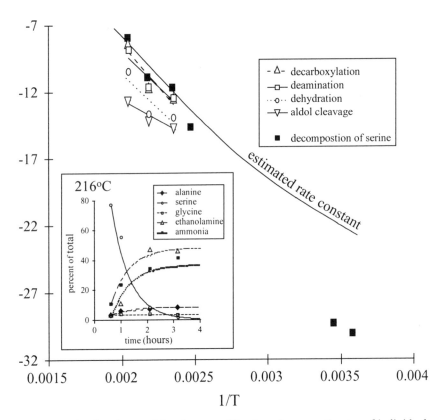

Figure 2. Predicted rate of Ser decomposition based upon estimates of individual reaction rates from high temperature experiments (an example is shown in the inset; all data from, *92*). Differences in the activation energies of different reactions leads to an curvilinear estimate of ln *k* vs. 1/T. However, comparison of the *overall* rates of decomposition of Ser in biominerals (*92, 93, 96, 97*) with experimental results suggests a much more linear Arrhenius relationship.

Hydrocarbon Generation. The ultimate products of decomposition, hydrocarbons (*103*), have received little attention in the past thirty years. The hydrocarbon products of complete defunctionalisation would be a range of C_1 - C_5 straight and branched chain alkanes, plus benzene and alkyl benzenes. Thompson and Creath (*104*) reported on the alkane distribution in a variety of fossil shells, which they attributed to complete defunctionalisation; certainly the distributions observed are remarkably similar to those predicted from the original amino acid composition. Most of these light hydrocarbons have multiple amino acid origins; *n*-pentane and i-pentane are of interest because the former derives only from lysine (Lys), the latter from either isoleucine (Ile) or leucine (Leu). In shells, the ratio of *n*- / *i*-pentane may reach values as high as 7.5 (*104*), whereas in sediments the ratio is close to 1 but does appear to rise slightly with maturity (*105*). In an experiment in which the protein ovalbumin was pyrolysed at different temperatures (*106*) the yield of pentane increases markedly with temperature, suggesting a reaction with a high activation energy. This in turn may explain the observed distribution of fatty acids in Cretaceous mollusc shells (*34*), the distribution of which was consistent with deamidation of only four amino acids, Asp, and the three pentane progenitors (Leu, Ile, Lys).

The most likely explanation for the elevated levels ratio of *n*- / i-pentane is that within shells the hydrocarbons derive from proteins, whilst in sediments they are not a significant source. As only one study has reported on the *i*-pentane/*n*-pentane ratio from an amino acid source (*104*) and no isotopic measurements have been made, it is not clear how useful this ratio will be in exploring the diagenesis of proteins, but it may be worthy of further investigation.

Conclusions

Despite much work on the detection of protein remnants in archaeological materials and fossils - which indicate that if biodegradation is excluded, macromolecular structure can survive many millennia, and amino acids for billions of years - the overall rates and fates of protein decomposition in the geosphere are poorly understood. This stems in part from the widespread use of biochemical approaches which may be at best semi-quantitative and worst inappropriate and from a limited knowledge of the decomposition of amino acids themselves. It also reflects the complex geochemistry of nitrogen containing heteropolymers in the geosphere, which have proved difficult to characterise. The importance of mineral surfaces as the sites of their concentration is a developing theme for carbon preservation and will certainly have a role to play in the geochemistry of nitrogen. Advances in spectroscopy and isotopic analysis outlined in this volume are offering new tools with which to unpick the Gordion Knot. However, the limit of our understanding can be expressed by the inability to conclusively disprove even the most extreme claims of the 'exceptional preservation' of ancient 'proteins'.

References

1. OBrien, S. J.; Wienberg, J.; Lyons, L. A. *Trends Genet.* **1997**, *13*, 393.
2. Bada, J. L. This volume.
3. Abelson, P. H. Carnegie Inst. Wash. Yb. **1954**, 53, 97.
4. Wang, X. C.; Lee, C. *Geochim. Cosmochim. Acta* **1995**, *59*, 1787.
5. Qian, Y. R.; Engel, M. H.; Macko, S. A.; Carpenter, S.; Deming, J. W. *Geochim. Cosmochim. Acta* **1993**, *57*, 3281.
6. Parkes, R. J.; Cragg, B. A.; Getliff, J. M.; Harvey, S. M.; Fry. J. C.; Lewis C. A.; Rowland, S. J. *Mar. Geol.* **1993**, *113*, 55.
7. Nguyen, R. T.; Harvey, H. R. This volume.
8. Schnitzer, M. In *Humic Substances in Soil, Sediment and Water*; Aitken, G. R.; McKnight, D. M.; Wershaw, R. L.; MacCarthy, P. Eds.; Wiley-Interscience: New York, **1985**; pp 303-325.
9. Bremner, J. M. In *Soil Biochemistry*; McLaren, A. D.; Peterson, G. H., Eds.; Marcell Dekker, **1967**; pp 19-66.
10. Keeney, D. R.; Konrad, J. G.; Chesters, G. *Water Pollution Control Fed.* **1970**, *42*, 411.
11. Stevenson, F. J.; Tilo, S. N. In *Advances in Organic Geochemistry*; Hobson, G. D., Speers, G. C., Eds.; Pergamon Press: Oxford, **1970**, pp 237-264.
12. Kemp, A. L. W.; Murchrochova, A. A. *Limnol. Oceanogr.* **1972**, *17*, 855.
13. Kemp, A. L. W.; Murchrochova, A. A. *Geochim. Cosmochim. Acta* **1973**, *37*, 2191.
14. Sowden, F. J.; Griffith, S. M.; Schnitzer, M. *Geochim. Cosmochim. Acta* **1977**, *42*, 1524.
15. Rosenfeld, J. K. *Amer J. Sci.* **1981**, *281*, 436.
16. Rosenfeld, J. K. *Limnol. Oceanogr.* **1979**, *24*, 1014.
17. Burdige, D. J.; Martens, C. S. *Geochim. Cosmochim. Acta* **1988**, *52*, 1571.
18. Haugen, J. E.; Lichtentaler, R. *Geochim. Cosmochim. Acta* **1991**, *55*, 1649.
19. Yamoto, S.; Ishiwatari, R. *Sci. Tot. Env.* **1992**, *117/118*, 279-292.
20. Whelan, J. K. Geochim. Cosmochim. Acta **1977**, 41, 803.
21. Tyson, R. V. *Sedimentary Organic Matter*; Chapman & Hall: London, **1994**, pp 1-615.
22. Knicker, H.; Kögel-Knabner, I. This volume.
23. Schimmelmann, A.; Wintsch, R. P.; Lewan, M. D.; DeNiro, M. J. This volume.
24. van Bergen, P. F.; Flannery, M. B.; Poulson, P. R.; Evershed, R. P. This volume.
25. Lesk, A. M., 1991, *Protein architecture : a practical approach*, Oxford, IRL Press, 287 pp.
26. Kielty, C. M.; Hopkinson, I.; Grant, M. E. In *Connective tissue and its heritable disorders. Molecular, genetic and medical aspects*; Royce, P. M.; Steinmann, B., Eds.; Wiley-Liss: New York, 1993, pp 103-147.

27. Schweitzer, M. H.; Marshall, M.; Carron, K.; Bohle, D. S.; Arnold, E. V.; Barnard, D.; Horner, J. R.; Starkey, J. R. *Proc. Natl. Acad. Sci. USA,*, **1997**, *94*, 6291.
28. Child, A. M.; Pollard, A. M. *J. Arch. Sci.,* **1991**, *19*, 39.
29. Oliyai, C.; Borchardt, R. T. *Pharm. Res.* **1993**, *10*, 95.
30. Le Berre, F.; Derenne, S.; Largeau, C.; Connan, J.; Berkaloff, C. In *Organic Geochemistry Advances and Applications in Energy and the Natural Environment*; Manning, D. A. C., Ed.; University Press: Manchester, pp 428-431.
31. Allard, B.; Templier, J.; Largeau, C. *Org. Geochem.* **1998** *in press*.
32. Tegelaar, E. W.; de Leeuw, J. W.; Derenne, S.; Largeau, C. *Geochim. Cosmochim. Acta* **1989**, *53*, 3103.
33. Larter, S. R.; Douglas, A. G. *Geochim. Cosmochim. Acta* **1980**, *44*, 2087.
34. Hare, P. E.; Hoering, T. C. *Carnegie Inst. Wash. Yb.* **1977**, *76*, 625.
35. Belluomini, G.; Branca, M.; Calderoni, G.; Schnitzer, M. *Org. Geochem.* **1986**, *9*, 127.
36. Eglinton, G.; Logan, G. *Phil. Trans. R. Soc. Lond. B* **1991**, *333*, 315.
37. Prager, E. M.; Wilson, A. C.; Lowenstein, J. M.; Sarich, V. M., *Science* **1980**, *209*, 287.
38. Bland, H. A.; van Bergen, P. F.; Carter, J. F.; Evershed, R. P. This volume.
39. McMenamin, M. A.; Blunt, D. J.; Kvenvolden, K. A.; Miller, S. E.; Marcus, L. F.; Pardi, R. R. *Quat. Res.* **1980**, *18*, 174.
40. Brenchley, J. E. *J. Ind. Microbiol. Biotech.* **1996**, *17*, 432.
41. Reference - fossil
42. Berman, A.; Addadi, L.; Weiner, S. *Nature* **1988**, *331*, 546.
43. Robbins, L. L.; Muyzer, G.; Brew, K. In *Organic Geochemistry Principles and Application*; Engel, M. H.; Macko, S. A., Eds.; Topics in Geobiology 11; Plenum Press: New York and London, 1993; pp 799-816.
44. Sykes, G. A.; Collins, M. J.; Walton, D. I. *Org. Geochem.* **1995**, *23*, 1059.
45. Collins, M. J.; Westbroek, P.; Muyzer, G.; de Leeuw, J. W. *Geochim. Cosmochim. Acta* **1992**, *56*, 1539.
46. King, K. Jr. In *Biogeochemistry of amino acids*; Hare P. E.; Hoering T. C.; King K. Jr., Eds.; Wiley: New York, **1980,** pp 377-391.
47. Krane, S. In *Dynamics of Connective Tissue Macromolecules*; Burleigh P. M. C., Poole A. R., Eds.; North Holland Publishing Company: Amsterdam, **1970**, pp 309-326.
48. Mayer, L. M. Geochim. Cosmochim. Acta **1994**, 58,1271.
49. Keil, R. G.; Tsamakis, E.; Fuh, C. B.; Giddings, J. C.; Hedges, J. I. *Geochim. Cosmochim. Acta* **1994**, *58*, 879.
50. Romanowski, G.; Lorenz, M. G.; Wackernagel, W. *Appl. Environ. Microbiol.* **1991**, *57*, 1057.
51. Ensminger, L. E.; Gieseking, J. E. *Soil Sci.* **1941**, *48*, 467.
52. Keil, R. G.; Montlucon, D. B.; Prahl, F. G.; Hedges, J. I. *Nature* **1994**, *370*, 549.

86

53. Nagata T.; Kirchman D. L. *Mar Ecol. Prog. Ser.* **1996**, *132*, 241.
54. Hedges, J. I.; Keil, R. G. *Mar. Chem.* 1995, *49*, 81.
55. Evershed, R. P.; Tuross, N. *J. Arch. Sci.* **1996**, *23,* 429.
56. Loy, T. H. *Science* **1983**, *220*, 1269.
57. Remington, S. J. *Science* **1994**, *266*, 298.
58. Garza-Ramos, G.; Degomezpuyou, M. T.; Gomezpuyou, A.; Yuksel, K. U.; Gracy R. W. *Biochemistry* **1994**, *33*, 6960.
59. Geiger, T.; Clarke, S. *J. Biol. Chem.* **1987**, *262*, 785.
60. Clarke, S. *Ann. Rev. Biochem.* **1993**, *54*, 479.
61. Brennan, M. *J. Biol. Chem.* **1989**, *264*, 20947.
62. Acharya, A. S.; Roy, R. P.; Dorai, B. *J. Biol. Chem.* **1991**, *10*, 345.
63. Fujisawa, R.; Kuboki, Y.; Sasaki, S. *Archiv. Biochem. Biophys.* **1985**, *243*, 619.
64. Curry, G. B. In *Molecular evolution and the fossil record*; Runnegar, B., Schopf, J. W., Eds.; Short courses in Paleontology Paleontological Society: Knoxville, Tennessee, Vol. 1, **1988**; pp 20-33.
65. Heidemann, E. *Fundamentals of leather manufacture*; Eduard Roether KG: Roetherdruck, Darmstadt, 1993.
66. Kristjansson, J. K.; Hreggvidsson, G. O. *World J. Microbiol. Biotechnol.* **1995**, *11*, 17.
67. Daniel, R. M.; Dines, M.; Petach, H. H. *Biochem. J.* **1996**, *317*, 1.
68. Tomizawa, H.; Yamada, H.; Wada, K.; Imoto, T. *J. Biochem.* **1995**, *117*, 635.
69. Volkin, D. B.; Staubli, A.; Langer, R.; Klibanov, A. M. *Biotechnol. Bioengineer.* **1991**, *37,* 843.
70. Lees, S. In *Calcified tissue: topics in molecular and structural biology*; Hukins, D. W. Ed.; London: MacMillan Press, **1989**, 153-173.
71. Muller, H. T.; Heidemann, E. *Das Leder* **1993**, *44,* 69.
72. Ulrich, M. M.; Perizonius, W. R. K.; Spoor, C. F.; Sandberg, P.; Vermeer, C. *Biochem Biophys. Res. Comm.* **1987**, *149*, 712.
73. Muyzer, G.; Sandberg , P.; Knapen, M. H. J.; Vermeer, C.; Collins, M. J.; Westbroek, P. *Geology* **1992**, *20*, 871.
74. Huq, N. L.; Tseng, A.; Chapman, G. E. *Biochem Int.* **1990**, *21*, 491.
75. Atkinson, R. A, Evans, J. S.; Hauschka, P. V.; Levine, B. A.; Meats, R.; Triffitt, J. T.; Virdi, A. S.; Williams, R. J. P. *Eur. J. Biochem.* **1995**, *232*, 515.
76. van Kleef, F. S. M.; de Jong, W. W.; Hoenders, H. J. *Nature* **1975**, *258*, 264.
77. Steinberg, S. M.; Bada, J. L. *J. Org. Chem.* **1983**, *48*, 2295.
78. Kahne, D.; Still, W. C. *J. Am. Chem. Soc.* **1988**, *110*, 7529.
79. Hill, R. L. *Adv. Prot. Chem.* **1965**, *20*, 37.
80. Tyler-Cross, R.; Schirch, V. *J. Biol. Chem.* **1991**, *266,* 22549.
81. Capasso, S.; Mazzarella, L.; Sorrentino, G.; Balboni, G.; Kirby, A. J. *Peptides* **1996**, *17*, 1075.
82. Mitterer, R. M.; Kriausakul, N. *Org. Geochem.* **1984**, *7*, 91.

83. Towe, K. M. In *Biogeochemistry of amino acids*; Hare, P. E., Hoering, T. C. King, K. Jr., Eds.; John Wiley and Sons: New York, **1980**, pp 65-74.
84. Hudson, J. D. *Geochim. Cosmochim. Acta* **1967**, *31*, 2361.
85. Gaffey, S. J. *J. Sediment. Petrol.* **1988,** *58*, 397.
86. Hare, P. E.; Miller, G. H.; Tuross, N. C. *Carnegie Inst. Wash. Yb.* **1975,** *74*, 609.
87. Qian, Y. R.; Engel, M. H.; Macko, S. A., Carpenter S.; Deming J. W., *Geochim. Cosmochim. Acta* **1993**, *57*, 3281.
88. Miller, G. H.; Beaumont, P. R. Jull, A. J. T.; Johnson, B. *Phil. Trans. Roy. Soc. Lond. B* **1992,** *337*, 149.
89. Müller, P. J. *Meteor. Forsch. Ergebn.* **1984**, *38*, 25.
90. Walton, D. In *Brachiopods;* Copper, P.; Juiso, J., Eds.; Balkema Press: Rotterdam, **1996**; pp 289-297.
91. Goodfriend, G. A.; Hare, P. E.; Druffel, E. R. M. *Geochim. Cosmochim. Acta* **1992**, *56*, 3847.
92. Vallentyne, J. R. *Geochim. Cosmochim. Acta* **1964**, *28,* 157.
93. Vallentyne, J. R. *Geochim. Cosmochim. Acta* **1968**, *32,* 1353.
94. Hare, P. E.; Mitterer, R. M. *Carnegie Inst. Wash. Yb.* **1967**, *65*, 362.
95. Totten, D. K.; Franklin, J. R.; Davidson, D.; Wyckoff, R. W. G. *Proc. Natl. Acad. Sci. USA* **1972**, 69, 784.
96. Bada, J. L.; Shou, M-Y.; Man, E. H.; Schroeder, R. A. *Earth Planet. Sci. Letts.* **1978**, *41*, 67.
97. Walton et al., unpublished data.
98. Bada, J. L.; Miller, S. L. *J. Amer. Chem. Soc.* **1969**, *91*, 3946.
99. Bada, J. L.; Miller, S. L. *J. Amer. Chem. Soc.* **1970**, *92*, 2774.
100. Robinson, A. B.; Rudd, C. J. *Curr. Topics Cell. Reguln.* **1974**, *8*, 247.
101. Csapó, J.; Capó-Kiss, Z.; Kolto, L; Némethy, S. *Anal. Chim. Acta* **1995**, *300*, 313.
102. Starke-Reed, P. E.; Oliver, C. N. *Archiv. Biochem. Biophys.* **1989**, *275*, 559.
103. Erdman, J. G. *Geochim. Cosmochim. Acta* **1961**, *22*, 16.
104. Thompson, R. R.; Creath, W. B. *Geochim. Cosmochim. Acta* **1966**, *30*, 1137.
105. Thompson, K. F. M. *Geochim. Cosmochim. Acta* **1979**, *43*, 659.
106. Phillippi, G. T. *Geochim. Cosmochim. Acta* **1977**, *41*, 1083.

Chapter 7

Protein Preservation During Early Diagenesis in Marine Waters and Sediments

Reno T. Nguyen and H. Rodger Harvey

Chesapeake Biological Laboratory, University of Maryland Center for Environmental Science, P.O. Box 38, Solomons, MD 20688

The fate of protein was followed during phytoplankton decay in flow-through laboratory incubations and with depth in the sediments of Mangrove Lake, Bermuda, which receives principally algal input. Two-dimensional electrophoresis and amino acid analysis suggest that while most proteins are degraded during early diagenesis, a significant fraction is preserved. Although >92% of the initial particulate nitrogen was lost during the oxic water column decay of the diatom *Thalassiosira weissflogii*, proteins remained a significant fraction of total amino acids and of the residual nitrogen (83% and 48%, respectively). Hydrolyzable amino acids associated with a >2 kDa molecular weight fraction accounted for 78 to 98% of total amino acids and 56 to 63% of total nitrogen in Mangrove Lake sediments. Two-dimensional electrophoresis revealed the preservation of few discrete, acidic protein species, some of which were common to both water column and sediment samples. Amino acid analysis of low (3.5-18 kDa), mid (18-43 kDa), and high (43-200 kDa) molecular weight protein fractions observed a shift towards high molecular weight proteinaceous material after extensive algal decay and with increasing sediment depth, compared to fresh algal material or recently deposited organic matter. These results suggest that only small amounts of discrete proteins survive early diagenesis, with most proteins retained in the residual organic matter as extensively modified and cross-linked, acidic species.

Among the three major biochemical classes (carbohydrate, lipid, and protein) synthesized by phytoplankton, proteins are typically the most abundant substances (*1*) and represent an important source of carbon and nitrogen to marine systems. The large amounts of protein in phytoplankton (up to 75% of the particulate nitrogen, *2*; and references therein) are generally thought to be rapidly recycled in the marine water column (*3,4*), and have seen limited examination in geochemical studies. Preservation of some fraction of proteins as the amino acids does occur, however, and hydrolyzable amino acids derived from proteins, low molecular weight peptides, and/or bound amino acid monomers may comprise a significant fraction (typically 40-60%) of particulate nitrogen in coastal and oceanic waters (5-9). Amino acids are preserved in smaller amounts in pelagic marine sediments (e.g., 10,11) and also appear in fossil shells which are millions of years old (12).

The potential preservation of protein in both dissolved and particulate organic matter pools has only recently been addressed. For example, dissolved protein reacted with glucose (13) or protein abiotically aged in sterile sea water (14) was found to be less readily assimilated by bacteria. Condensation reactions between sugars and amino acids (15) were suggested as responsible for the transformation of "labile" protein into "refractory" protein. It remains unclear, however, whether these types of reactions can occur at the low *in situ* concentrations of dissolved organic matter typical of natural systems. Studies of particulate proteins in the Pacific Ocean have shown that protein nitrogen may comprise 12 to 32% of particulate nitrogen in surface waters and 19 to 32% at 500 m depth (16). Dissolved proteins collected in the Pacific consisted of a relatively limited number of major protein species; one was tentatively identified as a 48-kDa porin P, a protein found in the outer membrane of Gram-negative bacteria (17,18). In experimental studies, proteins intimately associated with cell membranes were observed to be more slowly degraded by bacteria than unbound proteins (19). Hydrolyzable amino acids associated with a >2 kDa fraction have been found to comprise up to 75% of the residual nitrogen in phytodetritus (20), lending additional support to the potentially important contribution of proteins to refractory nitrogen.

Despite an increase in our understanding of proteins and their potential for preservation in marine systems, the mechanisms responsible for the resistance of proteins to degradation remain elusive. The first goal of this work was to follow early protein diagenesis in sediments of Mangrove Lake, Bermuda, a well-characterized, enclosed coastal lake which presently has brackish to saline waters (21,22). Terrigenous input in this system is minor, with sediment organic matter derived mainly from marine algae. Recent ^{15}N nuclear magnetic resonance (NMR) spectroscopic analysis indicates that organic nitrogen throughout the first 4000 yrs in this marine algal sapropel is dominated by the amide linkage (23). This signal is also the dominant one for protein-rich algae, suggesting the preservation of algal protein (23). Moreover, the absence of an increased signal assigned to ammonium or free amino groups indicates that substantial hydrolysis of proteins does not occur during early, sedimentary diagenesis in this lake. A second goal was to use laboratory incubations to link water column degradation with sediment incorporation as well as elucidate the mechanisms which might be responsible for protein preservation. A flow through laboratory system was used to track proteins during the simulated sedimentation of a marine diatom (after 4; 20), to compare diagenetic processes in the water column to those in sediments. Alternative protein extraction protocols were developed and used with high-resolution biochemical separation techniques to address the potential for protein preservation during early diagenesis.

Materials and Methods

Field Sample. Mangrove Lake, a small coastal lake located in Bermuda, presently has brackish to saline waters (21). Organic input to sediments is dominated by marine algae, and the sedimentation rate of 0.3 cm y^{-1} is high compared to other marine environments.

The sediments are anoxic and characterized by a low (<10%) mineral content (*21*). Sediment samples, spanning a range from recently deposited surficial (0.22 m) to 4000 years post-deposition (9.7 m), were initially collected by piston core (*21*) and were gifts from Patrick Hatcher (Pennsylvania State University, PA, USA).

Algal Culture. The diatom *Thalassiosira weissflogii* was cultured axenically in F/2 medium (*24*) prepared with 0.2 µm filtered and autoclaved seawater (salinity 14‰). Two parallel cultures were grown axenically under constant temperature (19°C) with a 12:12 L/D lighting regime provided by cool, white fluorescent bulbs. Upon reaching the late-stationary phase of growth, one culture was harvested for initial protein characterization, with the second subject to degradation experiments. The batch culture of *T. weissflogii* was harvested entirely by continuous-flow centrifugation (Sorvall Instruments, DE, USA), frozen quickly in liquid nitrogen, and stored at -70°C.

Experimental System. A flow-through system, which is described in detail previously (*4*), was used to simulate decomposition during sedimentation under oxic conditions. The system allowed cell culturing and decomposition by the natural microbial community to be conducted in the same vessel. Briefly, *T. weissflogii* was cultured axenically with constant aeration in a 42-L glass carboy under conditions similar to those described for the maintenance culture. Once the culture reached the late-stationary phase of growth as monitored by cell counts ($2.1\pm0.7 \times 10^5$ cells mL^{-1}), the decomposition experiment was started by enclosure of the vessel in darkness and initiation of flow through the system (designated as day 0). Inflow water was screened to 5 µm to exclude macro zooplankton grazers but allowed continual exposure to natural microbial consortia during degradation.

Sampling and Measurements. All Mangrove Lake sediment samples were received lyophilized and stored desiccated at 22-25°C until analysis. Total organic carbon was determined for the sediments after removal of carbonates by vapor phase acidification (*25*). Untreated subsamples were used for total nitrogen determination.

At defined time points during the diatom decomposition experiment, samples were withdrawn from the flow-through culture vessel for analysis. Particulate organic carbon (POC) and particulate nitrogen (PN) were quantified by standard methods (e.g., *26*), using material collected by filtration onto combusted (450°C, 4 h) glass-fiber filters (Whatman 25 mm GF/F). Precision for POC and PN averaged ±1% and ±2%, respectively. Dry weight was measured by duplicate filtering of a known volume through tared 0.2 µm polycarbonate filters (Gelman), followed by drying to constant weight (50°C) and correcting for salt retention. The vacuum for filtration was kept low (<100 mm Hg) to minimize loss of cell contents at higher vacuum pressures (*27*). Sampling precision was typically ±5%. For the analysis of particulate amino acids and proteins, particulate material was collected by centrifugation (20,000 x *g*, 4°C, 30 min) in 250 mL polycarbonate bottles. This *g*-force is sufficient to pellet cells, nuclei, cytoskeletons, and mitochondria, but not ribosomes, viruses, or soluble proteins (*28,29*). All samples were transferred to 50 mL polycarbonate centrifuge tubes for further concentration and eventual storage at -70°C until analysis. All values were corrected for sample removal and resulting dilution as the incubation progressed.

Analysis of Proteins and Amino Acids.

Sedimentary Protein Content. An alkaline extraction method (modified after Rausch (*30*)) followed by colorimetric assay was used for the determination of Mangrove Lake sedimentary protein content. One to 2 mg of lyophilized sediment was added to micro centrifuge tubes with 0.5 mL of 0.5 N NaOH. Samples were vortex-mixed and placed in a sonicator bath for 2 h to aid in rehydration. Extraction was continued at 80°C for 5 min in a water bath. Samples were then briefly centrifuged and sonicated with a probe (5 watts, 2 min). Samples were further extracted (80°C, 10 min) and quickly cooled to room temperature. After centrifugation (2000 x g, 5 min), the supernatants (0.45 mL) were transferred to new micro centrifuge tubes, and protein extracted a third time (80°C, 10 min). 50 μL of 0.5 N NaOH was added to the 0.45 mL extract, and samples were diluted to 1.5 mL with deionized water to yield a 0.17 N NaOH solution. Standards were prepared in the same final concentration of NaOH. 100 μL aliquots were then used for colorimetric protein determination with ribulose-1,5-diphosphate carboxylase (RuDPCase; a.k.a. RuBisCo) as standard (*2*). Protein concentrations were corrected using the protein-nitrogen mass: protein mass of RuDPCase (0.13±0.01; mean±SE; n=5) to express estimates as protein nitrogen.

Protein (>2 kDa Hydrolyzable) Amino Acids. As an alternative to the colorimetric method for determination of protein content, amino acids associated with a >2 kDa molecular weight (MW) fraction were quantified. The protocol is a major modification of the method described by Nguyen and Harvey (*2*) for the extraction and analysis of >2 kDa amino acids. The modified method is designed to minimize losses by employing a rigorous extraction scheme and eliminating glass-fiber filters which might lead to adsorptive loss. Milligram amounts of well-ground, lyophilized sedimentary or particulate material were weighed into micro centrifuge tubes followed by addition of 0.5 mL of cold (-20°C) 10% (w/v) trichloroacetic acid solution in acetone (with 0.1% β-mercaptoethanol), and sample extraction with sonication (5 watts, 1 min). Another 0.5 mL of the above solution was added to each tube, the sample vortexed and incubated at -20°C for 30 min with mixing at the midpoint. Samples were then centrifuged (16,000 x g, 2 min), and the resulting pellets, containing both cell debris and proteins, were resuspended in 1 mL of cold (-20°C) acetone (with 0.1% β-mercaptoethanol). After incubation at -20°C for 20 min, samples were centrifuged as described above. The acetone extraction was repeated, until the supernatant was clear. The pellets, containing cellular debris as well as proteins and polypeptides >2 kDa, were air dried, resuspended in 1.0 mL of a 25 μM solution of γ-methylleucine and quantitatively transferred to 4 mL precombusted amber vials. The samples were then dried in a centrifuge-evaporator (Savant Instruments, NY, USA). For acid hydrolysis, 0.5 mL of sequanal grade HCl (Sigma Chemical Co., MO, USA) was added and each vial was capped under nitrogen, and hydrolyzed at 150°C for 70 min. Samples were then transferred to micro centrifuge tubes, dried in a centrifuge-evaporator, and resuspended in 0.5 mL deionized water. Aliquots of ≥0.25 mL were transferred to 4 mL vials and derivatized as trifluoroacyl isopropyl esters, according to Silfer et al. (*31*), for analysis by gas chromatography/flame ionization detection (GC/FID) and gas chromatography/mass spectrometry (GC/MS). Separations were performed with a DB-5

(Hewlett Packard) fused-silica capillary column (60 m length, 0.32 mm i.d., 0.25 μm film thickness) using hydrogen (for GC/FID) or helium (for GC/MS) as the carrier gas. The GC temperature was programmed from 50°C to 85°C at 10°C/min, then to 200°C at 3.5°C/min, and finally to 300°C (5 min isothermal) at 10°C/min. Amino acid amounts were based on the internal standard (γ-methylleucine) and were also corrected for variable responses by comparing the areas of equimolar concentrations of γ-methylleucine and each amino acid in an external standard mix. The summation of anhydroamino acid residues (corrected for the nitrogen content of each amino acid) yielded an estimate of protein nitrogen. Precision for the analysis of individual amino acids was typically ±4%. Precision for summed amino acids averaged ±13% for duplicate samples.

Total Hydrolyzable Amino Acids. Sediment and samples collected during diatom decay were analyzed for total amino acids by GC/FID and GC/MS following hydrolysis of lyophilized samples (without prior fractionation) and derivatization as described above. This fraction includes proteins, polypeptides, peptides, and bound amino acids. Precision for summed amino acids averaged ±9% for duplicate samples. Mangrove Lake sediment samples were also analyzed for total amino acid composition using an o-pthaldialdehyde-high performance liquid chromatography (OPA-HPLC) method (20). Precision for individual amino acids analyzed by OPA-HPLC was typically ±10%. Lower precision (±20%) was occasionally found for the minor protein amino acids (histidine, methionine) and non-protein amino acids β-alanine and γ-aminobutyric acid.

Two-Dimensional Gel Electrophoresis. This technique offers the highest resolution separations available for proteins (for review see 28,32,33). The Bio-Rad Mini-PROTEAN II 2-D cell was used for the separation of proteins first by isoelectric focusing (charge) followed by sodium dodecyl sulfate (SDS) polyacrylamide gel electrophoresis (molecular weight). Samples were extracted as described above for the isolation of the >2 kDa proteinaceous fraction, except that an aqueous extraction solution was added (30 to 40 μL per mg dry weight of sample) to the air-dried pellets, to resolubilize proteins from the other pellet material. The extraction solution (8 M urea, 4% Triton X-100, 1.25% SDS, 5% β-mercaptoethanol, 60 mM Tris base; Bio-Rad Laboratories, CA, USA) and protocol are a slight modification of those developed by the Large Scale Biology Corporation (Rockville, MD, USA) for the analysis of plant proteins. An equivalent volume of a first-dimension sample buffer (9.5 M urea, 2% Triton X-100, 5% β-mercaptoethanol, 1.6% Bio-Lyte 5/7 ampholyte, 0.4% Bio-Lyte 3/10 ampholyte) was added to the extraction mix. The samples were mixed and centrifuged (16,000 x g, 5 min). Supernatant (~20 μL) containing approximately the same quantity of total protein (~60 μg) per sample was then loaded onto each first-dimension gel. First-dimension isoelectric focusing was accomplished in 6 cm polyacrylamide capillary gels containing 4:1 of Bio-Lyte (pH 5-7):Bio-Lyte (pH 3-10). Power conditions were 500 V for 10 min and 750 V for 4.5 h. Each capillary gel was carefully extruded into a 1.7 mL micro centrifuge tube and immediately frozen (-70°C) for storage without prior equilibration in Laemmli sample buffer (34). Avoiding the equilibration step prevented diffusive loss (32), which can be problematic with samples containing low amounts of protein. For second-dimension separation, first-dimension gels were thawed and quickly loaded onto 15% mini slab gels (7.5 cm x 8 cm) containing no

SDS. Following electrophoresis, slab gels were stained using Bio-Rad Silver Stain Plus which can detect a few nanograms of protein. Two-dimensional protein standards amended with partially purified spinach RuDPCase (Sigma Chemical Co.) were run in parallel isoelectric focusing gels for isoelectric point (pI; pH at which net charge is zero) calibration. The pI of the large subunit of RuDPCase was estimated by entering the RuDPCase sequence information for the green alga *Bryopsis maxima* (*35*) in the program Compute pI/MW. MW calibration was achieved with Bio-Rad broad-range standards loaded at 10 ng of each protein. Duplicate runs for proteins of late-stationary phase *T. weissflogii* yielded reproducible separations. No protein was detected in negative control samples (containing no lyophilized material). Other extraction and concentration protocols failed to yield samples sufficiently clean for two-dimensional separations; these methods included NaOH or detergent extraction followed by ultrafiltration; extraction with detergent followed by trichloroacetic acid precipitation.

Protein MW Fractions. Since transmission densitometry cannot be used to quantify very complex mixtures of proteins in electrophoresis gels, electro-elution was used to remove broad MW fractions of protein from the acrylamide matrix for subsequent amino acid analysis. Enough dry weight (2 to 5 mg) of a sample to yield 200 to 500 μg of total protein (estimated by amino acid analysis) was extracted as described above. Protein in the air-dried pellets was solubilized in 80 μL of Laemmli sample buffer. With phytodetritus samples, the buffering capacity of the Laemmli sample buffer was not sufficient to keep the pH at 6.8, and the pH of the mix was adjusted with 0.5 N NaOH. Samples were heated (95 °C, 10 min) and centrifuged (16,000 x g, 10 min). Vertical lane borders were drawn on the glass plates of 15% acrylamide slab gels which were cast with 10 sample wells per gel. Only two samples were loaded per gel, with each sample loaded in duplicate (25 μL per well), once on each half of the gel. Kaleidoscope prestained MW markers (Bio-Rad) provided MW calibration. An empty well surrounding each sample prevented cross-contamination, and each was loaded with Laemmli sample buffer to prevent band spreading. Following electrophoresis (200 V, 45 min), the glass plates were marked into three MW regions, mainly based on the mobilities of the prestained standards: low (3.5-18 kDa), mid (18-43 kDa), and high (43-200 kDa). The glass plates were carefully separated, and the gel, still bound to one marked plate, was cut accordingly. Gel pieces from sample lanes on one side of the gel were minced, transferred to micro centrifuge tubes, and frozen (-70 °C). The duplicate gel side was silver stained. If protein was detected in the silver stained gel half, electro-elution was performed on the MW fractions. A volatile protein buffer (50 mM ammonium bicarbonate, 0.1% SDS) was used to suspend the minced gel pieces for transfer to electro-elution sample tubes (Bio-Rad Model 422) with attached dialysis caps (3.5 or 12-15 kDa MW cutoff, depending on the fraction to be electro-eluted). Electro-elution was performed (10 mA/tube, 4 to 5 h) at the same time on all MW fractions of a particular sample. Electro-dialysis was performed using 50 mM ammonium bicarbonate halfway through a run, in order to remove SDS from the samples. The electro-eluted samples (0.4 to 0.8 mL) were transferred to 4 mL glass vials with γ-methylleucine as internal standard. The samples were dried in a centrifuge-evaporator to concentrate for amino acid analysis as described above. Although GC/FID had adequate sensitivity for detection of amino acids, it detected numerous non-amino acid compounds which were also eluted from the gels. For these fractions, GC with a nitrogen-phosphorus detector was used for quantification of amino acids. Blank values were obtained from gel pieces which had been cut out from lanes only loaded with sample buffer. Glycine was a significant component of the electrophoresis running buffer and thus its concentration was not included in calculation of summed amino acids. Precision for duplicate samples averaged ±7%, ±16%, and ±20% for low, mid, and high MW fractions, respectively.

Results

Bulk Organic Matter, Total Protein, >2 kDa Amino Acids, and Total Amino Acids.

Mangrove Lake Sediments. Sediments contained varied amounts of total organic carbon (25 to 41%) and total nitrogen (1.9 to 3.7%) based on dry weight (Table I). The atomic C:N ratio increased from 10.5 in surficial sediments to a maximum of 18.7 at 3.6 m depth and then declined to 13.9 at 9.7 m depth. Total amino acids comprised 60 to 74% of sedimentary nitrogen with depth (Figure 1). Amino acids in the >2 kDa fraction showed no significant variation with increasing sediment depth, comprising 56 to 63% of the nitrogen (Figure 1) and 78 to 98% (avg. 91%) of total amino acids. The concentration of "protein" extracted under the more traditional alkaline conditions showed good agreement with the concentrations of >2 kDa amino acids and total amino acids in surficial sediments, but decreased from 62% to 20% of sedimentary nitrogen with increasing sediment depth (Figure 1).

Diatom Decay. Organic carbon and nitrogen were rapidly lost from algal cells after death. First-order decay constants for POC and PN were not significantly different (11.5 y^{-1} and 11.8 y^{-1}, respectively; Figure 2). As a result, the atomic C:N ratio showed little change (6.0±0.3; mean±SE; n=7) during the decomposition process. In late-stationary phase *T. weissflogii* cells, amino acids in the >2 kDa fraction represented 71% of total amino acids and 54% of particulate nitrogen (day 0, Figure 2). By day 11, the contribution of >2 kDa amino acids to particulate nitrogen decreased to 29%, but still comprised 66% of total amino acid (Figure 2). By day 30, >2 kDa amino acids represented 91% of total amino acids and 48% of particulate nitrogen. At the end of the degradation period, only 8% of the initial particulate nitrogen remained. At this time, the >2 kDa amino acids accounted for 83% of the residual total amino acids and 48% of the residual particulate nitrogen (Figure 2).

Total Hydrolyzable Amino Acids in Mangrove Lake Sediments. No major changes were observed in the amino acid distribution with sediment depth (Table I). The major protein amino acids included aspartic acid, glutamic acid, glycine, and alanine, with each present at ≥8 mole % throughout the sediment depth examined. The ratio of acidic amino acids (aspartic acid, glutamic acid) to basic amino acids (arginine, histidine, lysine) also showed little variability with depth (2.4±0.2). The non-protein amino acids including

Table I. The distribution of sedimentary organic carbon, total nitrogen, and amino acids with depth in Mangrove Lake, Bermuda.

Depth (m)	%OC (w/w)	%N (w/w)	(C:N)$_a$	Amino Acid Mole %															Non-protein amino acids[‡]			
				Asp	Glu	Ser	His	Gly	Thr	Arg	Ala	Tyr	Met	Val	Phe	Ile	Leu	Lys	BALA	GABA	AABA	Orn
0.22	33.3	3.7	10.5	8.6	10	6.7	1.5	11	8.4	4.3	12	2.9	1.5	7.1	6.5	5.2	8.5	2.7	1.4	1.1	0.32	2.3
1.2	27.1	2.7	11.6	11	9.8	5.9	1.6	10	6.7	3.1	14	2.1	0.9	7.5	7.3	4.7	6.7	4.4	1.9	1.6	0.52	4.1
2.3	24.8	2.2	13.3	11	11	5.1	1.3	8.5	7.0	2.7	14	2.5	1.1	8.4	6.9	5.5	8.2	3.8	1.6	1.5	0.43	3.4
3.6	30.5	1.9	18.7	9.5	9.4	6.0	1.4	8.8	8.3	3.1	16	2.0	1.0	7.9	6.2	4.8	8.5	3.6	1.7	1.5	0.44	3.4
5.1	31.2	3.1	11.9	9.7	9.7	6.8	1.2	10	6.7	2.9	15	2.1	1.0	8.2	7.3	4.7	7.2	3.9	1.7	1.6	0.71	3.4
9.7	40.6	3.4	13.9	10	9.7	7.3	1.1	7.9	6.5	2.9	15	2.1	0.7	7.9	8.5	4.9	7.3	4.2	1.5	2.0	0.41	3.6

[‡]Abbreviations for non-protein amino acids: BALA, β-alanine; GABA, γ-aminobutyric acid; AABA, α-aminobutyric acid; Orn, ornithine.

g Amino Acid-N or Protein-N/g Total N

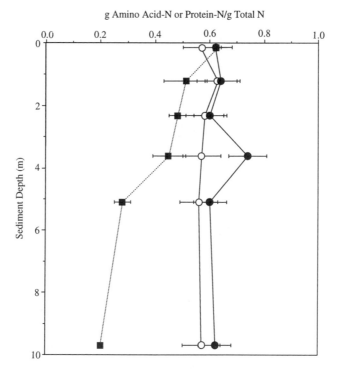

Figure 1. The contribution of amino acid- or protein-nitrogen to total sedimentary nitrogen for Mangrove Lake. Amino acid fractions include: (●) total hydrolyzable amino acids; and (○) >2 kDa hydrolyzable amino acids. For comparison, (■) colorimetrically-determined "protein" (mean±1 SD; n=2) following alkaline extraction is shown. Error bars for the amino acid data points represent the estimated uncertainty, based on the average coefficients of variation for 14 samples.

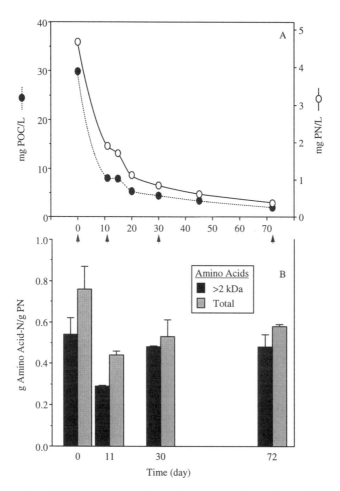

Figure 2. (A) Time course of particulate organic carbon (POC) and particulate nitrogen (PN) during the oxic water column degradation of the diatom *T. weissflogii*. (B) The contribution of amino acid-nitrogen to particulate nitrogen (mean±1 SD; n=2) for two amino acid fractions.

β-alanine, γ-aminobutyric acid, α-aminobutyric acid, and ornithine contributed a small percentage (0.3 to 4% each) to the total amino acids present (Table I).

Two-Dimensional Gel Electrophoresis and MW Distribution of Proteins.

Mangrove Lake Sediments. Only a few discrete protein species were visible at all sediment depths (Figure 3). All gels were characterized by an acidic (~pI 3.8-5), broad MW front of unresolved material which varied in staining intensity. For surficial sediments, four spots were scored: two proteins with a differing acidic pI (~6 and 6.3) but with the same MW of 16 kDa; and two proteins each with ~pI's of 3.8 and 5 and MW's of 33 and 37 kDa, respectively (Figure 3A). For sediments of 1.2 m depth, four spots could also be scored but these were all limited to a narrow MW range of 14-17 kDa (Figure 3B). Proteins "a" and "b" of 16 kDa MW, found at 0.22 m depth, appeared to have shifted to slightly more acidic proteins ("a*" and "b*") at 1.2 m depth (Figures 3A and 3B). Electrophoresis was performed for an extended period of time for the above samples, so <6.5 kDa proteins were not visible. Compared to surficial and 1.2 m deep sediments, which had four discrete proteins of MW's ≥14 kDa, only one spot of >14 kDa could be scored for the 5.1 m deep sediments (Figure 3C). This spot, designated "c", had a MW of 18 kDa and was located within the unresolved, acidic front region of the electrophoretogram (Figure 3C). Six other spots could be scored for the 5.1 m deep sediments, but these were all ≤6.5 kDa. For the 9.7 m deep sediments, eight spots of MW's between 18 and 26 kDa were scored, with all found in the unresolved acidic front region of the electrophoretogram (Figure 3D). In addition, five spots were ≤6.5 kDa. Some of the same protein species could be found in the two deepest sediment samples. For example, proteins "c", "d", "e", "f", "g", and "h" of MW's between 4 and 18 kDa were found at both 5.1 m and 9.7 m depths (Figures 3C and 3D). For surficial sediments, the MW distribution of proteins was 41%, 39%, and 20% for low (3.5-18 kDa), mid (18-43 kDa), and high (43-200 kDa) MW ranges, respectively (Figure 5A). At 3.6 m depth, the MW distribution shifted, with a decrease to 33% for low MW material and an increase to 26% for high MW material. At 9.7 m depth, the relative abundance of low MW material increased to 48%, while that for mid MW material decreased to 27%; the relative abundance of the high MW proteinaceous fraction showed no change compared to that for the same MW fraction at 3.6 m depth.

Diatom Decay. A large number of proteins (at least 103) in algal cells were present in the acidic pI range of 3.8 to ~6.5, with MW's highly distributed from 3.5 to 200 kDa (Figure 4A). The large (pI≈6.4) and small subunits of RuDPCase, which are dissociated under the denaturing electrophoresis conditions, were detected at low levels (Figure 4A). For late-stationary phase *T. weissflogii* cells, the MW distribution of proteins was 51%, 33%, and 16% for low (3.5-18 kDa), mid (18-43 kDa), and high (43-200 kDa) MW ranges (Figure 5B).

Compared to intact algal cells, only 18 visible spots were seen in the electrophoretogram by day 11, with the proteins found in the ~pI 4-6 and 3-23 kDa ranges (Figure 4B). By day 30, the number of spots scored decreased to 14, with proteins distributed from ~pI 4-6 and 3-35 kDa (Figure 4C). About 17 discrete protein species found in the ~pI 4-6 and 3-31

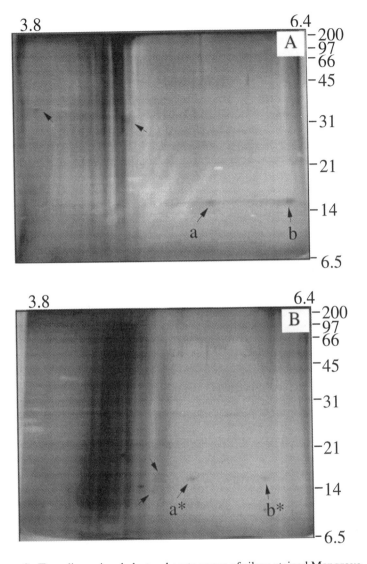

Figure 3. Two-dimensional electrophoretograms of silver-stained Mangrove Lake sedimentary proteins (equivalent total protein loadings) from (A) 0.22 m, (B) 1.2 m, (C) 5.1 m, and (D) 9.7 m depths. Spots labeled with identical letters are tentatively the same protein species. Letters followed by an asterisk indicate proteins which have been modified. Intense staining in ~pI 3-5 suggests a complex mixture.

Continued on next page.

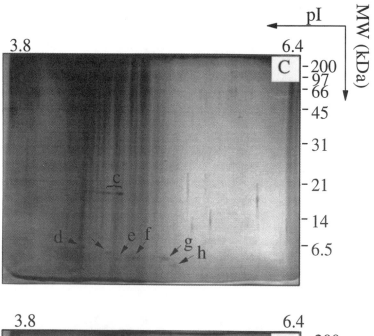

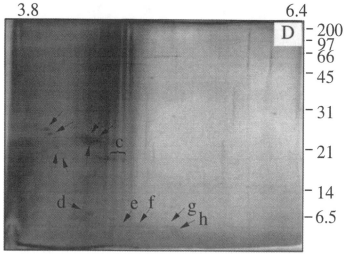

Figure 3. *Continued.*

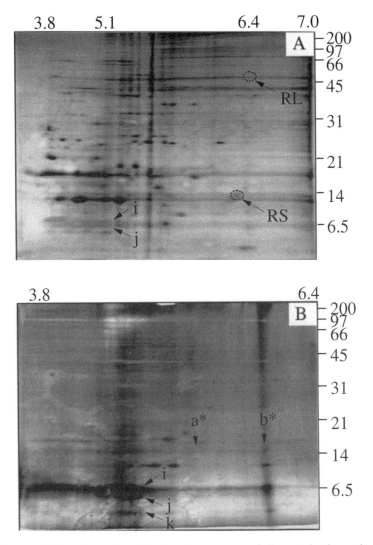

Figure 4. Two-dimensional electrophoretograms of silver-stained proteins (equivalent total protein loadings) at (A) day 0, (B) day 11, (C) day 30, and (D) day 72 of the oxic water column degradation of the diatom *T. weissflogii*. Spots shown with the same letters are tentatively identified as the same protein species. RuDPCase large and small subunits (RL, RS) are circled.

Continued on next page.

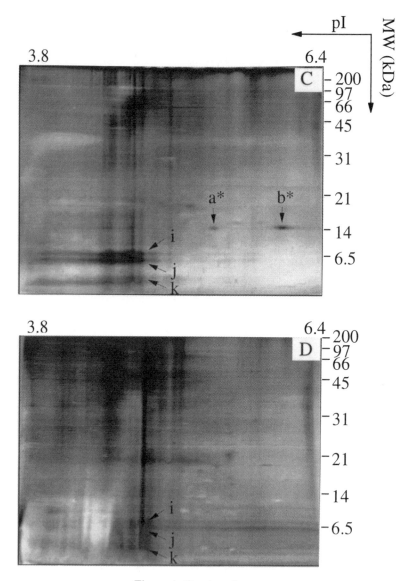

Figure 4. *Continued.*

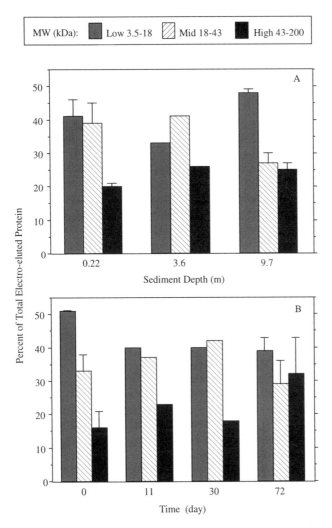

Figure 5. Molecular weight distribution of electro-eluted protein fractions in sediments and degraded algae. (A) recently deposited, surficial (0.22 m), intermediate age (3.6 m), and 4000 y old (9.7 m) Mangrove Lake, Bermuda, sediments; (B) late stationary-phase diatom (*T. weissflogii*) cells (day 0) and diatom detritus (days 11, 30, 72) from an incubation in an oxic water column. Endpoints are mean±1 SD (n=2).

kDa ranges were visible in the electrophoretogram for day 72 (Figure 4D). Two highly acidic species were visible in the degraded sample: one of ~pI 3.8 and 36 kDa MW; the other of ~pI 3.9 and 9 kDa MW (Figure 4D). Six spots, possibly negatively-stained proteins, were also visible on day 72. Proteins "a*" and "b*" found in the Mangrove Lake sediments were visible at days 11 and 30 of the diatom decay (Figures 4B and 4C). Proteins "i" and "j" were present throughout the diatom decay, although they were not always well resolved (Figure 4). Protein "k" could also be seen on days 11, 30, and 72. Many other protein species could not be easily traced throughout the decay experiment, and hence were not labeled. The acidic front seen for the Mangrove Lake sediments developed during the diatom decay experiment and was most pronounced by day 72 (Figure 4). The sizes of several proteins spots (e.g., "i" and "j") in the electrophoretograms appeared to increase during diagenesis, but is principally due to the fact that all gels were loaded with approximately the same quantity of *total* protein. Thus, relative amounts of a particular protein may increase (i.e., if it is selectively preserved), but absolute quantities of that particular protein may be similar to that in the earlier time point or shallower sediment horizon. By day 11 of the decay experiment, the MW distribution of "proteins" shifted, especially for low MW material (Figure 5B). MW distributions for days 11 and 30 were similar, with low MW "proteins" decreasing to 40%; mid and high MW "proteins" comprised ~40% and ~20%, respectively, of total electro-eluted "protein". By day 72, low MW "proteins" showed no change in relative abundance compared to days 11 and 30, while the relative contribution of mid and high MW "proteins" shifted to 29% and 32%, respectively.

Discussion

The incomplete hydrolysis of proteins observed in both laboratory incubations and field collections suggests that proteins are not as labile as commonly believed. The retention of relatively high MW material is demonstrated by amino acids associated with a >2 kDa fraction, which comprised up to half of the particulate nitrogen by the end of the 72-day diatom decay study (Figure 2). Similar results have been observed during decay studies with other algae such as dinoflagellates (20). Amino acids in high molecular weight material also account for 56 to 63% of the nitrogen in Mangrove Lake sediments. Sources of particulate or sedimentary nitrogen would include proteins, polypeptides, peptides, free amino acids, nucleic acids, amino-sugars, and lipid-N. With the possible exception of lipid-N, virtually all of these biochemical compounds are considered highly labile. The preservation of amino acid-N, whether from proteins, peptides, or bound amino acid monomers is significant, and suggests that while mineralization is an efficient process, some fraction of protein (or its high molecular weight fragments) escapes degradation and is retained.

The discrepancy in estimates of sedimentary protein using different extraction protocols is striking. With increasing Mangrove Lake sediment depth, the protein content estimated using the traditional alkaline extraction protocol decreased (Figure 1). A similar profile for protein in Mangrove Lake sediments was originally seen by Hatcher et al. (22). Estimates of protein in other coastal sediments also have been made following alkaline extraction but with subsequent protease treatment (e.g., *36-38*). These estimates, which represent protein that is available to enzymatic attack, comprise only a small fraction of the total nitrogen (<15%) and total hydrolyzable amino acid pools (<25%). In contrast to the alkaline extraction method, estimates of protein in Mangrove Lake sediments based on the recently developed, acid/acetone extraction protocol (trichloroacetic acid, acetone, HCl) remain essentially constant down core, and account for a large fraction of total nitrogen (~60%) in the oldest sediments (Figure 1). The newer protocol is expected to be more rigorous, with trichloroacetic acid and acetone separating relatively low MW (<2 kDa) material from higher MW material. Acid hydrolysis of the retained higher MW proteinaceous material results in the efficient release of amino acids whose concentrations may be used to estimate protein. The comparison of Mangrove Lake protein estimates using the acid/acetone and alkaline extraction methods suggests that proteins are increasingly difficult to extract down core using the alkaline method (Figure 1). Proteinaceous material *covalently* bound to the sedimentary organic matrix likely would not be released under the relatively mild alkaline conditions (i.e., 0.1 N to 2 N NaOH with incubation at room temperature or heating at <80°C) typically used. These results suggest that sedimentary material in older sediment sequences may be more cross-linked and that proteinaceous material represents a larger fraction of the sedimentary nitrogen than is commonly believed.

There is also evidence that the 6 N HCl hydrolysis may not completely hydrolyze proteinaceous material in decaying organic matter. Based on analysis of the hydrolysis residues by GC/MS (after TMAH treatment) and ^{15}N NMR, proteinaceous material was identified in the hydrolysis residue of deep Mangrove Lake sediments (*39*) and of soils amended with waste compost (*40*), suggesting that some peptide-like compounds are protected against acid hydrolysis. Generally, HCl hydrolysis is performed at ≥100°C for 12 to 24 h. The acid hydrolysis conditions that we employed are those of Cowie and Hedges (*41*): 150°C for 70 min; which is faster than, but has equivalent recoveries compared to other methods. Using HCl hydrolysis, ~40% of the total nitrogen in Mangrove Lake sediments cannot be accounted for by amino acids associated with a >2 kDa fraction. Although some of the sedimentary nitrogen may originate as inorganic nitrogen which has been incorporated into heterocyclic compounds (e.g., pyridines, indoles and pyroles), hydrolysis efficiency indicates that more of the unidentified nitrogen may be proteinaceous (*23*).

The lack of major changes in the amino acid distribution with depth in Mangrove Lake sediments (Table I) suggests that amino acids or original proteinaceous materials are not selectively lost during early diagenesis. The substantial contribution of "protein" (>2 kDa hydrolyzable amino acids) to the total hydrolyzable amino acid pool (avg. 91% down core; Figure 1) might help to explain why total amino acid distribution showed little variation with increasing sediment depth. The low relative abundance of non-protein amino acids in the Mangrove Lake sediments (Table I) also suggests minor diagenetic reactions of protein amino acids. Furthermore, kinetic studies of amino acid degradation in both oxic and anoxic water columns (*20*) and studies of early diagenesis of amino acids in coastal marine sediments (*42,43*) or abyssal plain sediments (*44*) suggest that selective attack on protein amino acids during degradation is unlikely.

Although most proteins found in biological materials are acidic (e.g., *32,45-47*), it is possible that some basic proteins could survive diagenesis. The pH gradient of the isoelectric focusing gels (established by the mixture of ampholytes) used in the present study did not allow the isolation of basic proteins. The contribution of basic proteins, however, is expected to be minor, based on two-dimensional electrophoresis and the analysis of amino acid composition during early diagenesis. In phytoplankton, the acidic amino acids comprise a larger fraction of the protein pool than do the basic ones; this composition exhibits minor changes during the degradation of the phytoplankton (*20*). The ratio of acidic to basic amino acids observed down core for Mangrove Lake sediments was >2, indicating that the majority of sedimentary proteins which are preserved have an acidic pI. The absolute quantity of acidic, amino acids is confounded by contributions of asparagine and glutamine which are deamidated to aspartic acid and glutamic acid under the 6 N HCl hydrolysis conditions. Nevertheless, the isoelectric focusing step of two-dimensional electrophoresis (Figures 3 and 4), which separates proteins mainly according to amino acid composition (i.e., the ratio of acidic to basic chemical groups), corroborates the amino acid data. A prior application of two-dimensional electrophoresis to investigate protein preservation includes one study (*48*) which revealed several high MW protein species with acidic to neutral pI's in the organic matrix of core-top and fossil plankton foraminifera.

The precise chemical and structural nature of the proteins observed in Mangrove Lake sediments and during diatom decay remains elusive. With the electrophoretic techniques we have employed in this study, one caveat is that the chemicals (i.e., SDS, Triton X-100, urea, and β-mercaptoethanol) found in the buffers denature proteins. In other words, large protein complexes normally held intact (by, e.g., hydrophobic forces, disulfide bonds, hydrogen bonds, dipolar interactions) will be unfolded and dissociated to their component polypeptide subunits. As a result, the MW's reported in this study may be minimal estimates and might not represent those of the proteins found in the natural environment. Conformational changes including the complete denaturation of proteins, however, might occur during diagenesis, as they may occur during the sorption of protein to surfaces (*49*). With biological samples, some proteins are difficult to solubilize or may remain associated with other proteins even in the presence of high concentrations of detergents and denaturing agents (*28,32*). This may explain some of the streaking associated with the protein separations (Figures 3 and 4). In addition, the incomplete denaturing or incomplete opening of condensed structures could result in protein-bound melanoidin that is retarded in the gel and thus does not give an accurate size indication (i.e., mobilities may not be equivalent to those of the MW standards). Silver typically binds to various chemical groups such as sulfhydryl and carboxyl moieties in macromolecules, although the detection method we applied (i.e., Bio-Rad Silver Stain Plus) does not stain highly glycosylated proteins or metalloproteins; these proteins would appear as white spots or bands on a yellowish background. White spots seen for day 72 of the diatom decay suggest that the condensation reaction may be an important diagenetic process. Negatively-stained species were not seen for Mangrove Lake sediments, and NMR spectroscopic analysis of the lake sediments (*23*) also suggests that condensation reactions are not important in that environment.

The sources of particulate proteins observed in this and previous studies may include both algae and bacteria and other members of the microbial community. The contribution of bacterial carbon to total POC in our flow-through incubations is generally small (*50*), but some bacterial proteins may persist compared to the more prevalent algal proteins which would allow the remaining pool to include both bacterial and algal sources. Interestingly, the two-dimensional electrophoretograms reveal the presence of several unique protein species at different time points or sediment depths. It is unclear whether these proteins are bacterial in origin or are algal proteins which have been modified. Slight modifications (e.g., acetylation or addition of charged carbohydrate groups) can produce an apparent shift in a protein's pI with no measurable changes (by electrophoresis) in its MW (*28,32*). Large modifications can produce observable shifts in both pI and MW. In that respect, two-dimensional electrophoresis is both a very powerful tool and one which requires cautious interpretation of results, especially for proteins in complex mixtures. Algal source variability during sedimentation in Mangrove Lake may provide another explanation for the observed differences in protein composition found in the electrophoretograms.

The analysis of proteinaceous material in the three different MW fractions (Figure 5) suggests that higher MW material is preferentially preserved during diagenesis. This shift may be a result of cross-linking of proteins with other proteinaceous material and/or with other macromolecular network. Interestingly, changes in MW distribution also were seen in the electrophoretograms for the diatom decay (Figure 4). These indicate a gradual change from well-resolved proteins in fresh material distributed over a wide MW range of~2 to 21 kDa range (Figure 4A), to intensely-stained, unresolved material in the 31-200 kDa range in highly degraded residue (Figure 4D). Similarities in the electrophoretograms for Mangrove Lake sediments at 1.2 m depth and for diatom detritus at day 30 (Figures 3B and 4C) suggest that organic matter at that sediment depth has been degraded to the same extent as organic matter which has already been processed to 18% of its original amount in the water column (Figure 2). One could speculate that the higher MW material seen in the two-dimensional gels of detritus simply could be cross-linked and condensed reservoirs of peptides and amino acids initially produced from diagenetic hydrolysis. Taken together, the amino acid analysis (Figures 1 and 2) and MW fraction analysis (Figure 5) suggest that the unresolved material (i.e., the acidic front seen in highly degraded samples) in the electrophoretograms is a complex mixture of proteinaceous material.

Most organic matter is preserved in coastal zones, which are characterized by relatively shallow water depths, high primary productivity, high sedimentation rates, and usually fine-grained sediments available for adsorption (see *51* for discussion). By the end of the 72-day diatom decay experiment, 8% of the initial particulate nitrogen remained, representing a value typical for the relative amount of particulate organic matter which ultimately settles onto coastal sediments. In contrast to many coastal environments, Mangrove Lake sediments are characterized by a low mineral content (<10% w/w) and low terrigenous input. In addition, organic matter cycling by benthic protozoa and meio- and macrofauna (*52*) is nonexistent due to the anoxic nature of these sediments. The total organic carbon and amino acid content and diagenetic trends for Mangrove Lake are also different from those observed for many coastal settings, and are likely attributed to the lake's unique depositional characteristics. For example, coastal sediments typically have

1 to 5% w/w organic carbon content (*38,53-55*) as compared to 25 to 41% w/w for Mangrove Lake sediments which do not show a consistent decrease in the carbon content with depth. In addition, amino acid nitrogen in other coastal sediments typically comprises less than half of total nitrogen and decreases or shows no consistent trend with depth (*38,42,43,54*), compared to amino acid nitrogen comprising ~60% of total nitrogen down core in Mangrove Lake sediments.

Adsorption to mineral surfaces (*56*) or within mineral pores (*57*) has been suggested as a mechanism to preserve organic matter in marine sediments. The present study indicates that proteins may be preserved in environments with a low abundance of minerals. Other mechanisms may be responsible for the preservation of proteinaceous material. For example, refractory cell wall material (e.g., highly aliphatic algaenans) might serve as a matrix which protects more easily degraded amide-containing molecules, as has been suggested by Derenne et al. (*58*). The major algal species found in Mangrove Lake is not expected to have algaenans, but other cell wall materials may impede degradation of the cellular constituents via encapsulation (*39*). Proteins intimately associated with cell membranes are degraded at slower rates than soluble proteins (*19*); this resistance may allow subsequent reactions to occur (e.g., condensation) which would further reduce their susceptibility to microbial attack. In addition, association with humic acid-type polymers in soils has been shown to reduce the biodegradation of ^{14}C-labeled amino acids and proteins (*59,60*). Although H_2S or polysulfide incorporation into functionalized biogenic lipids has been observed in anoxic environments (*61,62*), sulfur incorporation into proteins seems unlikely due to the lack of carbon-carbon double bonds in these macromolecules. However, proteins may be indirectly protected by sorption to macromolecular organosulfur compounds. The conformational stability of a protein would also determine its long-term preservation. Tightly folded proteins would be expected to be resistant to degradation and/or modifications, since their interior amino acid residues are not available to enzymatic attack and chemical reactions (*63*). If the preservation of proteins may occur in environments with low mineral content, enhanced preservation might occur in some mineral-rich environments, with the presence of individual functional groups in the proteins having an important effect on their partitioning behavior. The major amino acid groups found in Mangrove Lake sediments (Table I), acidic (aspartic and glutamic acids) and neutral (glycine, alanine), suggest that both ionic and hydrophobic interactions may play roles in sorption, e.g., to any carbonate surfaces and to other organic matter (*11*).

Based on these and previous results, it appears that the small fraction of proteins which is not mineralized in the marine water column or sediments has several potential fates during early diagenesis (Figure 6). A fraction of these proteins appears selectively preserved over short geological time scales. A second fraction may be hydrolyzed, albeit at slow rates. A third path for proteins is modification reactions (e.g., Schiff-base condensation, hydrophobic aggregations) with carbohydrate molecules, other proteins, or compounds to form large and insoluble, macromolecular complexes which are not readily degraded by bacteria. The potential pathways mentioned above are not mutually exclusive. For example, residual amino acids and peptides produced from the hydrolysis of proteins could then be sorbed to or cross-linked to other proteins or non-protein material (Figure 6). In addition, proteins which have been selectively preserved for a period of time might then be modified. Another diagenetic pathway is the enzymatic hydrolysis of proteins and

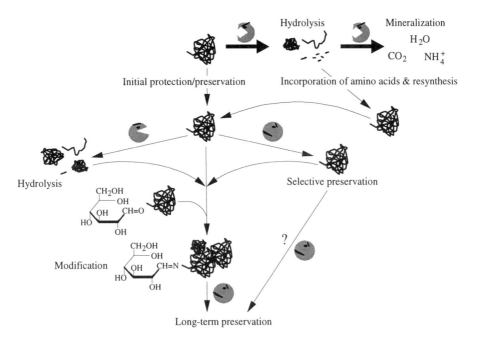

Figure 6. Conceptual diagram illustrating the potential fates of proteins during early diagenesis. Starting protein symbol represents proteins of a wide molecular weight distribution. Potential conformational changes in proteins are not shown.

peptides followed by the microbial assimilation of amino acids and their incorporation into proteins (*64,65*). These newly synthesized microbial proteins ultimately may enter the various diagenetic pathways illustrated in Figure 6. Therefore, diagenesis should not, or cannot, be viewed in simplistic terms. The two views of early diagenesis (depolymerization-recondensation or selective preservation) held by the geochemical community may both be operative. More work is needed to determine the types of linkages in the macromolecular organic material which survives diagenesis and the exact chemical nature of the proteinaceous material which is preserved.

Conclusions

The majority of algal proteins are lost during early diagenesis. Nevertheless, proteins (or their high molecular weight fragments), comprise a significant fraction of total amino acids and of the residual nitrogen which is retained in marine sediments of Mangrove Lake. Although several acidic protein species appear to be preserved, the contribution of proteins to the residual nitrogen arises largely from a complex mixture of proteinaceous material rather than the selective preservation of a few discrete moieties.

A shift towards higher molecular weight material was observed as diagenesis progressed, both in laboratory incubations of degrading algal cells and with increasing depth in Mangrove Lake sediments. This change in distribution could represent protein-protein aggregations, condensation reactions of carbohydrates and proteins, and/or extensive cross-linking of peptides and amino acids (initially produced from diagenetic hydrolysis) to non-proteinaceous material. The absence of minerals in the laboratory incubations and low abundance of minerals in field samples suggest that macromolecular organic matter also plays an important role in the preservation of proteinaceous substances during early diagenesis.

Acknowledgments

We thank Mary Haasch for two-dimensional electrophoresis equipment, Patrick Hatcher for the Mangrove Lake sediment samples, and Heike Knicker for helpful discussions. Comments by two anonymous reviewers significantly improved the final manuscript. This research was supported by the Donors to the Petroleum Research Fund of the American Chemical Society. Contribution No. 3031, University of Maryland Center for Environmental Science.

Literature Cited

1. Brown, M. R. *J. Exp. Mar. Biol. Ecol.* **1991**, *145*, 79-99.
2. Nguyen, R. T.; Harvey, H. R. *Mar. Chem.* **1994**, *45*, 1-14.
3. Hollibaugh, J. T.; Azam, F. *Limnol. Oceanogr.* **1983**, *28*, 1104-1116.

4.   Harvey, H. R.; Tuttle, J. H.; Bell, J. T. *Geochim. Cosmochim. Acta* **1995**, *59*, 3367-3377.
5.   Handa, N. In *Organic Matter in Natural Waters*; Hood, D. W., Eds.; Univ. of Alaska Institute of Marine Science Occasional Publication No. 1: 1970; pp 129-152.
6.   Siezen, R. J.; Mague, T. H. *Mar. Chem.* **1978**, *6*, 215-231.
7.   Lee, C.; Cronin, C. *J. Mar. Res.* **1982**, *40*, 227-251.
8.   Lee, C.; Cronin, C. *J. Mar. Res.* **1984**, *42*, 1075-1097.
9.   Lee, C.; Wakeham, S. G.; Farrington, J. W. *Mar. Chem.* **1983**, *13*, 181-194.
10. Maita, Y.; Montani, S.; Ishii, J. *Deep-Sea Res.* **1982**, *29*, 485-498.
11. Whelan, J. K.; Emeis K. In *Productivity, Accumulation, and Preservation of Organic Matter: Recent and Ancient Sediments*; Whelan, J. K.; Farrington, J. W., Eds.; Columbia University Press: New York, NY, 1992; pp 176-200.
12. Abelson, P. H. *Carnegie Inst. Washington Yearb.* **1954**, *53*, 97-101.
13. Keil, R. G.; Kirchman, D. L. *Limnol. Oceanogr.* **1993**, *38*, 1256-1270.
14. Keil., R. G.; Kirchman, D. L. *Mar. Chem.* **1994**, *45*, 187-196.
15. Hedges, J. I. *Geochim. Cosmochim. Acta* **1978**, *42*, 69-76.
16. Tanoue, E. *Deep-Sea Res.* **1992**, *39*, 743-761.
17. Tanoue, E.; Nishiyama, S.; Kamo, M.; Tsugita, A. *Geochim. Cosmochim. Acta* **1995**, *59*, 2643-2648.
18. Tanoue, E.; Ishii, M.; Midorikawa, T. *Limnol. Oceanogr.* **1996**, *41*, 1334-1343.
19. Nagata, T.; Fukuda, R.; Koike, I.; Kogure, K.; Kirchman, D. L. *Aquat. Microb. Ecol.*, in press.
20. Nguyen, R. T.; Harvey, H. R. *Org. Geochem.* **1997**, *27*, 115-128.
21. Hatcher, P. G. *The Organic Geochemistry of Mangrove Lake, Bermuda*; NOAA Prof. Paper, 1978; 10.
22. Hatcher, P. G.; Spiker, E. C.; Szeverenyi, N. M.; Maciel, G. E. *Nature.* **1983**, *305*, 498-501.
23. Knicker, H.; Scaroni, A. W.; Hatcher, P. G. *Org. Geochem.* **1996**, *24*, 661-669.
24. Guillard, R. L. (1972) In *The Culture of Marine Invertebrate Animals*; Smith, W. L.; Chaney, M. H., Eds.; Plenum: New York, 1972; pp 29-60.
25. Hedges, J. I.; Stern, J. H. *Limnol. Oceanogr.* **1984**, *29*, 657-663.
26. Bratbak, G. *Mar. Ecol. Prog. Ser.* **1987**, *36*, 267-276.
27. Dortch, Q.; Clayton, J. R. Jr.; Thoresen, S. S.; Ahmed, S. I. *Mar. Biol.* **1984**, *81*, 237-250.
28. Dunbar, B. S. *Two-Dimensional Electrophoresis and Immunological Techniques*; Plenum Press: New York, NY, 1987.
29. Voet, D.; Voet J. G. *Biochemistry*; Wiley: New York, NY, 1990; pp 102.
30. Rausch, T. *Hydrobiologia* **1981**, *78*, 237-251.
31. Silfer, J. A.; Engel, M. H.; Macko, S. A.; Jumeau, E. J. *Anal. Chem.* **1991**, *63*, 370-374.
32. O'Farrell, P. H. *J. Biol. Chem.* **1975**, *250*, 4007-4021.
33. Sinclair, J.; Rickwood, D. In *Gel Electrophoresis of Proteins: A Practical Approach*, 2nd Ed.; Hames, B. D; Rickwood, D., Eds.; IRL: London, 1990; pp 189-218.
34. Laemmli, U. K. *Nature* **1970**, *227*, 680-685.

35. Okada, M.; Okabe, Y.; Kono, M.; Nakayama, K.; Satoh, H. *Can. J. Bot.* **1991**, *69*, 1053-1061.
36. Mayer, L. M.; Schick, L. L.; Setchell, F. W. *Mar. Ecol. Prog. Ser.* **1986**, *30*, 159-165.
37. Mayer, L. M.; Macko, S. A.; Cammen, L. *Mar. Chem.* **1988**, *25*, 291-304.
38. Colombo, J. C.; Silverberg, N.; Gearing, J. N. *Mar. Chem.* **1996**, *51*, 295-314.
39. Knicker, H.; Hatcher P. G. *Naturwissenschaften* **1997**, *84*, 231-234.
40. Siebert, S.; Knicker, H.; Hatcher, M. A.; Leifeld, J.; Kögel-Knabner, I. 214th American Chemical Society National Meeting, Las Vegas, Nevada, 1997.
41. Cowie, G. L.; Hedges, J. I. *Mar. Chem.* **1992**, *37*, 223-238.
42. Rosenfeld, J. K. *Limnol. Oceanogr.* **1979**, *24*, 1014-1021.
43. Henrichs, S. M.; Farrington, J. W. *Geochim. Cosmochim. Acta* **1987**, *51*, 1-15.
44. Horsfall, I. M.; Wolff, G. A. *Org. Geochem.* **1997**, *26*, 311-320.
45. Bahrman, N.; de Vienne, D.; Thiellement, H.; Hofmann, J.-P. *Biochem. Genetics.* **1985**, *23*, 247-255.
46. Jungblut, P. R.; Seifert, R. *J. Biochem Biophys. Methods* **1990**, *21*, 47-58.
47. Wilkins, M. R.; Pasquali, C.; Appel, R. D.; Ou, K.; Golaz, O.; Sanchez, J.-C.; Yan, J. X.; Gooley, A. A.; Hughes, G.; Humphery-Smith, I.; Williams, K. L.; Hochstrasser, D. F. *Bio/Technology* **1996**, *14*, 61-65.
48. Robbins, L. L.; Brew, K. *Geochim. Cosmochim. Acta* **1990**, *54*, 2285-2292.
49. Taylor, G. T. *Limnol. Oceanogr.* **1995**, *40*, 875-885.
50. Harvey, H. R.; Macko, S. A. *Org. Geochem.* **1997**, *26*, 531-544.
51. Hedges, J. I.; Keil, R. G. *Mar. Chem.* **1995**, *49*, 81-115.
52. Deming, J. W.; Baross, J. A. In *Organic Geochemistry*; Engel, M. H.; Macko, S. A., Eds.; Plenum: New York, NY, 1993; pp 119-144.
53. Henrichs, S. M. (1993) In *Organic Geochemistry;* Engel, M. H.; Macko, S. A., Eds.; Plenum: New York, NY, 1993; pp 101-117.
54. Cowie, G. L.; Hedges, J. I. *Nature.* **1994**, *369*, 304-307.
55. Mayer, L. M. *Geochim. Cosmochim. Acta* **1994**, *58*, 1271-1284.
56. Keil, R. G.; Montluçon, D. B.; Prahl, F. G.; Hedges, J. I. *Nature* **1994**, *370*, 549-552.
57. Mayer, L. M. *Chem. Geol.* **1994**, *114*, 347-363.
58. Derenne, S.; Largeau, C.; Taulelle, T. *Geochim. Cosmochim. Acta* **1993**, *57*, 851-857.
59. Verma, L.; Martin, J. P.; Haider, K. *Soil Sci. Soc. Amer. Proc.* **1975**, *39*, 279-284.
60. Martin, J. P.; Parsa, A. A.; Haider, K. *Soil Biol. Biochem.* **1978**, *10*, 483-486.
61. Kohnen, M. E. L.; Sinninghe Damste, J. S.; ten Haven, H. L.; de Leeuw, J. W. *Nature* **1989**, *341*, 640-641.
62. Putschew, A.; Scholz-Böttcher, B. M.; Rullkötter, J. *Org. Geochem.* **1996**, *25*, 379-390.
63. Eglinton, G.; Logan, G. A. *Phil. Trans. R. Soc. Lond.* B **1991**, *333*, 315-328.
64. Law, B. A. In *Microorganisms and Nitrogen Sources: Transport and Utilization of Amino Acids, Peptides, Proteins, and Related Substrates*; Payne, J. W., Ed.; Wiley: New York, NY, 1980; pp 381-410.
65. Payne, J. W. In *Microorganisms and Nitrogen Sources: Transport and Utilization of Amino Acids, Peptides, Proteins, and Related Substrates*; Payne, J. W., Ed.; Wiley: New York, NY, 1980; pp 211-256.

Chapter 8

Early Diagenetic Transformations of Proteins and Polysaccharides in Archaeological Plant Remains

Helen A. Bland, Pim F. van Bergen[1], James F. Carter, and Richard P. Evershed[2]

Organic Geochemistry Unit, School of Chemistry, University of Bristol, Cantock's Close, Bristol BS8 1TS, United Kingdom

The chemical compositions of archaeological radish seeds (1400 years) and barley kernels (600 years), which have been held in a desiccated environment since deposition, and modern counterparts, have been studied to reveal changes occurring in their protein and polysaccharide composition. Pyrolysis-gas chromatography/mass spectrometry (Py-GC/MS) of barley storage tissue has revealed compositional changes in the dominant starch complex in the ancient specimens compared with their modern counterparts. For the radish storage tissue, pyrolysates are dominated by proteinaceous pyrolysis products, which show little difference in composition between the modern and ancient specimens. Analyses of volatiles released from the propagules upon crushing have revealed products of the advanced stages of the Maillard reaction between amino acids and sugars, most notably alkyl pyrazines and alkyl polysulphides. Despite the evidence for this reaction, the overall findings indicate that degradation of proteins and polysaccharides is limited.

The survival of organic matter into the archaeological and geological record is highly dependent on its original biomolecular composition and the environment into which it was deposited (1). However, the processes which govern the initial stages of diagenesis are yet to be fully understood. The major problems that exist include the chemically complex nature of the degradation products of biological matter and the lack of knowledge of the subtle but often significant physico-chemical differences between burial environments that influence preservation (1). For example, ancient mammalian soft-tissue remains have been recovered showing remarkable

[1]Present address: Organic Geochemistry Group, Faculty of Earth Sciences, Utrecht University, P.O. Box 80021, 3508 TA Utrecht, The Netherlands.
[2]Corresponding author.

morphological preservation from anoxic, acidic and waterlogged deposits (*2, 3*), where they appear to be preserved through a natural 'tanning' process, from permafrost (*4*), and locations of extreme aridity (*5*) where desiccation has resulted in natural mummification. In most instances, however, soft tissue is not preserved, and only bones and teeth survive (*6*). While it has been assumed that the morphological preservation will also be reflected in the component biomolecules, this is often not the case. For example, morphological preservation in bog bodies results from the unusual preservative properties of the peat environment for collagen (*3*, and references therein). Lipids are also preserved in such an environment (*7, 8*), however DNA is extensively degraded (*9*).

Biopolymers are most susceptible to attack and modification at the linkages between component moieties (for example, ester links between fatty acids in triacylglycerols, and acetal links between constituent sugars in polysaccharides, *1*). Whilst these linkages can be broken by a variety of processes, micro-organisms have been shown to be most highly pervasive and can metabolise most organic matter given the correct conditions. Environments showing extremes of pH, temperature or salinity restrict microbial activity and slow the otherwise rapid decay of detrital organic matter (*1*). The examples given above attest to this, and hence have allowed study of the physico-chemical transformations occurring during early diagenesis.

Ancient plant remains (of both geological and archaeological age) have been reported to survive through a range of processes including charring, partial mineralisation, and deposition in both desiccating and anoxic waterlogged environments (*10, 11*). The use of morphologically well-defined plant remains in the study of organic matter decay has an advantage in that the uncertainty raised by biomolecular input from multiple organic sources (for example, that occurring in studies of amorphous sedimentary organic matter) is minimised. Hence, the biomolecular composition of leaves, seeds, wood and other such morphologically well-defined entities can be studied in comparison with modern counterparts, enabling clearer understanding of compositional changes and transformations of individual classes of biomolecule occurring within the material. For example, the amplification of ancient DNA has been reported for desiccated radish seeds (1400 years old; *12*) and mummified maize seeds (ca.1000 years) uncovered from a tomb on the Peruvian coast yielded DNA strands of 90 and 130 base pairs (*13, 14*). Evidence for the survival of nucleic acids as rRNA was shown through molecular hybridisation and high performance liquid chromatography (HPLC; *14*). These studies show considerable fragmentation of the polynucleotide chains compared with modern DNA. Protein preservation in barley kernels (1000-3000 years) recovered from an Egyptian tomb has been studied (*15*). Gas chromatographic analyses, after hydrolysis, showed that amino acids were present in the ancient barley in a similar distribution to that found in modern samples, while electrophoresis showed that the proteins were at least partially degraded. Further evidence for the survival of proteinaceous moieties, and also that of ligno-cellulose, on a geological timescales has been reported using pyrolysis-gas chromatography/mass spectrometry (GC/MS) of water plant seed coats deposited under aqueous conditions (*11, 16*).

We have studied plant remains (seeds and fruits) from an archaeological site in Egyptian Nubia (Qasr Ibrîm), an area of extreme aridity extending across the border between modern day Egypt and Sudan. Once a hilltop fortress and trading centre overlooking the Nile, this settlement has provided large quantities of organic material exhibiting exceptional morphological preservation. The arid nature of the site means that microbial activity is greatly reduced, providing an excellent opportunity to study chemical and autolytic enzymatic degradation occurring in the initial stages of diagenesis.

Specimens from the site have already provided evidence for survival of nucleic acids, and purine and pyrimidine bases have been extracted from 1400 year old radish seeds and characterised using GC/MS/MS and HPLC/MS (*17, 18*). Extraction and amplification of ancient DNA from the same samples using PCR yielded DNA strands of 179 base pairs indicating the fragmentation of DNA through hydrolysis of the N-glycosyl bond leading to β-elimination and strand cleavage (*12, 19*). The transformations of proteins and ligno-cellulose in seed coats of these ancient radish propagules and the kernel walls of ancient barley (1400 years old) have been assessed using Curie-point pyrolysis-GC/MS (*20*). They showed high quality preservation of these macromolecules although some chemical alteration, including the loss of ester-linked hemicellulose and cinnamic acid moieties, was observed (*21*). Although studies of lipid preservation have also revealed severe retardation of degradative processes (*12, 20*), triacylglycerols have mostly, but not entirely, been hydrolysed to diacylglycerols and component free fatty acids. In radish (*Raphanus sativus*) seeds, triacylglycerol composition accounts for over 98% of total lipid in the modern specimens and less than 10% in the ancient, and in barley (*Hordeum vulgare*) kernels, triacylglycerol composition accounts for 76% in modern and 2% in ancient specimens. Although there is some loss of abundance in the mono- and polyunsaturated free fatty acids, $C_{18:2}$ and $C_{18:3}$ are still present in significant quantities, thus reflecting the retardation of oxidative processes. Furthermore, the distribution of sterol components remains essentially unchanged between the modern and ancient propagules (*12*).

In this contribution we report the results of the further application of pyrolysis (610°C)-GC/MS to investigate the quality of preservation of high molecular weight constituents in ancient radish (*Raphanus sativus*, 1400 years) and barley (*Hordeum vulgare*, 600 years), by comparison with their modern counterparts. These studies have been extended to include analyses of volatile degradation products trapped within the propagules (*22*). These volatile components were characterised by headspace concentration-GC/MS and provide new insights into the pathways involved in the transformation of proteins (note: the term 'protein', in this text, refers to all proteinaceous material whether it be proteins, peptides or amino acids) and polysaccharides in buried plant remains.

Experimental

Whole samples were powdered in a mortar and pestle with liquid nitrogen and the lipids extracted with chloroform/methanol solution (2:1 v/v, 10 ml x 15 mins sonication x 3). The solvent containing the total lipid extract (TLE) was then decanted from the remainder of the sample. Excess solvent was evaporated under a

116

stream of nitrogen, and the residue dried over silica. The outer walls of the extant and ancient propagules were removed from the internal tissue by dissection (*20*).

Pyrolysis-gas chromatography/mass spectrometry (Py-GC/MS). Py-GC/MS was performed using a Carlo Erba 4130 GC coupled to a Finnigan 4500 mass spectrometer (MS). A CDS (Chemical Data System) 1000 Pyroprobe pyrolysis unit was used with the both the injector and interface temperatures set to 250°C. Solvent extracted samples were weighed using a microanalytical balance (using 50 to 100 μg of sample) into quartz pyrolysis tubes (CDS 1000) and pyrolysed at 610°C (10 secs). Separation of pyrolysis products was achieved using a CPSil-5 CB fused silica capillary column (50 m, 0.32 mm i.d., film thickness 0.4 μm). The GC oven was programmed from 35°C (5 mins) to 310°C (15 mins) at 4°C min^{-1}. Helium was the carrier gas. The mass spectrometer was operated at 70 eV scanning over the range *m/z* 35-650 at 1 scan s^{-1} with emission current of 300 μA (full scan mode). The pyrograms of the ancient and modern specimens were normalised to the same vertical scale (total ion current) based on the weight of sample used. Compounds were identified by comparison of their mass spectra and retention times with those reported in the literature (*20, 23-33*) and in the NIST mass spectral library.

Headspace concentration-gas chromatography/mass spectometry (headspace-GC/MS). Whole weighed propagules were crushed in liquid nitrogen and the powder placed immediately in the glass headspace extraction apparatus. The apparatus was then immersed in a water bath kept at a constant temperature of either 35 or 40°C. Volatiles were collected on a Tenax cartridge by flushing the sample vessel with a stream of dry charcoal-filtered nitrogen for 30 mins (flow rate 25 ml min^{-1}). The cartridge was then removed and placed in a modified GC injection port, and the volatiles desorbed from the cartridge at 250°C and cryogenically focused at the head of the GC column. Volatile analysis of blank collections and whole propagules showed no detectable components when compared with crushed samples when normalised to the same vertical scale (total ion current).

Analyses were carried out using a Carlo Erba 4130 GC connected to a Finnigan 4500 MS. Compounds were separated on a CPSil-19 CB fused silica capillary column (50 m, 0.32 mm i.d., film thickness 0.2 μm). The GC oven was temperature programmed from 35°C (4 min) to 300°C (20 min) at a rate of 6°C min^{-1}. The MS was operated using the same parameters as described for pyrolysis-GC/MS (above).

Compounds were identified using both their mass spectra and retention times, comparing these with data in the literature (*34*) and from standard mixtures analysed concurrently. Isomers of alkyl pyrazines were distinguished by calculation of linear retention indices (LRIs) and compared with the LRIs of authentic compounds.

Results

Py-GC/MS. Analyses of the storage tissue of the ancient and modern radish and barley described below compliment our earlier work on the composition of external tissues (*20*). However, the storage tissue contains the majority of proteinaceous components (*35*).

Radish. The pyrolysate of solvent extracted storage tissue of the modern radish was dominated by 4-ethenyl-2,6-dimethoxyphenol (syringyl moiety; **t**), seen in Figure 1A, which is known to be a major pyrolysis product of dicotyledenous angiosperm lignin (*24*). Other characteristic lignin derived products include the unsubstituted (**p**) and ethyl substituted (**s**) 2,6-dimethoxyphenols and 4-ethenyl-2-methoxyphenol (guaiacyl moiety; **o**). Aside from the syringyl and guaiacyl derived products, the pyrolysate reveals numerous pyrolysis products which are known to derive from proteins (*25*): phenol [also known to derive from pyrolysis of lignin (*24*); **h**], 3 or 4-methylphenol (**i**), pyrrole (**b**), 2- and 3- methylpyrrole (**d, e**), and nitrogen-containing benzene derivatives including benzenacetonitrile (**j**), benzenepropanenitrile (**m**), indole (**n**) and methylindole (**q**). Also apparent are the 2,5-diketopiperazines with characteristic fragment ions of m/z 70, 154 and 194 (**v-y**) and diketodipyrrole (**u**), which originate from dipeptides and proteins (*32, 33*).

The pyrolysate of the ancient radish storage tissue indicates that the overall composition is essentially unchanged compared with the modern tissue (Figure 1B). Slight differences are seen in the relative abundance of some components; most significantly, the pyrolysate of the ancient storage tissue shows a marked decrease in abundance of the 4-ethenyl-2,6-dimethoxyphenol (**t**). However, two components which have yet to be characterised [(**k**) displaying significant fragment ions of m/z 127, 87, 57, 45, and (**r**) m/z 143, 114, 100, 87, 64, 53] are almost completely absent from the ancient specimens. The abundance of diketodipyrrole (**u**) and the 2,5-diketopiperazines (**v-y**) are greatly increased with respect to the rest of the pyrolysate and, interestingly, there is a slight increase in the overall abundance of these components when scaled with the modern pyrolysate.

In the pyrolysates of both ancient and modern radish storage tissue, polysaccharide pyrolysis products are of low relative abundance compared with those of proteinaceous origin. There is a significant contribution from 1,4:3,6-dianhydro-α-D-glucopyranose (**l**) in the modern specimen, however, this is depleted in the ancient storage tissue. Other polysaccharide pyrolysis products, including 2,5-dimethyl furan, were only identifiable by mass chromatography (m/z 96) but the low abundance and presence of co-eluting components hindered additional definitive identifications.

Barley. The pyrolysate of the modern barley storage tissue is dominated by a large fronting peak relating to 1,6-anhydro- β-D-glucopyranose [levoglucosan (**35**), Figure 2A] derived from the glucose moieties in the starch macromolecule that constitutes the storage tissue. There are also significant contributions from other polysaccharide pyrolysis products, including 2,3-dihydro-5-methylfuranone (**15**), 2-hydroxy-3-methyl-2-cyclopenten-1-one (**20**), 5-hydroxymethyl-2-tetra-hydrofuraldehyde-3-one (**27**), 1,4:3,6-dianhydro- α-D-glucopyranose (**29**) and 1,4-dideoxy-D-*glyco*-hex-l-enopyranose-3-ulose (**33**). A contribution from lignin is apparent from the various methoxyphenols, 4-ethyl-2-methoxyphenol (**32**) and 4-ethenyl-2-methoxyphenol (**34**). Nitrogen-containing pyrolysis products deriving

118

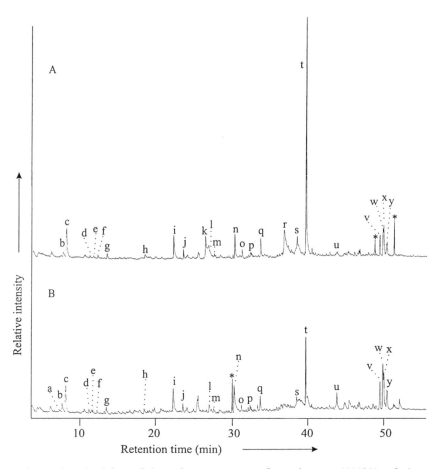

Figure 1. Partial total ion chromatograms of pyrolysates (610°C) of the insoluble storage tissue of (A) modern and (B) ancient radish (*Raphanus sativus*). Letters on peaks refer to compounds listed in Table I. Samples were weighed and pyrograms normalised to the same vertical scale (total ion current) based on weight of sample used. * = contaminant.

Table I Compounds present in the pyrolysates of ancient and modern radish storage tissue (Figure 1) detailing characteristic fragment ions (*m/z*) with their relative abundances. Ions underlined are the molecular ion and those in bold the base peak.

No.	Compound	Origin[1]	Characteristic fragment ions *m/z* (relative abundance)
a	Pyridine	Ala, AA/PS	79, 52 (55), 39 (60)
b	Pyrrole	Pro, Hyp, Glu	67, 41 (35), 40 (40)
c	Toluene	Phe	92 (70), 91, 65 (10), 63 (5)
d	2-methylpyrrole	Pro, Hyp	81 (70), 80, 53 (30), 40 (20)
e	3-methylpyrrole	Pro, Hyp	81 (70), 80, 53 (30), 40 (20)
f	Ethylbenzene	Phe	106 (30), 91, 51 (10),
g	Styrene	Phe	104, 103 (40), 78 (30), 51 (20)
h	Phenol	Tyr	94, 66 (20), 65 (20), 40 (40)
i	3(or 4)-methyl phenol	Tyr	108 (85), 107, 79 (15), 77 (20)
j	Benzenacetonitrile	Phe	117, 116 (40), 90 (40), 89 (25)
k	Unknown	~	~127 (60), 87, 53 (15), 45 (50)
l	1,4:3,6-dianhydro-α-*D*-glucopyranose	PS	144 (10), 98 (20), 69, 57 (55)
m	Benzenepropanenitrile	Phe	131 (30), 110 (10), 91, 65 (10),
n	Indole	Trp	117, 90 (35), 89 (25), 63 (10)
o	4-ethenyl-2-methoxyphenol	Lignin	150, 135 (65), 107 (20), 91 (10), 77 (20), 39 (5)
p	2,6-dimethoxyphenol	Lignin	154, 139 (45), 111 (25), 93 (25), 65 (25), 39 (30)
q	Methylindole	Trp	131 (55), 130, 103 (5), 77 (10)
r	Unknown	~	~143 (45), 114 (50), 100 (40), 87 (30), 64 (40), 53
s	4-ethyl-2,6-dimethoxyphenol	Lignin	182 (70), 167, 77 (10), 39 (10)
t	4-ethenyl-2,6-dimethoxyphenol	Lignin	180, 165 (30), 137 (20), 77 (10), 65 (5), 39 (5)
u	Diketodipyrrole	Hyp-Hyp	186, 138 (10), 93 (40), 65 (15)
v	2,5-diketopiperazine deriv.	Pro-AA	154, 125 (15), 86 (15), 70 (40), 55 (5), 41 (15)
w	2,5-diketopiperazine	Pro-Pro	194 (90), 166 (10), 138 (10), 96 (20), 70, 55 (10), 41 (30)
x	2,5-diketopiperazine deriv.	Pro-AA	194 (20), 167 (5), 154, 125 (10), 96 (5), 70 (55), 41 (15)
y	2,5-diketopiperazine deriv.	Pro-AA	~154, 125 (15), 96 (5), 70 (45), 55 (10), 41 (15)

[1]AA = amino acid (general or unknown); PS = polysaccharide pyrolysis product; AA/PS = amino acid/polysaccharide complex

120

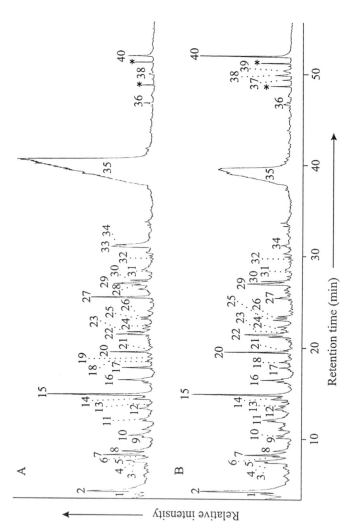

Figure 2. Partial total ion chromatograms of pyrolysates (610°C) of the insoluble storage tissue of (A) modern and (B) ancient barley (*Hordeum vulgare*). Numbers on peaks refer to compounds listed in Table II. Samples were weighed and pyrograms normalised to the same vertical scale (total ion current) based on weight of sample used. * = contaminants.

Table II Compounds present in the pyrolysates of ancient and modern barley storage tissue (Figure 2) detailing characteristic fragment ions with their relative abundances. Ions underlined are the molecular ion and those in bold the base peak.

No.	Compound	Origin[1]	Characteristic fragments ions m/z (relative abundance)
1	Acetic acid	St	$\underline{60}$ (65), **45**, 43 (95)
2	Hydroxypropanal	St	$\underline{74}$ (10), **43**, 42 (5)
3	Pyridine	Ala, AA/PS	$\underline{79}$, 52 (55), 39 (60)
4	Pyrrole	Pro Hyp Glu	**67**, 41 (35), 40 (40)
5	(3H) furan-2-one	St	$\underline{84}$, 55 (80), 39 (10)
6	Toluene	Phe	$\underline{92}$ (70), **91**, 65 (10),
7	Pyruvic acid, methyl ester	St	$\underline{102}$ (20), **43**
8	(2H) furan-3-one	St	**84**, 55 (25), 54 (80)
9	2,4-pentadienal	St	**82**, 54 (25), 53 (30), 39 (40)
10	2-furaldehyde	St	**96**, 95 (85), 39 (45)
11	2-(hydroxymethyl)furan	St	**98**, 97 (50), 81 (50), 70 (50), 69 (70), 39 (70)
12	(5H) furan-2-one	St	$\underline{84}$ (50), **55**, 54 (20)
13	2-methyl-2-cyclopenten-1-one	St	**96**, 68 (10), 67 (90), 53 (60), 41 (40)
14	2,5-dihydro-5-methylfuran-2-one	St	$\underline{98}$ (45), 97 (25), **70**, 69 (30), 55 (30), 43 (60)
15	2,3-dihydro-5-methylfuran-2-one	St	**98**, 70 (10), 69 (25), 55 (55), 43 (30), 42 (40)
16	5-methyl-2-furaldehyde	St	**110**, 109 (85), 81 (15), 53 (50), 43 (80)
17	4-hydroxy-5,6-dihydro-(2H)-pyran-2-one	He	**114**, 58 (50), 57 (25)
18	Phenol	Li (Tyr)	**94**, 66 (25), 65 (20), 40 (15)
19	3-hydroxy-2-methyl-2-cyclopenten-1-one	St	**112**, 84 (10), 71 (15), 69 (15), 53 (15), 42 (15)
20	2-hydroxy-3-methyl-2-cyclopenten-1-one	St	**112**, 84 (15), 83 (20), 69 (50), 56 (20), 55 (40)
21	2,3-dimethyl cyclopenten-1-one	St	$\underline{110}$ (85), 83 (15), **67**, 53 (35), 39 (45)
22	unknown (furanose)	St	~97 (5), 81 (5), 68 (10), 57 (70), **44**
23	2-(propan-2-one) tetrahydrofuran	St	$\underline{128}$ (15), **71**, 43 (35)
24	4-methylphenol	Li (Tyr)	$\underline{108}$ (75), **107**, 77 (20), 53 (20)
25	3-hydroxy-2-methyl-(4H)-pyranone	St	**126**, 98 (10), 97 (15), 71 (30), 55 (25), 43 (50)
26	unknown	~	~128 (20), 100 (10), 98 (15), 70 (20), **42**

Continued on next page.

Table III. *Continued*

No.	Compound (cont.)	Origin (cont.)	Fragment ions (cont.)
27	5-hydroxymethyl-2-tetrahydro-furaldehyde-3-one	St	<u>144</u> (tr), 85 (30), 70 (35), 69 (60), 57 (45), **43**, 41 (75)
28	3,5-dihydroxy-2-methyl-(*4H*)-pyran-4-one	St	**<u>142</u>**, 113 (10), 98 (5), 85 (20), 68 (15)
29	1,4:3,6-dianhydro-α-*D*-glucopyranose	St	<u>144</u> (10), 98 (20), **69**, 57 (55), 41 (35)
30	Unknown	~	~**154**, 125 (30), 70 (50)
31	4-ethenylphenol	Li (Tyr)	**<u>120</u>**, 119 (30), 91 (50), 65 (20), 51 (10), 39 (25)
32	4-ethyl-2-methoxyphenol	Li	<u>152</u> (65), **137**, 122 (5), 81 (25), 51 (50, 43 (45)
33	1,4-dideoxy-*D*-glycerohex-1-enopyranos-3-ulose	St	**<u>144</u>**, 97 (55), 87 (80), 69 (20), 57 (60), 43 (35)
34	4-ethenyl-2-methoxyphenol	Li	**<u>150</u>**, 135 (70), 107 (20), 77 (20), 51 (10), 39 (15)
35	1,6-anhydro-β-*D*-glucopyranose	St	<u>162</u> (tr), 73 (30), **60**, 57 (45), 43 (20), 42 (15)
36	2,5-diketopiperazine deriv.	Pro-AA	~**154**, 125 (30), 70 (50)
37	2,5-diketopiperazine deriv.	Pro-AA	**<u>154</u>**, 125 (10), 86 (15), 70 (50), 41 (15)
38	2,5-diketopiperazine	Pro-AA	<u>194</u> (30), **154**, 125 (10), 96 (10), 86 (15), 70 (85), 41 (20)
39	2,5-diketopiperazine deriv.	Pro-AA	**<u>154</u>**, 125 (15), 86 (5), 70 (35), 41 (10)
40	Hexadecanoic acid	Lipid	<u>256</u> (35), 213 (35), 185 (20), 129 (55), **73**, 43 (85)

[1] St = Starch; He = Hemicellulose; AA/PS = amino acid/polysaccharide complex

from proteins are also present, including: pyridine (**3**), pyrrole (**4**), indole (not shown) and benzenacetonitrile (presence of both confirmed by mass chromatography, *m/z* 117), and 2,5-diketopiperazines (**36-39**).

In the ancient barley there is a quantitative reduction in the overall yield of pyrolysate compared with that produced from the modern barley storage tissue (normalised as described in the experimental section, see figure 2), most noticeable in the reduction in abundance of 1,6-anhydro- β-*D*-glucopyranose (**35**). However, there is an almost complete absence of pyrolysis products relating to 4-ethenyl-2-methoxyphenol (**34**) and various polysaccharide pyrolysis products [5-hydroxymethyl-2-tetrahydrofuraldehyde-3-one (**27**), 1,4:3,6-dianhydro- α-*D*-glucopyranose (**29**) and 1,4-dideoxy-*D*-*glycero*-hex-1-enopyranos-3-ulose (**33**)], all of which were present in significant abundances in modern specimens. The pyrolysate is dominated by 2,3-dihydro-5-methylfuran-2-one (**15**) and hydroxypropanal (**2**), with hexadecanoic acid (**40**) appearing to have greatly increased in abundance (though this could be because of incomplete solvent extraction). There appears to be little change in abundances of the nitrogen-containing pyrolysis products in the ancient barley storage tissue compared with that seen in modern samples, especially the 2,5-diketopiperazines (**36-39**) which are more prominent in the ancient propagules.

Headspace Concentration-GC/MS

Radish. The volatile components released from crushing ancient radish seeds were dominated by sulphur-containing compounds; dimethyldisulphide (**1**), dimethyltrisulphide (**9**) and dimethyltetrasulphide (**26**) (Figure 3, Table III). These compounds have a very low odour threshold value (*36*) and are responsible for the pungent odour of the seeds when crushed. Also significant were alkyl pyrazines; methyl (**4**), dimethyl [2,5- (**6**) and 2,6- (**7**)], 2-ethyl-[5- (**11**) and 6- (**10**)]-methyl and trimethyl (**12**); and a notable contribution from 2,5-dimethyltetrahydrofuran (**18**). A series of 2-alkanones and *n*-alkanals were also present, encompassing the carbon number range C_6 to C_9.

Barley. The composition of the volatile components of the crushed ancient barley differed significantly from that of radish; most noticeable being the absence of the dominant sulphur compounds seen in the radish. Instead, the alkyl pyrazines are the more abundant components, showing the same range of isomers present in ancient radish volatiles together with more highly substituted components including, 2-ethyl-[3,5- (**15**) and 3,6- (**14**)]-dimethyl, 2,5-diethyl (**16**), 2-propyl-(5 or 6)-methyl (**17**), 2,6-diethyl-3-methyl (**22**) and 2-propyl-dimethyl (**23**) pyrazines. A series of 2-alkanones and *n*-alkanals is again present and appears to be relatively more abundant than in the volatile components of the crushed ancient radish.

Discussion

Py-GC/MS — Radish. 4-ethenyl-2,6-dimethoxyphenol (**5**), the most abundant compound in both the ancient and modern radish storage tissue, is a characteristic pyrolysis product of

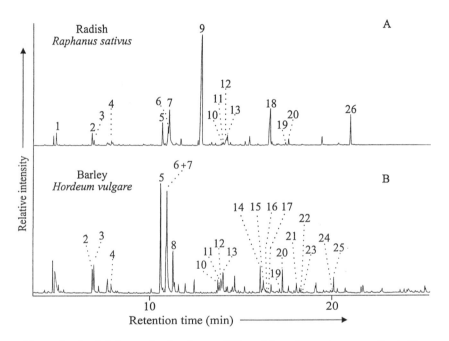

Figure 3. Partial desorption headspace-GC total ion chromatograms of volatile compounds released from ancient (A) radish seeds and (B) barley kernels upon crushing. Numbers on peaks refer to individual compounds listed in Table III.

Table III The volatile components of ancient radish and barley detailing characteristic fragment ions, with their relative abundances, and linear retention indices (LRIs, see experimental section). Ions underlined indicate molecular ion and those in bold the base peak.

No.	Compound	Ancient Radish	Ancient Barley	Characteristic fragment ions (m/z)	LRI CPSil 19
1	Dimethyldisulphide	✓	x	94 (70), 79 (45), 64 (10), 47 (30), **45**	805
2	2-hexanone	✓	✓	100 (10), 85 (5), 70 (5), 58 (40), **43**	
3	n-hexanal	✓	✓	100 (tr), 82 (10), 72 (15), 57 (40), 56 (60), **44**	
4	Methylpyrazine	✓	✓	**94,** 67 (90), 53 (35), 42 (35), 39 (75)	905
5	2-heptanone	✓	✓	114 (5), 85 (5), 71 (15), 58 (55), **43**	
6	2,5-dimethylpyrazine	✓	✓	**108**, 81 (10), 42 (85), 39 (35)	988
7	2,6-dimethylpyrazine	✓	✓	**108**, 93 (5), 81 (5), 67 (5), 42 (50), 39 (35)	990
8	n-heptanal	x	✓	114 (tr), 70 (5), 55 (20), **43**	
9	Dimethyltrisulphide	✓	x	**126**, 111 (35), 79 (70), 64 (50), 47 (60), 45 (80)	1042
10	2-ethyl-6-methyl-pyrazine	✓	✓	**122** (60), **121**, 94 (10), 56 (10), 42 (5), 39 (20)	1072
11	2-ethyl-5-methyl-pyrazine	✓	✓	122 (65), **121**, 94 (10), 56 (15), 39 (20)	1075
12	Trimethylpyrazine	✓	✓	**122**, 121 (35), 81 (15), 42 (70)	1078
13	2-octanone	✓	✓	128 (5), 85 (5), 71 (10), 58 (65), **43**	
14	2-ethyl-3,6-dimethyl-pyrazine	x	✓	136 (75), **135**, 108 (10), 56 (10), 42 (25)	
15	2-ethyl-3,5-dimethyl-pyrazine	x	✓	136 (75), **135**, 108 (10), 56 (25), 42 (15)	
16	2,5-diethyl-pyrazine	x	✓	136 (90), 135 (95), **121**, 108 (10), 56 (20), 39 (35)	1158
17	2-propyl-5 (or 6)-methyl-pyrazine	x	✓	136 (15), 121 (20), **108**, 94 (5), 39 (25)	
18	2,5-dimethyl-tetrahydrofuran	✓	x	100 (10), 85 (70), **56**, 41 (80)	
19	2-nonanone	✓	✓	**142** (5), 71 (15), 58 (85), **43**	
20	n-nonanal	✓	✓	142 (tr), 127 (5), 114 (5), 98 (15), 74 (40), 57 (80), **41**	
21	C5-pyrazine	x	✓	**150**, 149 (70), 135 (80), 121 (25), 94 (70), 56 (65), 39 (95)	
22	2,6-diethyl-3-methyl-pyrazine	x	✓	**150** (70), **149**, 135 (10), 122 (15), 107 (5), 53 (35), 39 (45)	
23	2-propyl-dimethyl-pyrazine	x	✓	150 (5), 135 (15), **122**, 107 (10), 39 (20)	
24	2-decanone	x	✓	156 (5), 98 (5), 85 (5), 71 (20), 58 (80), **43**	
25	n-decanal	x	✓	**156** (tr), 128 (5), 112 (15), 82 (25), 70 (30), 57 (70), **41**	
26	Dimethyltetrasulphide	✓	x	**158** (85), 111 (10), 94 (20), 79 (90), 64 (45), 47 (50), **45**	1258

126

angiosperm lignin (24, 26). The presence of this component, together with other lignin derived products (n, o, r), indicates that a lignin-like biopolymer is a structural component of the internal radish tissue. Lignin, being one of the most resistant organic macromolecules known (1, 37), is predominantly associated with cellulose in woody tissues, cereal stalks and grasses, and the external walls of plant cells (24). Although, the lignin structure varies with different organs and taxa, in each case the pyrolysate is usually dominated by four 2-methoxyphenol (guaiacyl) or 2,6-dimethoxyphenol (syringyl) derivatives; unsubstituted, 4-methyl, 4-ethenyl and 4-(2-E-propenyl) (11). In the case of radish, the pyrolysate appears to reveal a rather unusual form of this macromolecule characterised by only one major pyrolysis product. Furthermore, lignification causes inherent hardening and the presence of a lignin-type macromolecule within the storage tissue rather than the seed coat would have to be highly localised or, conversely, delocalised to ensure that germination was not impeded. The decrease in abundance of the 4-ethenyl-2,6-dimethoxyphenol in the ancient propagules is consistent with the trends seen previously during fossilisation of dicotyledonous lignin in seed coats of samples preserved in aquatic deposits, in particular, preferential degradation of 2,6-dimethoxyphenols compared with 2-methoxyphenols (38).

Other major pyrolysis products seen in the modern radish storage tissue derive from protein moieties (amino acid and peptides). Pyrrole and methyl substituted analogues have been shown to be formed upon pyrolysis of proline, hydroxyproline and glutamine (25), benzene derivatives (toluene, ethylbenzene and styrene) and benzonitrile compounds (benzenacetonitrile and benzene-propanenitrile) are produced from pyrolysis of phenylalanine (25, 31), phenolic compounds (phenol and methyl phenol) can derive from tyrosine (25), and indole and methylindole are formed from tryptophan (25). These data agree with previous reports that radish storage proteins (including napins) contain a high proportion of proline and glutamic acid (35, 39-40) with lower but still significant proportions of cysteine, alanine and arginine (35, 39-40).

The abundance of proline in radish storage tissue is further confirmed by the presence of diketopiperazine; this component, deriving from pyrolysis of a proline-proline dipeptide moiety (w), gives a characteristic ion of m/z 194 and base peak of m/z 70 (32). Two other 2,5-diketopiperazine derivatives shown (v and y) have similar mass spectra to those reported for proline-arginine (Pro-Arg) and proline-valine (Pro-Val) dipeptide derived pyrolysis products (3, 32). Diketodipyrrole (u) has been reported as forming from hydroxyproline alone and dipeptide (Hyp-Hyp) moieties. This indicates the presence of this amino acid in the storage tissue of radish and hence supports the proposed contribution of hydroxyproline to the pyrrole pyrolysis products (3, 23, 25). Hydroxyproline is known as a major constituent of glycoproteins in plant cell walls, which are thought to play a structural role (41).

Comparing the data obtained during this work with data published on the macromolecular constituents of seed coats of these modern and ancient radish samples (20), it can be said that the pyrolysis products relating to amino acids are consistent between the external and internal tissues, including the diketodipyrrole and diketopiperazines. The phenolic macromolecule (20) associated with the seed coat is known to display antifungal properties and may have contributed to the excellent preservation of protein derived moieties in the storage tissue. The pyrolysates of both ancient and modern radish seed coats (20) display significantly more polysaccharide pyrolysis products than those of the storage tissue. Since the separation of seed coat from storage tissue is very effective, it is assumed that only trace quantities of contamination occur between the two tissues.

Py-GC/MS — Barley. The pyrolysate of the modern barley storage tissue is consistent with data reported for the storage tissue of cereals, showing a high proportion of polysaccharide derived products (starch; 42). The pyrolysate is very similar to that produced from cellulose (27), reflecting the similarity in the macromolecular constituents between this polysaccharide and starch (amylose, α-1,4 linked glycosyl units, and amylopectin, α-1,6 linked glucosyl units, for starch and β-1,4-linked D-glucosyl units for cellulose; 42). The decrease in abundance of 1,6-anhydro- β-D-glucopyranose (**35**) and other polysaccharide pyrolysis products, in the ancient sample (Figure 2), agrees with the preferred pathways of degradation for monocotyledonous lignin-cellulose complexes in aquatic depositional environments (38), namely removal of polysaccharides. The contribution from lignin can be rationalised as either being 'contamination' from the kernel wall, or more likely, arising from macromolecular material involved in compartmentalising the starch grains. The reduction in the abundance of 4-ethenyl-2-methoxyphenol, most probably deriving from ferulic acid, is comparable to the preferential degradation of this phenolic acid (where it is ester-linked to polysaccharides) in other depositional settings (20).

The appearance of protein derived moieties in both the modern and ancient barley samples is not surprising as the storage tissue is the principle reserve of nitrogen and sulphur prior to germination (35). The presence of these compounds contrasts with the pyrolysate of the kernel wall which showed no detectable protein markers (20). Pyridine is reported to form from pyrolysis of both α and β-alanine (43) and peptide/polysaccharide complexes, whilst the origins of pyrrole, toluene and phenol have been described in the discussion of radish storage tissue (see above). The protein group, prolamins, which constitute up to 50% of N-containing compounds in modern barley storage tissue, are high in proline and glutamine, and low in lysine and threonine (35). These data therefore support the contribution of proline and glutamic acid towards pyrrole, and also of proline toward the 2,5-diketopiperazine derivatives detected in the radish.

In comparing the biomolecular composition of barley storage tissue and kernel walls, the compositional differences seen in the pyrolysis products reflects their differing functions. The storage tissue is comprised predominantly of starch, the principle reserve of carbohydrates in plants (42), whilst the kernel wall displays a typical angiosperm lignin-cellulose complex which forms a strong protective layer around the interior of the propagule (20). As mentioned above, the storage tissue of both ancient and modern barley samples contains significantly more pyrolysis products relating to proteinaceous material than the kernel wall (20).

Headspace Concentration-GC/MS — Radish. The sulphur and nitrogen-containing compounds identified in the volatiles of ancient radish are formed through the Maillard, or browning, reaction (*22, 36*). This term encompasses reactions occurring between amino acids, peptides, proteins and amines with reducing sugars, aldehydes and ketones (*44*). The initial stage involves condensation to form a Schiff base, then cyclisation to an *N*-substituted glycoylamine. The next step, the Amadori rearrangement, involves the transition from an aldose to a ketose sugar derivative and up to this point the reaction is reversible. At more advanced stages, flavour and odour compounds are formed (usually heterocyclic compounds; *34*), as are brown nitrogenous polymers that give the reaction the alternative name of the 'browning' reaction (*44*). The alkyl polysulphides are one of the classes of compound formed in these later stages via a process known as Strecker degradation of sulphur containing amino acids (cysteine and methionine) which also yields hydrogen sulphide.

Of most significance, however, are the alkyl pyrazines since they are highly diagnostic products of the Maillard reaction (*36*). Alkyl pyrazines are extremely uncommon natural products [existing in some pheromones (e.g. *45*) or in coffee beans (*46*)]. They, too, are thought to form via condensation of dicarbonyls with an amino compound via Strecker degradation to form α-aminoketones, which then self condense or condense with other aminoketones to produce a dehydropyrazine that is oxidised to the alkyl pyrazine (*47*). The alkyl substituents of the pyrazines derive from the carbonyl group rather than the amino acid so it is not possible to hypothesise as to their precise amino acid origin. However, it should be noted that two of the most abundant amino acids in radish, proline and hydroxyproline, do not produce aminoketones and Strecker aldehydes by this reaction (to form pyrazines) as they contain a secondary amino group in a pyrrolidine ring (they form pyrrolidines, piperidines and pyrrolizines which were not seen in these ancient propagules; *36, 48*).

Although this reaction is best recorded in heated foodstuffs, in model systems the reaction has also been shown to progress at a severely reduced rate at lower temperatures (*48*). Laboratory degradation (accelerated ageing) experiments on soybeans have detected formation of initial Maillard reaction products using fluorescence (*49*), however, our samples have provided the first unequivocal evidence for this reaction occurring and progressing to the advanced form as a decay process in buried ancient plant remains (*22*). The well-preserved lignified seed walls have encapsulated the volatiles within, hence enabling confirmation of this reaction by characterisation of low molecular weight compounds, rather than the polymeric products which are also formed in the advanced stages of the Maillard reaction (*22*).

The C_6 to C_{10} *n*-alkanals are known to form from enzymatic oxidative attack (*50*) across double bonds in unsaturated fatty acids, whereas the 2-alkanones occur through free radical oxidative attack of the same bonds. This supports previous work (*12, 20*) showing evidence for oxidation of the unsaturated fatty acid components of lipids in the seeds from Qasr Ibrîm.

Headspace Concentration-GC/MS — Barley. The absence of the alkyl polysulphide compounds amongst the volatile components of the crushed ancient barley could indicate fewer sulphur-containing amino acids in the barley storage protein compared with radish. Since dimethyltrisulphide has been shown to be present in minute quantities in modern samples of freshly malted barley (*51*), by crushing a greater quantity of the propagules or using selected ion monitoring to increase the sensitivity of the GC/MS analysis, these components may be detected. The greater variety and abundance of highly substituted alkyl pyrazines in the volatile composition of crushed ancient barley can be attributed to the high abundance and variety of carbohydrate compounds in this taxon compared with radish (*42*).

Conclusions

Modern and ancient samples of radish (1400 years) and barley (600 years) have been analysed by pyrolysis-GC/MS (610°C) and headspace concentration-GC/MS in order to elucidate the chemical transformations occurring in macromolecular constituents (especially proteins and polysaccharides) upon desiccation. The pyrolysate of modern radish storage tissue is dominated by pyrolysis products of amino acids most probably present as partially degraded proteins. There was also a significant contribution from 4-ethenyl-2,6-dimethoxyphenol, which cannot be explained at this time. Pyrolysis of the ancient radish storage tissue showed that 4-ethenyl-2,6-dimethoxyphenol was greatly depleted in abundance which is consistent with literature on preferential side chain oxidation (*20*). There was very little difference in the pyrolysis products of proteinaceous moieties between the modern and ancient radish storage tissue. In contrast to the radish, the pyrolysate of modern barley storage tissue was dominated by pyrolysis products of starch. Many of these compounds showed a significant decrease in overall abundance in the ancient sample, which most probably relate to alterations of the starch macromolecule occurring over time.

Characterisation of volatile components of ancient radish and barley by headspace concentration-GC/MS revealed that the decay of proteins and polysaccharides in buried plant matter involved the Maillard reaction. These analyses have revealed a significantly easier method of detecting this reaction than characterisation of heteropolymeric melanoidins. It has been possible to study this reaction in these ancient propagules as the volatiles are preserved by encapsulation within the propagules It is clear, from these data, that desiccation retards the degradation of proteins and polysaccharides and yet supports the preservation of their decay products.

Acknowledgments

NERC is thanked for grant to RPE (GR3/9578) and NERC studentship to HAB. The Egyptian Exploration Society and Dr Peter Rowley-Conwy are gratefully acknowledged for donation of ancient samples. Dr Donald Mottram is thanked for provision of alkyl pyrazine standards. The use of NERC Mass Spectrometric facilities are also acknowledged.

Literature cited

1. Eglinton, G., Logan, G.A. *Phil. Trans. R. Soc. Lond.* B **1991**, *333*, 315-328.
2. Evershed, R.P. *Archaeometry* **1990**, *32*, 139-153.

3. Stankiewicz, B.A., Hutchins, J.C., Thomson, R., Briggs, D.E.G., Evershed, R.P. *Rap. Comm. Mass Spec.* **1997**, *11*, 1884-1890.

4. Lowenstein, J.M., Sarich, V.M., Richardson, B.J. *Nature Lond.* **1981**, *291*, 409-411.

5. Gülaçar, F.O., Buchs, A., Susini, A. *J. Chromatog.* **1989**, *479*, 61-72.

6. Koch, P.L., Fogel, M.L., Tuross, N. In *Stable Isotopes in Ecology and Environmental Science;* Lajtha, K., Michener, R.H., Ed.; Methods in Ecology; Blackwell Scientific Publications: Oxford, UK, 1994, pp 63-92.

7. Evershed, R.P., Connolly, R.C. *Naturwissenschaften* **1988**, *75*, 143-145.

8. Evershed, R.P., Connolly, R.C. *J. Archaeol. Sci.* **1994**, *21*, 577-583.

9. Hughes, M.A., Jones, D.S., Connolly, R.C. *Nature* **1986**, *323*, 208.

10. Brown, T.A., Allaby, R.G., Brown, K.A., Jones, M.K. *World Archaeol.* **1993**, *25*, 64-73.

11. van Bergen, P.F., Collinson, M.E., de Leeuw, J.W. *Anc. Biomol.* **1996**, *1*, 55-81.

12. O'Donoghue, K., Clapham, A., Evershed, R.P., Brown, T.A. *Proc. R. Soc. Lond.* B **1996**, *263*, 541-547.

13. Rollo, F., Amici, A., Salvi, R., Garbuglia, A. *Nature* **1988**, *335*, 774.

14. Rollo, F., Venanzi, F.M., Amici, A. *Genet. Res.* **1991**, *58*, 193-201.

15. Shewry, P.R., Kirkman, M.A., Burgess, S.R., Festenstein, G.N., Miflin, B.J. *New Phytol.* **1982**, *90*, 455-466.

16. van Bergen, P.F., Collinson, M.E., Briggs, D.E.G., de Leeuw, J.W., Scott, A.C., Evershed, R.P., Finch, P. *Acta Botanica Neerlandica* **1995**, *44*, 319-342.

17. O'Donoghue, K., Brown, T.A., Carter, J.F., Evershed, R.P. *Rap. Comm. Mass Spec.* **1994**, *8*, 503-508.

18. O'Donoghue, K., Brown, T.A., Carter, J.F., Evershed, R.P. *Rap. Comm. Mass Spec.* **1996**, *10*, 495-500.

19. Lindahl, T. *Nature* **1993**, *362*, 709-715.

20. van Bergen, P.F., Bland, H.A., Horton, M.C., Evershed, R.P. *Geochim. Cosmochim. Acta* **1997**, *61*, 1919-1930.

21. Iiyama, K., Lam, T.B.T., Stone, B.A. *Phytochemistry* **1990**, *29*, 733-737.

22. Evershed, R.P., Bland, H.A., van Bergen, P.F., Carter, J.F., Horton, M.C., and Rowley-Conwy, P.A. *Science* **1997**, *278*, 432-433.

23. Stankiewicz, B.A., van Bergen, P.F., Duncan, I.J., Carter, J.F., Briggs, D.E.G., Evershed, R.P. *Rap. Comm. Mass Spec.* **1996**, *10*, 1747-1757.

24. Galletti, G.C., Bocchini, P. *Rap. Comm. Mass Spec.* **1995**, *9*, 815-826.

25. Chiavari, G., Galletti, G.C. *J. Anal. Appl. Pyrolysis* **1992**, *24*, 123-137.

26. Ralph, J., Hatfield, R.D. *J. Agric. Food Chem.* **1991**, *39*, 1426-1437.

27. Pouwels, A.D., Eijkel, G.B., Boon, J.J. *J.Anal. Appl. Pyrolysis* **1989**, *14*,237-280.

28. Faix, O., Meier, D., Fortmann, I. *Holz als R. Werk.* **1990**, *48*, 351-354.

29. Faix, O., Fortmann, I., Bremer, J., Meier, D. *Holz als R. Werk.* **1991**, *49*, 213-219.

30. Faix, O., Fortmann, I., Bremer, J., Meier, D. *Holz als R. Werk.* **1991**, *49*, 299-304.

31. Tsuge, S., Matsubara, H. *J. Anal. Appl. Pyrolysis* **1985**, *8*, 49-64.

32. Smith, G.G., Reddy, G.S., Boon, J.J. *J. Chem. Soc. Perkin Trans. II* **1988**, 203-211.

33. Ratcliff, M.A., Medley E.E., Simmonds, P.G. *J. Org. Chem.* **1974**, *39*, 1481-1490.

34. Vernin, G., Parkanyi, C. In *Chemistry of Heterocyclic Compounds in Flavours and Aromas;* Vernin, G., Ed., Ellis Horwood, Chichester, 1982, pp.151-207.

35. Shewry, P.R. *Biol. Rev.* **1995**, *70*, 375-426.

36. Mottram, D.S. *Amer. Chem. Soc. Symp. Ser.* **1994**, *543*, 104-126.

37. Tegelaar, E.W., de Leeuw, J.W., Derenne, S., Largeau, C. *Geochim. Cosmochim. Acta* **1994**, *53*, 3103-3106.

38. van Bergen, P.F., Goñi, M., Collinson, M.E., Barrie, P.J., Sinninghe Damsté, J.S., de Leeuw, J.W. *Geochim. Cosmochim. Acta* **1994**, *58*, 3823-3844.

39. Monsalve, R.I., Menéndez-Arias, L., Gonzàlez de la Peña, M.A., Batanero, E., Villalbe, M., Rodríguez, R. *J. Exp. Bot.* **1994**, *45*, 1169-1176.

40. Laroche, M., Aspart, L., Delseny, M., Penon, P. *Plant Physiol.* **1984**, *74*, 487-493.

41. Dey, P.M., Brinson, K. In *Advances in Carbohydrate Chemistry and Biochemistry*; Tipson, R.S. and Horton, D., Eds., Academic Press: London, 1984, Vol. 42; pp. 266-383.

42. John, P. *Biosynthesis of the Major Crop Products*; Wiley Biotechnological Series, John Wiley and Sons, Chichester, 1992.

43. Schulten, H.R., Sorge, C., Schnitzer, M. *Biol. Fertil. Soils* **1995**, *20*, 174-184.

44. Ellis, G.P. *Adv. Carbohyd. Chem.* **1959**, *14*, 63-134.

45. Evershed, R.P., Morgan, E.D., Cammaerts, M.C. *Naturwissenschaften* **1981**, *67*, 374-376.

46. Semmelroch, P., Grosch, W. *Lebensm.-Wiss. u.-Technol.***1995**, *28*, 310-313.

47. Weenen, H., Tjan, S.B., de Valois, P.J., Bouter, N., Pos, A., Vonk, H. *Amer. Chem. Soc. Ser.* **1994**, *543*, 142-157.

48. Mauron, J. *Prog. Fd. Nutr. Sci.* **1981**, *5*, 5-35.

49. Sun, W.Q., Leopold, A.C. *Physiol. Plant.* **1995**, *94*, 94-104.

50. Gunstone, F.D., Harwood, J.L., Paisley, F.B. *The Lipid Handbook,* The University Press, Cambridge, 1986.

51. Beal, A.D., Mottram, D.S. *J. Agric. Food Chem.* **1994**, *42*, 2880-2884.

Chapter 9

Preservation of DNA in the Fossil Record

Hendrik N. Poinar

Zoological Institute, University of Munich, PF 202136 D-80021, Munich, Germany

All organisms store their genetic makeup in either the DNA or RNA
polymer, hence the preservation of Deoxyribonucleic Acid (DNA) in
the geosphere and its retrieval from fossil organisms has intrigued
the molecular biologist for many years. An ancient sequence would
enable the scientist to place an extinct species in a phylogenetic
context, to look at the diversity of extinct animal and human
populations or even to trace disease associated bacteria or viruses
back through time. The DNA molecule, although stronger than
RNA, has a limited lifespan estimated at 10^4 years in most
geological settings. The advent of the Polymerase Chain Reaction
(PCR) has enabled the retrieval of small amounts of DNA and in
essence opened fossils to DNA analysis. Most research involving
ancient DNA revolves around extraction, PCR amplification and
sequencing of DNA from the remains of extinct animals and plants.
There has been little work on the actual factors involved in the
preservation of DNA. A more thorough understanding on the
environments where this molecule is likely to be preserved would
help molecular biologists focus their search.

Molecular Preservation

Preservation in general can be thought of on many different levels. There is
preservation on the large scale, i.e. the environmental conditions throughout burial
history, for example the amount of leaching at a site, the bacterial flora and the
temperature. On a smaller scale, within the organism, the bone as a source of
ancient DNA as opposed to the skin (more susceptible to direct attack), or the teeth
which may house the most protection (but will have quantitatively less DNA than a
bone per dry weight). And finally at the molecular level within the cell, the nucleus
(DNA protected by histones) as opposed to the mitochondria (DNA is naked but
may have many more copies per cell than nuclear DNA). So when one asks the
question where is DNA best preserved, the answer is not an easy one. One has to
take into consideration the preservational site, the type and molecular makeup of an
organism, and the specific tissue to target for DNA isolation.
     Presumably the two most important factors for long-term DNA preservation
are the immediate environment (temperature, humidity, aridity) in which an
organism is preserved and the bacterial flora surrounding it (*1*). Autolytic

degradation in dying tissues may likely cleave nuclear DNA at the linker regions between histones into small (200bp) fragments (it is unknown what happens to mitochondrial DNA). These processes may be dependent upon the rate at which tissues dry. In cold climates such as permafrost, drying may be rapid and therefore preservation best. If these processes can be halted in certain environments, than DNA may be preferentially preserved. Environments where tissues can be protected from bacterial degradation (acidic bogs and peats, amber, permafrost) seem to house some of the most pristine specimens. Aside from these, degradation will depend upon the temperature, pH and the availability of water (Hydrolysis) and oxygen (Oxidation). These factors will determine the extent and speed at which DNA breakdown will take place. A constant burial temperature and relative humidity will likely play a major role in preservation.

Unfortunately for the molecular biologist, the RNA and DNA polymers are relatively weak and unstable in comparison to other macromolecules (Table I) and their survival is suggested to be limited to 10^4 years in most environments. This instability may be a design 'flaw' which, however, enables its deconstruction and reconstruction by repair enzymes, so that the roughly 100,000 insults which occur daily to this molecule can be fixed and we may live (2, 3).

Ancient DNA

The first report on ancient DNA came from the extinct equid, Quagga, a museum skin some 140 years old (4). This was still in the pre-PCR era and it was not until the invention of PCR that the field exploded with reports from fossil material. The oldest reproduced sequences to date stem from 50,000 year old permafrost remains of the woolly mammoth (5). In Table II are a selection of fossils from which DNA

Table I : Compounds, their Susceptible Bonds and Preservation Potential.

Compound Class	Susceptible Bonds	Susceptible Groups	Preserving Potential*
Nucleic acids: RNA DNA	Phosphate esters; Glycosidic,; 5,6- C,C bond of pyrimdines	Heterocyclic rings Amino, Methyl	- +
Proteins	Peptide bond	Side chain functionalites Chiral center	-/+
Carbohydrates	Acetal C-O	Hydroxy, amide	+
Lipids glycolipids	Ester, Ether, C-O Amide, C-N	Hydroxyl, carboxyl ester	++
Cutin,	Ester, Ether, C-O	Hydroxyl, carboxyl	+++
Lignin	Ether, C-O	Methoxyl aromatic rings	++++

- to ++++ (weakest to strongest) a rough estimate on the preservation potential.
* Preservation potential for an unaltered molecule. [Adapted from ref.1].

has been recovered. The maximum base pairs amplified are indicated, along with a description on the possible reasons for the unusually good preservation. Additional information on the preservation of other macromolecules from the same site are also listed.

In recent years there have been several reports based on samples that date 17 to 120 million years old (5, 6, 7, 8, 9, 10). These contradict the theoretical longevity of the DNA molecule (10^4 to 10^5 yrs) and have either been shown to be a contaminant or unable to be reproduced, these will not be discussed here. A complete review of all published works on ancient DNA is not the scope of this review, but rather to understand from a selection of the works, the type of DNA damage that is seen in ancient samples.

Polymerase Chain Reaction

The Polymerase Chain Reaction (PCR) is a unique enzymatic process that can be thought of as a genetic xeroxing machine with the ability to exponentially increase the number of copies. The PCR works in repeated cycles composed of three steps. The first step is a denaturing step (90-95°C) where the double stranded DNA is teased apart (hydrogen bonds between bases broken) so that the base sequence is open for copying. In the second step (40-60°C) synthesized DNA molecules (15-30bp in length) known as primers, designed for complimentary conserved regions of genes, anneal to their respective targets on the denatured strands (depending on how well they match). In the final step (72°C) an enzyme known as *Taq* (*Thermus aquaticus*) DNA polymerase sits on the annealed primer, and like the engine of a train on its track, the polymerase moves down the denatured DNA strand (the template) reading its bases and adding complimentary bases to the growing primer. This cycle, made up of these three steps is repeated on average 40 times and an exponential increase in copies results. There are pros and cons of course to every method used, especially when used with fossil material. Benefits of the PCR machine are its potential sensitivity, under the right conditions a single molecule can be amplified to generate millions, this is much more sensitive than most analytical methods such as (High Pressure Liquid chromatography (HPLC) or Gas Chromatography Mass Spectroscopy (GCMS). It is selective; by intuitive primer design one can select for bacterial, vertebrate or plant DNA thereby making it a very discriminatory method. Through use of an internal standard, known as a competitive construct, one can quantitate the number of endogenous DNA molecule fragments of a requisite sequence exist in the ancient tissues (21). Of course there are problems with PCR as well. Due to its sensitivity, modern contamination can be a big problem and in the case of human remains it becomes difficult to identify an ancient endogenous sequence from a modern contaminant, as was the case with the 'Ice Man' (11). PCR depends upon the polymerase's reaction to a degraded template which may be quite different from when it is offered an undamaged modern equivalent. Certain DNA base lesions resulting from various forms of damage will act as polymerase blocks and thus will prevent the amplification of damaged templates, so a negative PCR may be the result either of no DNA remaining in the fossil or the DNA being damaged beyond amplifiability. Other oxidative and hydrolytic DNA lesions can be bypassed by the polymerase and result in miscodings (Table III and Figure 1). Individual base substitutions seen in clones from a single PCR product may be the result of these processes and could possibly provide insight into the types of damage occurring in ancient DNA.

Damaged DNA base lesions that block the enzyme in PCR may induce jumping of the partially extended product to another strand, where the enzyme template complex will continue to extend (12). This results in recombination products during the PCR which can lead to erroneous results especially when the

Table II. DNA Retrieval from Some Specimens, their Environment and Additional Preservation Information.

SOURCE ,NAME Scientific, and common	AGE Kyrs	TISSUE type	SITE	DNA bp	PRESERVATIONAL COMMENTS	ADDITIONAL INFO	REFS.
Thylacinus cynocephalus Marsupial wolf	0.08	Muscle	Museum	411	museum dried specimen no treatment, natural drying	proteins, antigenic response	50, 51, 52
Equus quagga Quagga	0.140	Muscle	Museum, Mainz Germany	117*	preserved with salts	proteins, antigenic response	3, 53
Zea mays Corn	.980	Seeds	Huari Tomb, Peru	130	dessication and Seed coating as protection	NA	54
Megalapteryx didinus Moa	3.5	Bone Tissue	Mt. Owen New Zealand	438 B 150 T	cooler temps higher elevation?	low race	17, 18
Homo sapiens Ice Man	5.0	Tissue	Tyrolean Alps Austria Italy?	394	glacier <0°C freeze dried (dessicated) low bacteria	low race	11
Homo sapiens Windover samples	7.0	Brain	Windover Pond Windover Florida	200	waterlogged, acidic, or alkaline, anoxic, high humic acids, demineralization	protein, no DNA controversy? fatty acids. degraded TAGs	36, 38, 55
Mylodon darwinii Ground sloth	13.0	Bone	Cave, Ultima Esperanza Patagonia ,Chile	140	cold cave ,magnesium sulfate (Epsom salts, hygroscopic drying) low bacterial content	hide preservation hist. excellent low race	28, 18, 57
Smilidon fatalis Sabre tooth tiger	14.0	Bone	LaBrea Tar pits Los Angeles, CA	200	anaerobic asphalt, oils and salts involved in dessication. Low humidity, low bacteria	protein& chitin preservation collagen, varying aa race	57, 60, 61
Equus hemionus Selerikan horse	27.0	Bone	Tundra Fairbanks, Alaska	140	permafrost <0°C, freeze dried (dessicated) (humics)	low race Hist= 5, %N = 4.80	18, 30 63
Mammut primigenius Wolly mammoth	50.0	Tissue	Khatanga, Siberia	200	permafrost <0°C, freeze dried dessicated tissues	protein, lipids albumin good histology ,low race	5, 18, 52, 62, 63

DNA preservation in tissues of varying ages and locations. Maximum length PCR product or Cloned* DNA product indicated in base pairs (bp). Histology (Hist) and Percent Nitrogen (%N) based on ref 63 . Abbreviations :TAG = Triacylglycerides; Low race = low amino acid (aa) racemization, REFS = References.

Table III. Some DNA Base Damages and their Miscoding by *Taq* DNA Polymerases

Base	Lesion	Result	Base pair; Mutation
Adenine	Depurination	Apurinic site	A : T ---> T : A
Guanine		Apurinic site	G : C ---> T : A
Cytosine		Uracil	C : G ---> T : A
Adenine	Deamination	Hypoxanthine	A : T ---> G : C
Guanine		Xanthine	G : C ---> G : C
5MetCytosine		Thymine	C : G ---> T : A
Cytosine			C : G ---> T : A
Guanine	Oxidation	Many oxidative	G : C ---> T : A
Uracil		lesions	5OHU : G ---> T : A
Guanine	Methylation	O6 Methylguanine	G : C --->A : T
Thymine		O4 Methylthymine	T : A ---> C : G

[Adapted from ref. 3].

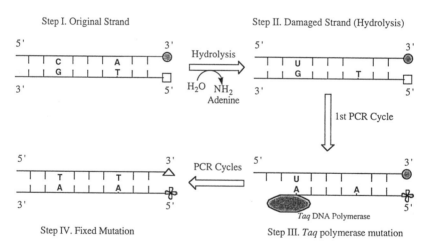

Figure 1. DNA damage (Cytosine (C) deamination and Adenine (A) depurination) and mutation fixation by PCR. G, Guanine; T, Thymine; U, Uracil.

amplification begins with a few template molecules. In cases where the amplification may begin with many templates jumping PCR products will rarely if at all, be seen unless many clones from these products are sequenced.

Nucleic Acid Structure

The single stranded DNA molecule is a polymer composed of 2'-deoxyribose sugar units linked via phosphate ester groups. To every sugar a purine (A,G) or pyrimidine (C,T) base is attached at the 1' position by a glycosidic bond(C-N) (Figure 2). In vivo in its hydrated B form the DNA molecule is supported by interactions between water molecules in the major grooves and the bases and sugars. DNA polymers are subject to multiple forms of damage including hydrolytic, oxidative, alkylation and condensation reactions which limit the molecules survival in the geosphere (Figure 2). In vitro hydrolysis of the naked DNA molecule suggests a limited life span of 10^4 years in most environments (*13, 14*). The RNA polymer is single stranded, has a ribose instead of a deoxyribose and the bulkier 2' OH group which is more reactive . The phosphodiester bond in DNA is some 200 times more stable than in RNA, making RNA preservation in the geosphere less likely (*2*). In general it is unlikely that RNA will be preserved for periods of time that make its study of equal importance as DNA.

Hydrolytic Damage to DNA

It is thought that hydrolytic damage is the most significant form of damage for the DNA molecule in terms of its preservation in the geosphere (*2*). With this in mind, the most sensitive sites to hydrolysis in the DNA molecule are the phosphate ester bonds linking the deoxyribose sugars together (Figure 2 and Table IV) (*15*). Once a strand is cleaved, its preservation may depend upon its ability to remain in the double stranded (ds) form (length and number of hydrogen bonds and cations holding it together) to bind to a surface or become encapsulated, thereby protecting the suceptible bases from further attack. Aside from direct cleavage of phosphoester bonds the least stable bond in DNA is the glycosidic bond (C-N) between the bases and the ribose sugars. Protonation of the base (making it a better leaving group) will in turn lead to cleavage of the glycosidic bond (depurination or depyrimidination). This will lead to an Apurinic or Apyrimidinic site (AP site) which in turn will distort the DNA molecule and subsequently lead to chain cleavage through a ß-elimination (Figure 2). In aqueous solution the phosphate esters cleave at a rate 2x faster than depurination, the purine bases (Guanine is released about 1.5x faster than adenine) are released at a rate approximately 20x that of the pyrimidines and subsequent ß-elimination leading to chain cleavage is about the same rate as depurination (Table IV). Surprisingly there is only about a 4 fold decrease in the rate of depurination for naked dsDNA as compared to single stranded (ss) DNA (*2, 15*).

Finally hydrolysis in the form of deaminations can take place on cytosine, adenine and guanine which will result in uracil, hypoxanthine and xanthine, respectively. Cytosine and adenine will result in miscoding during the PCR (Table III). For example deamination of cytosine will result in uracil. In the first cycle the polymerase will place an A opposite the uracil and in the next cycle a T will be placed in the opposite strand so a C:G base pair has the potential to become a T:A base pair (Table III and Figure 1). Hypoxanthine and xanthine will preferentially pair with cytosine so adenine deamination has the potential to be mutagenic while guanine deamination does not. Cytosines are much more sensitive to deaminations when in the single stranded DNA form (half life some 200 years), than in the double stranded form (half life 30,000 years) (*2*).

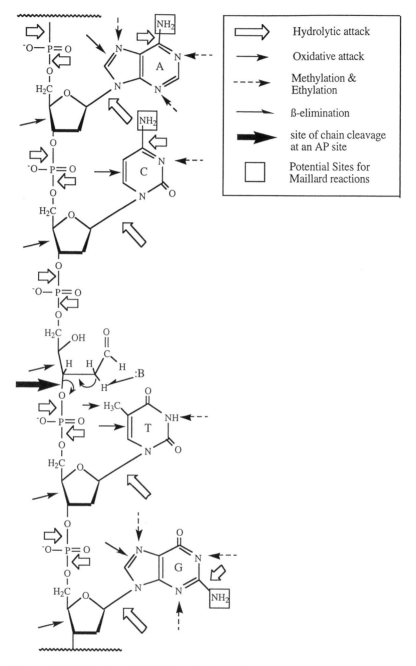

Figure 2. DNA Strand (5'-3') indicating sites of hydrolytic, oxidative, methylation, ethylation, and possible maillard reaction attack, along with Apurinic site(AP site) and site of chain cleavage following ß-elimination. Adenine (A), Cytosine (C), Thymine (T), and Guanine (G). (Adapted from [2])

Table IV: Time in Hours Needed for one Event (pH 7.4, 37°C)

Damage	ssDNA (2x10^{6}bp)	dsDNA(2x10^{6}bp)
Depurination	2.5	10
Depyrimidation	50	200
Deamination of C	2.8	700
Chain break after depurination	?	10
Chain break direct	?	2.5

Hydrolytic Damage Seen in Fossil Material

Ancient DNA is present in small fragments, this is most likely a direct result of hydrolysis of the N-glycosidic bond leading to base loss (AP site) followed by strand cleavage through ß-elimination and direct hydrolysis of the phosphoester bonds. However, it could also be autolytic degradation. Ethidium bromide stained gels and electron microscopy showed that among several specimens of different origin and ages (4 to 5000 years) the predominant size distribution of the DNA consisted of fragments some 100-500 bp in length (16). Considering the maximum amplification length achieved in most of the publications on ancient DNA thus far, this range has been shown to be a general feature (Table II). Even though in some cases, differences in length could be attributed to different tissue types (17) (Table II). In general the individual burial conditions of a specimen seem to be the crucial component (18) However, it should be noted that individuals buried under seemingly similar conditions can be strikingly different in their DNA preservation (19, 20). The fact that there is an inverse correlation between the length and efficiency of amplification products has been widely used as one criterion for authenticity, in the ancient origin of a DNA sample (16, 21). Amplifications not complying with this common feature have usually been identified as resulting from contaminants of microbial origin (21) with PCR products from previous amplifications and/or human DNA (Höss, and Poinar unpublished results).
    Only one publication to date has mentioned the possibility of deaminations playing a role in post mortem change (22). Considering the rather slow rate of deamination in double stranded DNA this is not surprising.

Oxidative Damage to the DNA Molecule

Even under "dry" conditions, DNA can still be subject to oxidative damage. Both the DNA backbone and its bases experience free radical induced damage. It is envisaged that most of the oxidative damage to DNA is mediated by hydroxyl radical attack (23, 24). The presence of metal ions (which may be bound to the DNA) may catalyze the formation of these reactive species. Almost 100 different free radical induced damages have been identified (23, 24). The DNA's double stranded nature does afford some protection of the bases from free radical attack. However the 3',4' bond of the deoxyribose is subject to attack and can lead to a strand break (2). DNA pyrimidines are both sensitive to free radical attack especially at the fragile 5,6- double bond (25). The imidazole ring in purines is another site of oxidative attack and it may lead eventually to ring fragmentation (23). To the molecular paleontologist oxidative lesions such as ring fragmentation will affect the ability to retrieve DNA by blocking the polymerase in the PCR. Some oxidative lesions, such as 8-oxo-G formation (like cytosine deamination) will result in a miscoding during replication in the PCR (Table III).

Lipid peroxides decompose to give a wide range of carbonyl containing compounds which may react with and damage DNA, for example, malondialdehyde may react with the amino group containing bases (A,C,G) (23). It has been suggested that by preventing lipid peroxidation, and hence the resulting highly reactive byproducts, DNA preservation may be enhanced (26).

Oxidative DNA Damage Seen in Fossil Material

Recently gas chromatography combined with mass spectrometry (GC/MS) has been used to look at the type and amount of oxidative base damage in DNA extracts from 11 different ancient tissues (27). The eleven samples used in the study comprised five specimens where DNA could be reliably amplified and sequenced according to criteria published elsewhere (21, 28), as well as six samples where no DNA could be amplified. With the exception of a 50 year old horse sample, all of the determined DNA sequences have been published (5, 28,-30). All of the specimens that yielded DNA sequences, with the exception of the 50 year old horse, stem from Arctic or subantarctic regions. Seven oxidatively derived base lesions could be detected in at least one of the samples and three were abundant enough to be identified in all samples. These modifications are the two oxidized pyrimidines, the thymine derived 5-hydroxy-5-methylhydantoin (5-OH-MeHyd) and the cytosine derived 5-hydroxyhydantoin (5-OH-Hyd), as well as the oxidized guanine base 8-hydroxyguanine (8-OH-Gua). The fact that these lesions were abundant enough to be detected in all of the samples is consistent with the observations that 5-OH-MeHyd is one of the major degradation products of thymine in gamma-irradiated DNA (31) and may have enhanced stability within the DNA molecule (32). 8-OH-Gua is the predominant oxidation product of guanine upon hydroxyl attack (24). These data are also consistent with the results of an older study (16) which showed by enzymatic analysis that approximately 1 in 10 pyrimidines were oxidatively damaged.

In summary, the hydantoin lesions were 3-30 times more abundant in the samples that didn't yield DNA sequences than in the ones that did. Secondly, no such difference could be seen in the case of purine 8-OH-Gua. Selection of good over bad templates takes place during the extension step in the PCR. As 8-OH-Gua is a mutagenic lesion that does not block *Taq* polymerase, DNA molecules carrying this lesion are expected to be amplified. The hydantoins on the other hand are likely to be blocking lesions which prevent the extension of *Taq* polymerase.

Modification of DNA by Reducing Sugars (Maillard reactions)

In 1984 Bucala et al showed that DNA, like proteins, can undergo the Maillard reaction where a primary amine will condense with a carbonyl group to from a schiff base and eventually with further condensations form advanced glycosylation end products (33). The primary amine of the three bases adenine, cytosine and guanine can interact with the carbonyl of a reducing sugar. Maillard reaction products have been found in long lived proteins such as collagen and lens crystallins (33). In addition it was shown that the DNA adducts generated from addition of lysine and sugars (glucose 6-phosphate) to DNA were insensitive to reduction by sodium borohydride, and therefore extremely resistant to breakdown (34). Recently a guanine glycosylation product has been determined (35). It may very well be that in fossil material the DNA may be actively crosslinked to such products and protected for longer periods of time than the naked molecule.

Maillard Reaction in Ancient Material

A common observation made during most DNA extractions from ancient material is the brown color of the extract. When an aliquot is run out on an agarose gel which is then subsequently examined under UV light, a bluish smear in the low molecular weight region is seen (16). If such extracts are not purified by various silica binding methods (30), then the PCR will be inhibited. No work has been published on quantitating these products in fossil material. However simple fluorescence spectroscopy at wavelengths characteristic for Maillard products, of various extracts with high brown color, indicate the possible presence of these products (H.Poinar unpublished observation).

There are likely to be many more reactions taking place as the fossil material ages however nothing to date has been published.

What is Left: Amount of Endogenous DNA in Fossil Material

Once broken into small fragments by hydrolytic and oxidative damage, as with small peptides, the DNA may likely leach from the organism and become part of the overall organic debris, such as humic acids and eventually kerogen. Very little work has been done on attempting to quantitate the endogenous DNA extracted from fossil material. In most cases reported, the DNA content of the fossil material was very much lower than that of modern equivalents. For example, the muscle tissue from a 140 year old Quagga yielded about 1% of what can be retrieved from modern animal muscle (4). Brain tissues from Windover mummies also had only about 1% of what a modern brain contains (36), and with a Quantitative PCR method applied to the Tirolean Ice Man's DNA, only a few picograms of DNA per gram of tissue remained (11). In addition to lower amounts of total DNA, early observations indicated that ancient DNA extracts usually contained a large percentage of contaminating DNA, which required discriminative quantification methods. The fact that ancient DNA might be too damaged to be detected by some methods (e.g. hybridization and Polymerase Chain Reaction) further complicated the situation. All publications reporting quantification of ancient DNA used either quantitative PCR (11) or hybridization with DNA from extant related species (28, 37-39). Table V lists a few samples from which the endogenous DNA content has been estimated through hybridizations.

Table V: Amount of DNA Retrieval (ug/g tissue) and Percent Endogenous (% endo.) for Various Ancient Remains.

Sample Scientific (common)	Age Kyrs	Location	DNA (ug/g tissue)	% endo.	Ref
Homo sapiens (Mummies)	1.5	Qilakitsoq Greenland	?	0.001	59
Mammutus primegenius (Woolly mammoth)	4.0	Magadan Siberia	2 - 5	5	39
Mammutus primegenius (Woolly mammoth)	10.0	Yuribei Siberia	8	2 - 3	39
Mylodon darwinii (Ground sloth)	13.0	Patagonia Chile	3	0.1	28
Mammutus primegenius (Woolly mammoth)	50.0	Khatanga Siberia	5	0.1	37

Johnson and colleagues (*39*) concluded by hybridization of a mammoth DNA extract with modern Asian elephant DNA, that from 8μg of total mammoth extract (from 1g of tissue) only 2-3 % hybridized. More recently it was shown by dot blot analysis of ancient DNA from a ground sloth and a woolly mammoth that only some 0.1% of the total extract is of ground sloth and mammoth origin (*28, 37*). Although one has to allow for shortcomings in the accuracy of these methods when applied to ancient DNA, these results indicate that only a small fraction of ancient DNA extracts is truely endogenous. The exact origin of the "foreign" DNA is still unknown; however, likely possibilities include both bacterial and fungal DNA. The abundance of microbial DNA in ancient extracts has been shown by southern blots (*40*) and by PCR where strong amplification products could be seen when bacterial and fungal specific primers were used (*21*). The age of the exogenous contamination is not known, whether from bacterial colonization of the tissues shortly after death, slow attack overtime or more recent inhabitation of tissues.

Quality of Ancient DNA by Determing Amino Acid Racemization

An indirect way to study the extent of hydrolytic damage to DNA in ancient tissue samples is to study the extent of racemization of ancient proteins within the fossil (*18*) (For more information on Amino acid racemization see the chapter by J. Bada in this volume). Amino acid racemization, the conversion of L- amino acids to D-enantiomers, is influenced by water, temperature, heavy metal ion chelation and pH, similar to factors which affect DNA hydrolysis. The rate of aspartic acid racemization in enamel proteins (at roughly 37°C and pH 7.4) is 2.62×10^{-11} sec^{-1} (*41*)and that of DNA depurination at pH 7.4 and 37°C is 3.0×10^{-11} sec^{-1} (*42*)are very similar. In fact, plotting points from both reactions over a wide temperature range shows the similarities of the two rates. Ancient samples from a range of ages and environments were analyzed for their extent of aspartic acid racemization. A rough cut off value of 0.08 (the ratio of D- to L- enantiomer of aspartic acid) was found (*18*). Samples with lower D/L ratios yielded DNA authenticated via a host of criteria, and those with higher ratios yielded no detectable endogenous DNA. Of the 9 fossil samples that yielded DNA and had low levels of racemization, 7 stemmed from cold regions. Based on calibrated racemization rates for aspartic acid in dated bones, a D/L aspartic acid ratio of 0.10 would limit the preservation of endogenous DNA to a few thousand years in warm climates such as Egypt and to 10^5 years in cold regions such as Siberia. Thus million year old samples such as dinosaur bones (65ma), and leaf compression fossils (17ma), had only trace quantities of amino acids, which were extensively racemized. Moreover the D/L ratios of all three amino acids analyzed (asp ala and leu) did not follow the predicted patterns of racemization, i.e. D/L asp > D/L ala $\geq$ D/L leu, therefore suggesting that the amino acids were mostly of contaminating origin (*43*). However, amber samples are unique in that their amino acids have very low levels of amino acid racemization. The anhydrous nature of the amber matrix may be involved in the long term preservation of these pristine amino acids.

Desiccation and DNA Preservation

In cases where the DNA molecule can successfully escape the bacterial onslaught and 'dry' (the DNA molecule requires water or something in its place as support) it may be preserved for longer periods of time (most likely dependent upon the speed of drying as well). As it dries it may actively crosslink to proteins, itself or carbohydrates to maintain its support. It may bind to apatite in bone or tooth, or even condense with glycosylation products perhaps protecting it from further insult.

It has been demonstrated, however, by direct experimentation that drying DNA and even desiccation tolerant bacteria under vacuum (not likely what happens in the environment) will cause the DNA to distort from the fully hydrated B form to the less hydrated A form inducing single stranded breaks along with protein crosslinks (*44*). Thus limiting the half-life of DNA in desiccating environments (*44, 45*). The rate of depurination, however, in dried DNA drops to some 20x that of the hydrated form (*46*). It is unknown what becomes of the tightly hydrogen bonded water supporting the DNA molecule as the cells dry, or if this water will promote DNA damage. Desiccation tolerant plants, arthropods and bacteria have developed unique ways to cope with dry environments. Lotus seeds from Pulantien Manchuria (carbon date of 1,288 ± 271yr BP), found within an ancient lake bed must have well preserved DNA, since they have been germinated (*47*). It has been suggested that their pericarp is impervious to water and that the high levels of ascorbic acid and glutathione (27mg/g dry weight) in the seeds keep radicals at low levels, playing a vital role in their long term preservation. Bacterial endospores have an amazing set of ways to cope with desiccation. Their DNA is heavily condensed, water levels are much lower than in an active cell, reactive species (ATP) are absent and radical formation is inhibited (*48, 49*). In summary the DNA molecule can be sequestered from damage and protected for longer periods of time than what in vitro chemistry predicts, if organisms such as these can remain in suspended animation.

What Have We Learned

In brief, we know the following facts about truly ancient DNA (aDNA). The endogenous DNA that remains is present as small fragments, usually below 500bp in length. The aDNA is oxidatively damaged, most frequently at the pyrimidine bases which then act as polymearse blocks in the PCR. The overall DNA content of fossil material is low and mostly originates from exogenous sources such as bacterial and fungal.

As can be seen from Table II, there is no correlation between the age of the samples and the preservation and retrieval of DNA. The preservation is strictly a function of the molecular makeup of the organism and environment in which the sample was preserved. In general, samples from colder regions such as Alaska and Siberia are the best preserved, suggesting that the single most important factor for longterm DNA preservation, and perhaps preservation in general, may be the temperature during burial. For example, the almost complete sample overlap of the racemization (*18*) and the GC/MS study (*27*) allows us to assess if hydrolytic and oxidative damage are influenced by different factors in tissue preservation. In both studies a correlation between the amount of damage, the amplifiability and the burial temperature was found. Cold environments can be expected to provide better conditions for molecular preservation by lowering the rates of reactions, such as racemization and depurination. Chemical reaction rates in general decrease by a factor of 3-4 times for every 10 °C drop in temperature.

Conclusion

The DNA molecule is a rather labile molecule which has limited persistence in most geological environments. Its preservation in the fossil record is a function of the molecular makeup and structure of an organism and the environment in which it is preserved particularily in regard to the availability of water and oxygen. Certain organisms that have developed ways to cope with extreme conditions of desiccation and cold may be better suited for longterm DNA survival.

144

Acknowledgments

I would like to thank Debi Cacesse for corrections, constant help and things more important than science, Geoffrey Eglinton for providing stimulating discussions on the possible preservation of DNA in various medias. Matthias Hoss for constructive criticisms and friendship, Hans Zischler for laboratory help and fatherly advice, George Poinar for encouragment and constant support, Jeff Bada for help and direction, and Svante Pääbo for active discussions and excellent working facilities, and the rest of the Ancient DNA group in Munich.

Literature Cited

1. Eglinton,G., G. Logan, *Phil. Trans. R. Soc. Lon. B* **1991**,*333*, 315.
2. Lindahl, T.*Nature* **1993**,*362*, 709.
3. Lindahl, T. *Phil. Trans. R. Soc. Lon. B* **1996**,*351*,1529.
4. Higuchi, R.; Bowman, B.; Freiberger, B.; Ryder, O. A.; Wilson, A. C., *Nature* **1984**,*312*, 282-284 .
5. Höss, M.; Vereshchagin, M. K.; Pääbo, S., *Nature* **1994**,*370*, 333 .
6. Cano, R. J.; Poinar, H. N.; Roubik, D. W.; Poinar, G. O. J., *Med.Sci.Res.* **1992**,*20*, 619-622 .
7. Cano, R. J.; Poinar, H. N.; Pieniazek, N. J.; Acra, A.; Poinar, G. O. J., *Nature* **1993**,*363*, 536-538 .
8. DeSalle, R.; Gatesy, J.; Wheeler, W.; Grimaldi, D., *Science* **1992**,*257*,1933-1936 .
9. Golenberg, E. M.; Giannasi, D. E.; Clegg, M. T.; Smiley, C. J.; Durbin, M.; Henderson, D.; Zurawski, G., *Nature* **1990**,*344*, 656-658 .
10. Woodward, S. R.; Weyand, N. J.; Bunnell, M., *Science* **1994**,*266*, 1299-1232 .
11. Handt, O.; Richards, M.; Trommsdorff, M.; Kilger, C.; Simanainen, J.; Georgiev, O.; Bauer, K.; Stone, A.; Hedges, R.; Schaffner, W.; Utermann, G.; Sykes, B.; Pääbo, S., *Science* **1994**,*264*, 1775-1778 .
12. Pääbo, S., D.M. Irwin, A.C.Wilson *J. Biol.Chem.* **1990**, *265*, 4718.
13. Lindahl, T.,*Nature* **1993**,*365*, 700 .
14. Pääbo, S., A. C. Wilson, *Current Biology* **1991**,*1*, 45.
15. Shapiro. R., in *Chromosome Damage and Repair;* eds. E. Seeberg, K. Kleppe; Plenum Publishing Corporation, England, 1981;3-18
16. Pääbo, S., *Proc. Natl. Acad. Sci. USA* ,**1989**,*86*, 1939-1943 .
17. Cooper, A., C. Mourer-Chauvire, G.K. Chambers, A.von Haeseler, A.C. Wilson, A. C., S. Pääbo, S., *Proc. Natl. Acad. Sci. USA.* **1992**,*89*, 8741.
18. Poinar, H. N.; Höss, M.; Bada, J. L.; Pääbo, S., *Science* **1996**,*272*, 864-866 .
19. Handt, O.; Krings, M.; Ward, R.; Paabo, S., *Am. J. Hum. Gen.*.**1996**,*59*, 368-376 .
20. Hagelberg, E., J.B., Clegg, in *Molecules through time;* eds. G. Eglington Curry, G.B.; Royal Society of London, London, 1991, 399.
21. Handt, O.; Höss, M.; Krings, M.; Pääbo, S. *Experientia* **1994**,*50*, 524-529 .
22. Higuchi, R. H.; Wrischknik, L. A.; Oakes, E.; George, M.; Tong, B.; Wilson, A. C., *J. Mol. Evol* **1987**,*25*, 283-287 .
23. Halliwell, B., O. I. Aruoma, *FEBS* **1991**,*281*, 9.
24. Dizdaroglu, M.;*Mut. Res.***1992**,*275*, 331.
25. Hayatsu, H., *Biochemistry* **1996**,*119*, 391-395 .

26. Matsuo, S.; Toyokuni, S.; Osaka, M.; Hamazaki, S.; Sugiyama, T., *Biochem.and Biophys. Res. Com.***1995**,208, 1021-1027 .
27. Höss, M.; Jaruga, P.; Zastawny, T. H.; Dizdaroglu, M.; Pääbo, S., *Nucl. Acid.Res.***1996**,24, 1304-1307 .
28. Höss, M.; Dilling, A.; Currant, A.; Pääbo, S., *Proc. Natl. Acad. Sci. USA* **1996**,93, 181-185 .
29. Taylor, P. G., *Mol.Biol.Evol.***1996**,13, 283-285 .
30. Höss, M.; Pääbo, S.*Nuc. Acid. Res.* **1993**,21, 3913-3914 .
31. Breimer, L. H.; Lindahl, T., *Biochemistry* **1985**,24, 4018-4022 .
32. Teoule, R.; Bert, C.; Bonicel, A., *Rad. Res.***1977**,72, 190-200 .
33. Bucala, R.; Model, P.; Cerami, A., *Proc. Natl. Acad. Sci. USA* ,**1984**,81, 105-109 .
34. A. Lee, A. Cerami, *Mut. Res.* **1987**,*179*, 151.
35. Papoulis, A.; Al-Abed, Y.; Bucala, R., *Biochemistry* **1995**,34, 648-655 .
36. Doran, H. G.; Dickel, D. N.; Ballinger, W. E.; Agee, O. F.; Laipis, P. J.; Hauswirth, W. W., *Nature* **1986**,323, 803-806 .
37. Höss, M., Ph.D. Thesis, Ludwig-Maximilians University, Munich, Germany (1995).
38. Hauswirth, W. W.; Dickel, C. D.; Lawlor, D. A. in *Ancient DNA;* eds. H. B. a. H. S; Springer Verlag, New York, 1994; 104-121
39. Johnson, P. H.; Olson, C. B.; Goodman, M., *Comp. Biochem. Physiol.* **1985**,81B, 1045-1051 .
40. Sidow, A.; A.C., W.; Pääbo, S. in *Molecules through time;* eds. G. Eglington Curry, G.B.; Royal Society of London, 1991; 429-433
41. Masters, P.; Bada, J., *Proc. Natl. Acad. Sci. USA* ,**1975**,72, 2891-2894 .
42. Lindahl, T.; Nyberg, B., *Biochemistry* **1972**,11, 3610-3618 .
43. Bada, J. L.; Kvenvolden, K. A.; Peterson, E., *Nature* **1973**,245, 208-210.
44. Dose,K.,A. Bieger-Dose, O. Kerz, M. Gill, *Origins of Life and Evolution in the Bioshpere* **1991**,*21*, 177.
45. Dose, K.; Bieger-Dose, A.; Labusch, M.; M.Gill, *Adv. Space Res.* **1992**,12, 221-229 .
46. Greer, S.; Zamenhof, S., *J. Molec Biol.* **1962**,4, 123-141 .
47. Shen-Miller, J.; Mudget, M. B.; W.Schopf, J.; Clarke, S.; Berger, R., *Am. J. Bot.***1995**,82, 1367-1380 .
48. Setlow, P., *J. Bacteriology* **1992**,174, 2737-2741 .
49. Potts, M.,*Microbiological Reviews* **1994**,58, 755.
50. Krajewski,C., A. C. Driskell, P. R. Baverstock, M. J. Braun, *Phil. Trans. Roc. Soc. Lon. B* **1992**,250, 19.
51. Thomas, R. H., W. Schaffner, A. C. Wilson, S. Pääbo, *Nature* **1989**,*340*, 465.
52. J. R. Lowenstein, V. M. Sarich, B. J. Richardson, *Nature* **1981**,*291*, 409.
53. Lowenstein, J. M; Ryder, D.A. *Experientia* **1985**,*41*, 1192.
54. Rollo,F.; Venanzi, F.M. *Genetical Research* **1991**,*58*, 193.
55. Hughes, M. A.; Jones, D.S.; Connoly, R.C. *Nature* **1986**,*323*, 208 .
56. Evershed,R. P.; Connoly, R.C.*Naturwissenchaften* **1988**,*75*, 143. .
57. Sutcliffe,T. *On the Track of Ice Age Mammals* (Dorset Press, Dorchester, 1985).
58. McMenamin, M. A. S.; Blunt, D.S.; Kvenvolden, K.A.; Miller, S.E.; Marcus, R. R. Pardi., *Quaternary Research* **1982**,*18*, 174.
59. Neilsen, L.E.; Engberg,J.;Thuesen, I. in *Ancient DNA;* eds. H. B. a. H. S; Springer Verlag, New York, 1994;104.
60. Jancziewski, D. N.; Yuhki, N.; Gilbert, D.A.; Jefferson, G.T.; O'Brien, S.J. *Proc. Natl. Acad. Sci. USA* **1992**,*89*.9769

61. Stankiewicz, B.A.; Briggs, D.E.G; Evershed, R.P, Duncan, I.J.*Geochim. Cosmochimic.Acta*, in the press.
62. Prager, E.M, Lowenstein, J.M.; Sarich, V.M. *Science* **1980**, *209*, 287.
63. Cooper,A; Poinar, H.N.; Pääbo,S.; Radovcic, J.; Debenath, J.; Caparros,M.; Barroso-Ruiz, C.; Bertranpetit, J.; Nelson-Marsh, C.; Hedges, R.E.M.; Sykes. B.*Science* **1997**,*277*,1021.

CHITIN: THE "FORGOTTEN" MACROMOLECULE

Chapter 10

Native and Modified Chitins in the Biosphere

Riccardo A. A. Muzzarelli and Corrado Muzzarelli

Center for Innovative Biomaterials, Faculty of Medicine, University Via Ranieri 67, IT-60100 Ancona Italy

Chitin, as abundant as cellulose in the biosphere, is present in countless living organisms, which continuously synthesize and degrade it enzymatically for nutritional, morphogenetic and defensive/aggressive purposes. Chitin is promptly metabolized in sediments, thus fossil chitin is not frequently encountered. Chemically modified chitins are important in light of their biochemical significance; their environmental friendly behavior permits to propose some of them for technological applications in the areas of crop protection, cosmetics, textiles and packaging.

The essential informations on the role of chitin, and its chemical derivatives, in the biosphere are briefly discussed in this chapter. The applications of chitin in the manufacture of a variety of biodegradable commercial products are also taken into consideration.

Chitin Structure

Chitin is a highly ordered copolymer of 2-acetamido-2-deoxy-β-D-glucose and 2-amino-2-deoxy-β-D-glucose and differs from other polysaccharides by the presence of nitrogen in its structure. At least ten Gigatons (1.10^{13} Kg) of chitin are synthesized (and degraded) each year in the biosphere (*1*). Chitin is widely distributed as a skeletal material among invertebrates in different crystallographic form (i.e. α, β). Orthorhombic α-chitin is found in the calyces of hydrozoa, the eggshells of nematodes and rotifers, the radulae of mollusks and the cuticles of arthropods. The monoclinic β-chitin exists in the shells of brachiopods and mollusks, the cuttlefish bone, the squid pen, the pogonophoran and vestimentiferan tubes, as well as in the *Aphrodites* chaetae and in the spines of some diatoms. While α-chitin exhibit antiparallelism of the chitin chains, β-chitin is characterized by parallel chains as a result of unidirectional biosynthesis and concomitant crystallization. β-Chitin also has less numerous hydrogen bonds as compared with α-chitin. Chitin in fungal walls varies in degree of crystallinity, degree of covalent bonding to other wall components, mainly glucans, and degree of acetylation (*2, 3*).

The chiral nematic order of chitin crystallites in arthropod cuticles has been described in analogy with cellulose in plant cell walls and collagen in skeletal tissues. It becomes macroscopically evident when microfibrillar fragments of purified crustacean chitins are prepared in 3M HCl at 104°C. The resulting colloidal suspensions spontaneously self-assemble in a chiral nematic liquid crystalline phase and reproduce the helicoidal organization that characterizes the cuticles (*4*). Circular dichroism of Congo red bound to chitin films reveals a cholesteric structure, having an organization similar to chitin naturally occurring in the cuticle (*5*).

The highly ordered structure of chitin is not immediately perceived due to the fact that it is generally encountered in combination with, tanned to various extent, proteins and with inorganic components, most commonly $CaCO_3$. The subject of the preservation of the native structure during the chemical isolation of chitin has been recently discussed by Bade (*6*).

Chitosan, obtained by deacetylation of chitin from a crab tendon, possesses a crystal structure showing an orthorhombic unit cell comprised of four glucosamine units, where two chains pass through the unit cell with an antiparallel packing arrangement. The main hydrogen bonds are O3···O5 (intramolecular) and N2···O6 (intermolecular) (*7*).

Chitobiose, O-(2-amino-2-deoxy-β-D-glucopyranosyl)-(1-4)-2-amino-2-deoxy-D-glucose, can be considered the structural unit of native chitin (*7*). The latter observation is based on the fact that chitinases produce the dimer (a point of similarity to cellulases), and that dimer is widespread in N-linked glycoproteins, even in organisms that do not synthesize chitin.

Chitin Synthesis

Chitin is synthesized following a common pathway, which includes the polymerization of *N*-acetylglucosamine from the activated precursor UDP-GlcNAc. The synthetic pathway includes the action of chitin synthases, which in fungi are transmembrane proteins that accept substrate UDP-*N*-acetylglucosamine at their cytoplasmic faces and vectorially feed nascent chitin into the extracellular matrix. In crustacean, the Golgi complex is directly involved in synthesis and secretion of chitin (*8*). In this process, the nitrogen comes from glutamine. The regulation, biochemistry and genetics of fungal chitin synthases are well known. The following reaction shows chitin synthesis:

$$UDP\text{-}GlcNAc + (GlcNAc)_n \rightarrow (GlcNAc)_{n+1} + UDP$$

Fungal chitin synthases are found as integral proteins of the plasma membrane and in chitosomes. A divalent cation (Mg(II)) is required for enzyme activity, but neither primers nor lipid intermediates are required. The enzyme is allosteric and is activated by the substrate and other effectors, notably free GlcNAc. The byproduct of the enzyme activity, uridine diphosphate (UDP) is strongly inhibitory to chitin synthase, but it may be metabolized readily to uridine monophosphate by a diphosphatase (9).

The chitin is modified to impart it the structure required by the particular tissue, via crystallization, deacetylation, cross-linking to other biopolymers, and quinone tanning. The result of the latter reactions is formation of very complex structures, endowed with high mechanical and chemical resistance, capable of exceptional performances from the physiological standpoint. In a living organisms chitin production is finely tuned with chitin resorption, in order to permit morphogenesis and growth. Therefore delicate equilibria exist between various enzymatic systems with opposite activities.

Detection and Localization of Chitin

Chitin can be easily identified by X-ray diffraction and infrared spectroscopy in various kinds of organic material when present in sufficient amounts. Solid state NMR spectroscopy offers the possibility of obtaining information on conformation, crystal structure and polymorphism of chitins. The localization of chitinous tissues within biological materials requires most elaborated methods, essentially based on molecular recognition. In this section information is presented on some recent developments.

Enzymatic Methods. Chitinases, specific for chitin and devoid of hydrolytic effects on any other polysaccharides, hydrolyze chitin to chitobiose and chitotriose, following hydrolysis of the latter, using *N*-acetylglucosaminidase, to *N*-acetylglucosamine. Because organic and inorganic constituents such as quinone-tanned proteins, glycoproteins and $CaCO_3$ hinder the enzymatic process, the determination of chitin has to be preceded by chemical isolation. *N*-Acetylglucosamine is determined colorimetrically (10).

Enzyme linked immunosorbent assay (Elisa) allows the determination of chitin and chitosan in nanomol to micromol quantities. This method is based on the use of a chitin antibody obtained from a rabbits injected with purified chitin and a secondary antibody, the goat-anti-rabbit IgG covalently linked to peroxidase. Calibration curves have to be obtained from chitin having identical physico-chemical properties.

Assay of Fungal Chitin. The vast majority of fungi (e.g. Ascomycotina, Basidiomycotina, Deuteromycotina and Mastigomycotina) have walls containing chitin and glucans or mannans, whereas Zygomycotina fungi contain both chitin and chitosan. In contrast, higher plants do not contain chitin in their cell walls. Therefore, a specific assay of chitin, estimates the amount of mycelium from a fungus, and detects whether a parasite or a symbiont is present in the tissue of a host plant. Thus, the degree of infection of plants can be estimated as well as the mycelium in mycorrhiza.

The colorimetric assay involves two different stages: the glucosamine residues are deaminated by nitrous acid to produce 2,5-anhydromannoses having a free aldehyde group on C1. These compounds then react with MBTH, a specific reagent for aldehydes, and give a blue color in the presence of $FeCl_3$ (*10*).

Molecular Recognition. WGA-Au labeling using the lectin wheat germ agglutinin and colloidal gold as a marker is the most reliable and versatile method for chitin localization. However, in studies of vertebrate tissues, treatment of the sample with neuraminidase is necessary prior to WGA-Au labeling, as well as the preliminary removal of protein and the use of siliconized glassware. Labeling is feasible with material embedded in a conventional epoxy resin that can be incubated with colloidal gold. In investigations of fungal cell walls, the glutaraldehyde-cacodylate fixed specimens are postfixed with OsO_4; ultrathin sections are incubated in the WGA-Au solution. WGA-fluorescent dye conjugates may also be used.

Lipo-chitooligomers and the Agricultural Use of Chitosan

Rhizobia are the N-fixing bacteria, which induce the deformation of the root hairs of leguminous plants. These bacteria invade the roots by means of the infection thread (i.e. newly formed tube) and induce the formation of the nodule (a specialized organ) where they multiply and reduce atmospheric N to ammonia which is later utilized by the plant. This relationship (i.e. symbiosis) between plants and the rhizobia is very specific (*11*).

The agents responsible for the formation of root nodules are lipo-chitooligomers. They consist of an oligosaccharide backbone of β-1,4-linked *N*-acetyl-D-glucosamine carrying a fatty acyl group on the nitrogen atom of the non-reducing end unit. Rhizobia produce complex mixtures of lipo-chitooligomer species, where the number of units varies between 3 and 5. The fatty acyl moieties seem to reflect the composition of the fatty acyl pool present as components of the phospholipids (*12, 13*).

Lipo-chitooligomers exert various unspecific effects on cultured plant cells that can be measured by electrophysiological techniques. For example, the transient alkalinization of suspension cultures of tomato cells occurring within 5 min after addition of lipo-chitooligomers to the culture medium (*14*). Lipo-chitooligomers represent a new substrate class for plant chitinases.

The mycorrhyza-induced chitinases are presumably able to release oligosaccharide elicitors from chitinous arbuscular mycorrhizal cell walls, which in effect may stimulate plant genes implicated in symbiotic process (*15*). In this light it can be easily perceived that the applications of chitin/chitosan in agriculture are not solely based on its high N-content, but rather on its biological significance. Chitosan appears to have multiple influences directed toward the improvement of the reproductive phase. At present, chitosan's consistent performance as a yield enhancer proved to be very reliable. Treatment of rice seeds with chitosan, for example, substantially increases root development (14 %

greater than control), tillering (7.7 %) and pollination. Furthermore, rice plants are more productive (5–20 %) and commercial rice fields in California and Louisiana has overproduced controls by 450-1000 kg/ha (16).

Chitin Isolates

In the areas of fisheries, textiles, food and ecology, scientists and industrialists were prompted to upgrade chitin in order to exploit a renewable resource and to alleviate waste problems. Today chitins and chitosans from different animals are commercially available.

The shells of crabs, shrimps, prawns and lobsters are used for the industrial preparation of chitin. The isolation includes two steps: demineralization with HCl, and deproteination with aqueous NaOH. Lipids and pigments may also be extracted. These operations are mainly empiric and vary with the type of mineralized shells, seasons and the presence of different crustaceans in the catch.

Chitin isolates differ from each other mainly by the degree of acetylation (typically ca 0.90), the elemental analysis (typical N content ca 7 %, N/C ratio 0.146 for fully acetylated chitin), molecular size and polydispersity. The average molecular weight of chitin in vivo is probably in the order of 1 MDa, but chitin isolates have lower values due to partial random depolymerization occurring during the chemical treatments and depigmentation steps. The average molecular weight for commercial chitins are 0.5 - 1.0 MDa. Polydispersity may vary depending on such treatments as powder milling and blending of various chitin batches.

The solubility of chitin is remarkably poorer than that of cellulose, due to its high crystallinity, supported by hydrogen bonds mainly through the acetamido groups. It can be only partially (up to 5 %) dissolved in ethanol containing calcium chloride, dimethylacetamide containing 5-9 % lithium chloride (DMAc+LiCl), and N-methyl-2-pyrrolidinone+LiCl.

Chitosans

Chitosan is a continuum of progressively deacetylated chitin. In general, chitosans have a N content higher than 7 % and a degree of acetylation lower than 0.40. The removal of the acetyl groups is a harsh treatment usually performed with concentrated NaOH solution (either aqueous or alcoholic). Protection from oxygen, with a nitrogen purge or by addition of sodium borohydride to the alkali solution, is necessary in order to avoid undesirable reactions such as depolymerization and generation of reactive species. Determination of intrinsic viscosity, average molecular weight (M_w) and polydispersity indices (M_w/M_n) indicates negligible depolymerization of the acid-soluble fractions when deacetylation is performed under nitrogen at 75°C (degree of acetylation 0.20 - 0.52) (17).

Chitosan can be brought into solution by the formation of salts due to the presence of a prevailing number of 2-amino-2-deoxyglucose units. It can also be chemically depolymerized by the reaction with nitrous acid. The latter process is selective, rapid and easily controlled. Nitrosating species attack the glucosamine, but not the N-acetylglucosamine moieties, and consequently cleave the anydroglycosidic linkage. The rate-limiting step is nitrosation of the unprotonated amine by protonated nitrous acid (*18*). Hydrogen peroxide can also be conveniently used to depolymerize chitosan. Oligomers are prepared at ambient temperature with no side-product formation by fluorolysis of chitin and chitosan (*19*).

Chitosan can be obtained from fungi, easily cultured on a simple nutrient (*20*). It can be isolated from the cell wall of Mucorales by the extraction with either acetic acid or alkali. The final molecular weight is in the order of 500 KDa and the degree of acetylation is around 0.10 (*21*).

Chitin and Chitosan Derivatives, Their Commercial Application

Chitin has been often considered as an intractable biopolymer due to the difficulties with its dissolution and chemical reactions. However, as soon as the molecular association is prevented or suppressed, chitin lends itself to many reactions affording a wide choice of modified chitins (Figure 1). Moreover, during the past few years a number of novel solvents have been proposed and reactions in homogeneous or nearly homogeneous media have been carried out (*22*). Because deacetylation is also feasible in various ways, chitosans and modified chitosans have also become available. Indeed, the reactions of the primary amino group, and primary and secondary hydroxyl groups can be easily performed with chitosan. The chemical modifications of chitin and chitosan, carried out under mild conditions (to protect glycoside and acetamido linkages) yield more soluble polymers. The latter have higher biodegradability in animal bodies and their physical properties are of interest for applications in solid state or in solution.

Chitin treated with NaOH yields alkali chitin, which further reacts with 2-chloroethanol to yield O-(2-hydroxyethyl) chitin, known as glycol chitin. This material was probably the first chitin derivative to find practical use and to be recommended as a substrate for lysozyme. The reaction of alkali chitin with sodium monochloroacetate gives water-soluble O-carboxymethylchitin sodium salt.

Chitosan can be reacetylated with acetic anhydride to obtain water-soluble partially reacetylated chitin (*23*). The Schiff reaction between chitosan and aldehydes or ketones gives the corresponding aldimines and ketimines, which are converted to N-alkylderivatives on hydrogenation with cyanoborohydride. Chitosan acetate salt can be converted into chitin upon heating (*24*).

N-Carboxymethyl Chitosan. Water-soluble *N*-carboxymethyl chitosan can be obtained by using glyoxylic acid (*25*). *N*-Carboxymethylchitosan from crab and shrimp chitosans is obtained in water-soluble form by proper selection of the reactant ratio, i.e. with equimolar quantities of glyoxylic acid and amino groups.

154

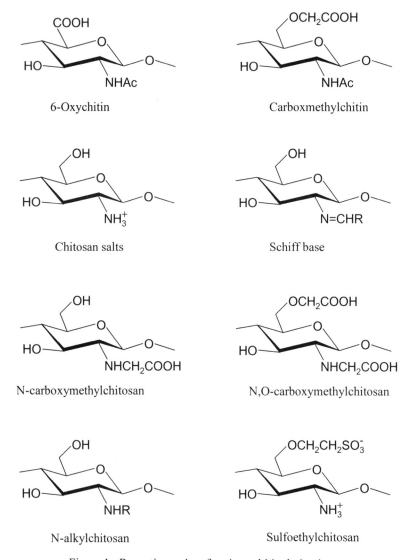

6-Oxychitin

Carboxmethylchitin

Chitosan salts

Schiff base

N-carboxymethylchitosan

N,O-carboxymethylchitosan

N-alkylchitosan

Sulfoethylchitosan

Figure 1. Repeating units of various chitin derivatives.

The product is in part *N*-mono-carboxymethylated (0.3), *N,N*-dicarboxymethylated (0.3) and *N*-acetylated depending on the starting chitosan (0.08 - 0.15) (*26*).

The application of *N*-carboxymethylchitosan is quite wide and important. Transition metal ions are chelated by *N*-carboxymethylchitosan more effectively than plain chitosan with the formation of colored and generally insoluble products, depending on pH values (*27*). The chelating ability of the cross-linked *N*-carboxymethylchitosan can be put to advantage for the removal of transition metal ions from brines, drinking water and nuclear plant effluents (e.g. trace Cu, Cr, Fe could be isolated from sodium fluoride brines; *28*). Another study reported on the separation of palladium and platinum (*29*).

N-Carboxymethylchitosan is a valuable functional ingredient of cosmetic hydrating creams in view of its durable moisturizing effect on the skin (*30*). Carboxymethylated chitins and chitosans are currently also used to preserve excised flowers and fruits.

Glutamate Glucan. This chitosan derivative, obtained from 2-oxoglutaric acid under reducing conditions, has extended conformationally mobile side chains, whatever the protonation state of the secondary amine. Instead, both the charge state of ionizable groups and chirality of the C atom in the side chain strongly affect the structural features (*31*). Complexes with Cu(II) and Fe(III) are stable even at pH as high as 10, the feature differentiating this derivative from insoluble 'plain' chitosan (*32*). In the sorption of uranium, pH and metal concentration are the only significant parameters, and the exceptional capacity is in excess of 500 mg/g at pH 5-6. This high value is attributed to the chelation process in addition to other mechanisms such as surface precipitation adsorption. The initial chelate promotes the formation of inorganic aggregates on polymer flakes.

Hydrophobic Chitosans. Hydrophobic, water-soluble polymers are a new class of industrially important macromolecules. Hydrophobic derivatives of chitosan can be easily obtained from long chain acyl chlorides and anhydrides and are intended to mimic the endotoxins (*33*).

Chitosans with Methoxyphenyl Functions. The methoxyphenyl aldehydes such as vanillin, *o*-vanillin, syringaldehyde and veratraldehyde react with chitosan under normal and reducing conditions and impart insolubility and other properties characteristic to chitosan. The films obtained from veratraldehyde are insoluble, biodegradable and mechanically resistant (Table I), and are suitable as special, biodegradable packaging (*34*).

N-**Ethylglycoside Branched Chitosans**. Formylmethyl glycosides obtained from reductive ozonolysis of allyl glycosides are reductively alkylated to chitosan. The glycosides of D-glucopyranose, D-galactopyranose, 2-acetamido-2-deoxy-D-glucopyranose, β-D-glucuronic acid and β-D-lactose are also incorporated by this method. Varying the reactant ratio, providing an array of water soluble products controls the degree of substitution.

Table I. Characteristics of veratraldehyde-chitosan films (ASTM 882)

Film	Thickness, µm	Tensile strength, kg	Elongation, %
Veratraldehyde chitosan	50	9.8	9
Low density polyethylene	50	1.5	300
Polypropylene	30-50	10.5	30
Polystyrene	30-50	10 ca.	10 ca.

Modified from ref. *34*.

Glycolipids from Chito-oligomers. Lipid A, a component of endotoxins, in the type analogues are synthesized using chitobiose and lower oligomers (which differ from lipid A in the type of glycosidic linkage). These modified chitosans possess low toxicity, immunostimulating and antitumor activities, leading to induction of interleukin-1 and tumor necrosis factor and augmentation by 140-180 % of the mean life of mice with carcinoma.

Tyrosine Glucan. These derivatives were inspired by the chemistry of insect cuticle tanning *in vivo* and are used to obtain stable and self-sustaining gels (a modified chitosan synthesized with 4-hydroxyphenylpyruvic acid) in the presence of tyrosinase (*35*). Similar gels are obtained from 3- and 4-hydroxybenzaldehyde, and 3,4-dihydroxybenzaldehyde, however all of them are hydrolyzed by lysozyme, lipase and papain. Mixtures of collagen + chitosan + tannin, when exposed to catalytic action of tyrosinase, give partially crystalline, hard, mechanically resistant and scarcely wettable materials. Phenoxyacetate is used in the production of penicillin and is often recycled; in order to remove *p*-hydroxylated derivatives of this precursor, tyrosinase is used followed by adsorption of the quinone species on chitosan (*36*).

Highly Cationic Chitosans. Trimethyl chitosan is prepared from trimethyl iodide (*37*). When submitted to 6-O-carboxymethylation, an intermediate is available for an amide bond formation with a suitable amine.

Anionic Chitins. The regioselective oxidation of C-6 position in chitins leads to formation of 6-oxychitins, that can be transformed into salt, ester and/or amide forms. Because this reaction is much simpler than the deacetylation and requires less labour and thermal energy, it is expected that 6-oxychitins will be protagonists in the exploitation of chitin resources during the forthcoming years (*38*).

Polyurethane-Type Chitosans. Chitins of various origins reacting with excess 1,6-diisocyanatohexane in DMAc+LiCl solution (upon exposure to water vapor for 2 days) yield flexible and opaque materials. The latter products are characterized by insolubility in aqueous and organic solvents, remarkable crystallinity, high N/C ratio (0.287) and relatively high degree of substitution (0.29), but no thermoplasticity. Chitosan, similarly treated under heterogeneous conditions in anhydrous pyridine, yields reaction products with a lower degree of substitution (0.17). Spherical chitosan particles (Chitopearl, trade name of Fuji Spinning Co., Shizuoka, Japan) are produced by this chemical approach and are suitable for chromatographic purposes and as enzyme supports (*39*). Microencapsulation of lactic acid bacteria based on the crosslinking of chitosan by 1,6-diisocyanatohexane has been performed as well (*40*).

Biodegradation of Chitins

The most common biodegradation pathway involves chitinases, produced by nearly all types of organisms, which hydrolyze bonds randomly (endo) or cleave dimers from the non reducing end of the chain (exo). Accompanying β-*N*-acetylglucosaminidases split the dimers and some of them also remove terminal *N*-acetylglucosamine from the polymer. Based on amino acid sequences, it seems that there are two evolutionary origins of these enzymes, i) family 19 plant chitinases that produce α-anomers (inverting), and ii) family 18 chitinases of plants and all other chitinolytic organisms producing β-anomers (retaining).

Chitinases fulfill nutritional, morphogenetic, defensive and aggressive roles. Many bacteria and fungi in soil and water environments can utilize chitin as a sole source of carbon. Nutritional chitinases are produced by protozoa and invertebrates that digest fungi or other invertebrates. Fish and other animals use chitinases mainly as food processing enzymes capable of dismantling the chitinous structures of their prey, thus allowing access to the more nutritious meat (*41-43*).

Morphogenetic chitinases are agents for assembling, dissembling or modifying cell and body structures. Many resting cysts of protozoa and eggs of invertebrates utilize chitin as the major protective material of their cases, allowing the hatching organism to use chitinases to aid its emergence. The role of chitinases in cleavage of the chitinous links between body and the exoskeleton is well documented. The latter process is especially evident during the development of arthropods, where the exoskeleton must be shed (and partially resorbed) at regular intervals.

The role of chitinases is also evident in the world of fungi. The release of mature endospores of the human pathogenic fungus *Coccidioides immitis*, takes place as a consequence of the digestion of a chitinous segmentation wall under the action of chitinases. Moreover, the enzyme-controlled autolysis facilitates spore release in the ink-cap mushroom *Coprinus* spp. In the yeasts, *Saccharomyces cerevisiae* and *Candida albicans*, for example, buds are cleaved from the mother cells by the action of chitinases, the septum being made of chitin (*41-43*).

Constitutive defensive chitinases in plants act on chitinous pathogens, especially fungi, to produce potent chito-oligomer elicitors, which induce massive production of further chitinases with antifungal properties.

Pathogens of chitinous invertebrates aid the penetration of the host exoskeleton, provide nutrients and aid in the release of progeny pathogens. Aggressive chitinases are virulence factors for bacteria parasitic on chitinous organisms, causing for example exoskeleton lesions of crabs.

Malarial and trypanosome parasites use chitinase to penetrate the chitin-bearing peritrophic membranes formed around the blood meals of their insect vectors. Chitinase is formed during the maturation of ookinetes, the invasive form of *Plasmodium gallinaceum*, which penetrates the mosquito peritrophic membrane (*44, 45*).

Human enzymatic activities capable to degrade exogenous chitin and derivatives include lysozyme, *N*-acetylglucosaminidase and the newly reported human chitinase. These enzymes are to aid in defense against pathogens.

Chitosans appear to be unexpectedly vulnerable to a range of hydrolases such as proteases (e.g. pepsin, bromelain, ficin and pancreatin), which display lytic activities that surpass those of chitinases and lysozyme. Amylase, lipases, cellulases, hemicellulases and other enzymes are also unspecifically active on chitosans (*46*). This is possibly due to the simplicity of the hydrolytic mechanism.

The marine environment is the major pool of chitin producers. The large annual production of this polysaccharide in the oceans have to be balanced by its consequent degradation. Detailed processes of chitin biodegradation in this environment are discussed later in this volume (*47*).

Fossil Chitin

Whilst chitinous materials are relatively resistant to degradation under certain conditions (suspension in sea water, desert sand) (*48*), they are promptly degraded in other environments (*47, 49, 50*).

Detection of chitin in fossils is not frequent, and in fact rather rare. There are some reports of fossil chitin in pogonophora, in insect wings from amber, and in beetle elytra from terrestrial sediments, but fossils of Crustacea were found to contain only traces of chitin, and no chitin-like microfibrils were detected by electron microscopy (*43, 51-54*). Coleopteran fossils (Lismore, Canada) were recovered from a buried peat and their chitin content determined, following isolation of glucosamine used later for isotopic study of D/H as a paleoclimate tool (*54, 55, 56*).

Although the terrestrial arthropods, with preserved cuticle fragments, have a fossil record that reach back to Silurian (*ca* 420 Myr), they are usually mineralized rather than organically preserved. Thus, chitin is generally not preserved as such, but some of the most spectacular examples of soft part preservation involve replication in calcium phosphate, since the apatite minerals inhibit the decay of organic compounds. Although proteins have a short survival time even within the $CaCO_3$ crystals, it is possible that phosphate salts provide protection from degradation (*57*). Other spectacular examples of fossilization include arachnids from Rhynie (Scotland), which were found preserved by precipitation of silica (*58*), whereas millipeds occur as calcified remains (*59*). Furthermore, the fossil cuticles, when organically preserved, often reveal lack of chitin but presence of other organic compounds such as aliphatic polymer (*60-62*). Fate of chitin in sedimentary strata is discussed later in this volume (*53*).

Conclusions

The nitrogen present in chitin imparts unique properties to this class of polysaccharides. Chitins are widely distributed in the biosphere, mainly as the strengthening material of the exoskeletons of many organisms, but also are a part of such delicate structures and tissues as tendons of insects and crustaceans. There are many derivatives of chitin, found either in biosphere or produced industrially, however chitin and its deacetylated polymer, chitosan, have the highest biological and commercial importance.

The chitins nitrogen is recycled by biodegradation operated by microbial genera which exist in just about every conceivable environment. In fact, chitin is susceptible to the action of specific hydrolases that are amply used for nutrition, self-defence, as well as for aggression. Chitin nitrogen can be therefore obtained from another organism either by digestion of the latter or by transfer, for instance from a mycorrhizal fungus to a plant.

Similarly, goods manufactured from chitin are environmentally important items, such as the chitosan solutions used to coagulate proteins and to preserve cereal seeds, or sophisticated biomaterials such as wound dressings and dietary ingredients. In most cases, the enzymatic hydrolysis of the chitin is important, because it can mobilize some oligomeric species having particular biochemical significance, or because a chitin-bearing material is more environmentally acceptable than a corresponding artificial polymers. Thus, it becomes evident that chitins are endowed with profound biochemical significance.

Acknowledgments

The authors are grateful to Mrs Maria Weckx for assistance in the retrieval of the bibliographic references. Work performed with Fondi MURST Quaranta Percento, and under the auspices of Agenzia Spaziale Italiana, Roma.

Literature Cited

1. Jeuniaux, C.; Voss-Foucart, M.F. *Biochem. Syst. Ecol.* **1991**, *19*, 347-356.
2. *Chitin in Nature and Technology;* Muzzarelli, R.A.A.; Jeuniaux, C.; Gooday, G.W., Eds.; Plenum: New York, NY, 1986.
3. Muzzarelli, R.A.A. In *The Polysaccharides;* G.O. Aspinall, Ed.; Academic Press: New York, 1985, Vol. 3.
4. Revel, J.F.; Marchessault, R.H. *Int. J. Biol. Macromol. 1993*, *15*, 329-335.

5. Bianchi, E.; Ciferri, A.; Conio, G.; Marsano, E. *Mol. Cryst. Liq. Cryst. Letters* **1990**, *7*,111-115.

6. Bade, M.L. In *Applications of Chitin and Chitosan,* M.F.A. Goosen, Ed.; Technomic: Basel, 1997; pp 57-78.

7. Yui, T.; Kobayashi, H.; Kitamura, S.; Imada, K. *Biopolymers* **1994**, *34*, 203-208.

8. Horst, M.N.; Walker, A.N. In *Chitin Enzymology*; R.A.A. Muzzarelli, Ed.; Atec: Grottammare, Italy, 1993; Vol. **1**; pp 109-118.

9. Adams, D.J.; Causier, B.E.; Mellor, K.J.; Keer, V.; Milling, R.; Dada, J. In *Chitin Enzymology;* R.A.A. Muzzarelli, Ed.; Atec: Grottammare, Italy, 1993; Vol. **1**; pp 15-25.

10. *Chitin Handbook,* Muzzarelli R.A.A.; Peter M.G., Eds.; Atec: Grottammare, Italy, 1997.

11. Lerouge, P. *Glycobiolog.* **1994**, *4*, 127-134.

12. Cedergreb, R.A.; Lee, J.; Ross, K.L.; Hollingsworth, R.I. *Biochem.* **1995**, *34*, 4467-4477.

13. Spaink, H.P. *Critic. Rev. Plant. Sci.* **1996**, *15*, 559-582.

14. Staehelin, C.; Granado, J.; Muller, J.; Wiemken, A.; Mellor, R.B.; Felix, G.; Regenass, M.; Broughton, W.J.; Boller, T. *Proc. Natl. Acad. Sci. USA* **1994**, *91*, 2196-2200.

15. Slezac, S.; Dessi, B.; Dumas-Gaudot, E. In *Chitin Enzymology;* Muzzarelli, R.A.A., Ed.; Atec: Grottammare, 1996; pp 339-347.

16. Freepons, D. In *Applications of Chitin and Chitosan;* M.F.A. Goosen, Ed.; Technomic: Basel, 1997; pp 129-139.

17. Ottoy, M.H.; Varum, K.M.; Smidsrod, O. *Carbohydr. Polym.* **1996**, *29*, 17-24.

18. Allan, G.G.; Peyron, M. *Carbohydr. Res.* **1995**, *277*, 257-272.

19. Defaye, J.; Gadelle A.; Pedersen, C. *Carbohydr. Res.* **1994**, *261*, 267-277.

20. Muzzarelli, R.A.A.; Tanfani, F.; Scarpini, G.F. *Biotechnol. Bioengin.* **1980**, *22*, 885-896.

21. Rane, K.D.; Hoover, D.C. *Food Technol.* **1993**, *7*, 11-33.

22. Terbojevich, M.; Carraro, C.; Cosani, A.; Marsano, E. *Carbohydr. Res.* **1988**, *180*, 73-86.

23. Hirano, S.; Yoshida, S.; Takabuchi, N. *Carbohydr. Polym.* **1993**, *22*, 137-140.

24. Toffey, A.; Samaranayake, G.; Frazier, C.E.; Glasser, W.G. *J. Appl. Polym. Sci.* **1996**, *60*, 75-85.

25. Muzzarelli, R.A.A.; Delben, F.; Tomasetti, M. *Agro-Food Industry High Tech.* **1994**, *5*, 35-39.

26. Rinaudo M.; Le Dung, P.; Gey, C., Milas, M. *Int. J. Biol. Macromol.* **1992**, *14*, 122-128.

27. Delben, F.; Muzzarelli, R.A.A. *Carbohydr. Polym.* **1989**, *16*, 221-232.

28. Muzzarelli, R.A.A.; Weckx, M.; Filippini, O.; Sigon, F. *Carbohydr. Polym.* **1989**, *16*, 293-306.

161

29. Inoue, K.; Yoshizuka, K.; Baba, Y. *New Dev. Ion Exch. Proc. Int. Conf. Ion Exch.* **1991**, 543-548.

30. Lapasin, R., Stefancic S.; Delben, F. *Agro-Food Ind. High Tech.* **1996**, *7*, 12-17.

31. Muzzarelli, R.A.A.; Zattoni, F. *Int. J. Biol. Macromol.* **1986**, *8*, 137-142.

32. Chiessi, E.; Paradossi, G; Venanzi, M.; Pispisa, B. *J. Inorg. Biochem.* **1992**, *46*, 109-118.

33. Desbrières, J.; Martinez, C.; Rinaudo, M. *Int. J. Biol. Macrom.* **1996**, *19*, 21-28.

34. Muzzarelli, R.A.A.; Ilari, P. *Carbohydr. Polym.* **1994**, *23*, 155-160.

35. Payne, G.F.; Sun, W.Q. *Appl. Environ. Microbiol.* **1994**, *60*, 397-401.

36. Payne, G. F.; Chaubal, M.W.; Barbari, T.A. *Polymer* **1996**, *37*, 4643-4648.

37. Muzzarelli, R.A.A.; Tanfani, F. *Carbohydr. Polym.* **1985**, *5*, 297-307.

38. Muzzarelli, R.A.A. *It. Patent Appl.* AN97062, 31 Oct. 1997.

39. Yoshida, H.; Nishihara, H.; Kataoka, T. *Biotechnol. Bioengin.* **1994**, *43*, 1087-1093.

40. Grobouillot, A.R.; Champagne, C.P.; Darling, G.D.; Poncelet, D.; Neufeld, R.J. *Biotechnol. Bioengin.* **1993**, *42*, 1157-1163.

41. Gooday, G.W. In *Biochemistry of Cell Walls and Membranes in Fungi;* Kuhn, P.J.; Trinci, A.P.J.; Jung, M.J.; Goosey, M.W.; Copping, L.G.; Eds.; Springer-Verlag, 1990; pp 61-79.

42. Gooday, G.W. In *Molecular Aspects of Chemotherapy;* Borowski, E.; Shugar, D., Eds.; Pergamon Press: Oxford, UK, 1990; pp 175-185.

43. Gooday, G.W. In *Advances in Microbial Ecology;* Marshall, K.C., Ed.; Plenum Press: Oxford, UK; 1990, Vol. 11; pp 378-430.

44. Gooday, G.W. In *Chitin Enzymology;* Muzzarelli, R.A.A., Ed.; Atec: Grottammare, Italy, 1996; Vol. 2; pp 125-134.

45. Lehane, M.J. *Annu. Rev. Entomol.* **1997**, *42*, 525-550.

46. Yalpani, M.; Pantaleone, D. *Carbohydr. Res.* **1994**, *256*, 159-175.

47. Poulicek, M.; Gail, F.; Goffinet, G. This volume.

48. Hirano, S.; Koishibara, Y.; Kitaura, S.; Taneko, T.; Tsuchida, H.; Murae, K.; Yamamoto, T. *Biochem. System. Ecol.* **1991**, *19*, 379-384.

49. Poulicek, M.; Jeuniaux C. *Biochem. System. Ecol.* **1991**, *19*, 385-394.

50. Poulicek, M.; Machiroux, R.; Toussant, C. In *Chitin in Nature and Technology;* Muzzarelli, R.A.A.; Jeuniaux, C.; Gooday, G.W. Eds.; Plenum: New York, NY, 1986; pp 523-530.

51. Miller, R.F. *Biochem. System. Ecol.* **1991**, *19*, 401-412.

52. Miller, R.F.; Voss-Foucart, M.F.; Toussaint, C.; Jeuniaux, C. *Palaeogeogr. Palaeoclim. Palaeocol.* **1993**, *103*, 133-140.

53. Stankiewicz, B.A.; Briggs, D.E.G.; Evershed, R.P.; Miller, R.F.; Bierstedt, A. This volume.

54. Miller, R.F.; Morgan, A.V. *Atlantic Geology* **1991**, *27*, 193-197.

55. Miller, R.F. *Atlantic Geology* **1994**, *30*, 65-66.

56. Schimmelmann, A.; Miller, R.F.; Leavitt, S.W. *AGU Geophys. Mag.* **1993**, *78*.

57. Briggs, D.E.G. *Palaios* **1995**, *10*, 539-550.
58. Trewin, N. *Trans. R. Soc. Edinb.* **1994**, *84*, 433-442.
59. Shear, W.A.; Gensel, P.G.; Jeram, A.J. *Nature* **1996**, *384*, 555-557.
60. van Bergen, P.F.; Collinson, M.E.; Briggs, D.E.G.; de Leeuw, J.W.; Scott, A.C.; Evershed, R.P.; Finch, P. *Acta Bot. Neerl.* **1995**, *44*, 319-342.
61. Briggs, D.E.G.; Kear, A.J. *Palaios* **1994**, *9*, 431-456.
62. Stankiewicz, B.A.; Briggs, D.E.G.; Evershed, R.P. *Energy Fuels* **1997**, *11*, 515-521.

Chapter 11

Chitin Biodegradation in Marine Environments

Mathieu Poulicek[1], François Gaill[1], and Gerhard Goffinet[3]

[1]Animal Ecology and Ecotoxicology Laboratory and [3]General Biology and Ultrastructural Morphology Laboratory, Zoological Institute, Liège University, 22, Quai Van Benéden, B-4020 Liège, Belgium
[2]Marine Biology Laboratory, 7 Quai Saint Bernard, University "Pierre and Marie Curie", F-75005 Paris, France

After cellulose and lignin, chitin is surely one of the most abundant biopolymers in the biosphere. However the amounts of chitin actually produced and accumulated in different ecosystems are difficult to evaluate accurately: according to Gooday (*1*), "*the annual production of chitin is enormous, but just how enormous is difficult to say*". A reasonable estimation for both annual production and the steady state amount of chitin in the biosphere is of the order of 10^{12} to 10^{14} kilograms (*1-3*). The estimated high biomass and production of chitin in marine ecosystems, together with its relatively recent commercial applications (for reviews see *4-10*), led to extensive studies of this biopolymer for the last twenty years.

This polysaccharide is known to be one of the most decay resistant of all biomacromolecules. It is generally firmly associated with proteins *via* different kinds of cross-linkages (e.g. catechol and quinonic sclerotization, histidyl links, disulfur bridges; *11, 12*) and the chitin-protein complexes are often mineralized (calcium carbonate with different mineralogies, silicate, etc.). These associations determine the rates and pathways of degradation in natural environments.

From an ecological point of view, the chitin biopolymer plays a key role in many biogeochemical studies as it contains both N and C, and so forms a link between the two most important cycles: rates of chitin production and biodegradation will affect C and N stocks and availability. Moreover, the high chelating capacities of this polysaccharide (in its native and deacetylated form, chitosan) affect the distribution and transfer of heavy metals and other elements through the whole water column and in surficial sediments (*6, 13, 14*).

Chitin in Marine Environment: Distribution and Abundance

Chitin biosynthesis appears to be a primitive property of the eukaryotic cell (*15-17*). Different types of unicellular organisms produce chitin: diatoms (*18-20*), Rhizopoda, Foraminifera, Cnidosporidia, ciliates, and others (*21-25*). Yeasts, molds and fungi are also well-known chitin producers (*26, 27*). This biosynthetic capacity is completely lacking in procaryotic Archaea, bacteria and Cyanobacteria.

Chitin is widely distributed in the animal kingdom as a supporting molecule in skeletal structures (e.g. tubes, cuticles, shells, eggs; *15-18*), mainly of epidermal origin. Chitin is used by many diploblastic organisms like cnidarians (mainly Hydrozoa but not by corals) and, in the protostomian lineage, triploblastic classes as bryozoans, brachiopods, nematodes, mollusks, pogonophorans, vestimentiferans, polychaete worms and, of course, arthropods. The main exceptions are platyhelminths and nemerteans. On the other hand, in the deuterostomian lineage (echinoderms, Enteropneusta, Pterobranchia, urochordates, ... that supposedly leads to vertebrates) chitin is absent, except in tunicates (peritrophic membrane, *28*), and in the cuticular layer of a fish fin (*29, 30*).

Three different crystallographic structures were described in chitinous material differing in the number of chains in the unit cell and their organization. In β-chitin, each crystallite contains only one chitin chain, all chains running parallel. In the α-chitin, there are two chains per unit cell, running in antiparallel directions. In γ-chitin (the existence of which is subject to debate), the crystallite should be formed by three chains, the central one running antiparallel between the two adjacent ones (*16, 31*). The distribution of the three crystallographic forms does not appear to be taxonomically related and the three forms can even coexist within the same organism as in the squid *Loligo* (beak in α-chitin, pen with β-chitin and stomach cuticle in γ-chitin) or the brachiopod *Lingula* (*16*). According to Rudall (*31*), β-chitin and γ-chitin are associated with collagen-like structures or with neighboring collagen-secreting tissues, while α-chitin replaces collagen-like structures.

The various chitinous structural tissues show striking differences in macromorphology, fine structure, chemical composition and physical characteristics. In living systems, chitin occurs in most cases in the form of microfibrils 2.5 to 3.0 nm in diameter, clearly demonstrated recently by electron microscopists. 18 to 25 individual molecular chains are arranged in two or three rows to form a single crystalline rod, with proteins bound at the periphery (*32-35*). In arthropod cuticles, the chitin-protein supramolecular organization first highlighted the presence of twisted liquid crystalline orders in biological structure (*35*). This model was subsequently extended to other biological polymers. The chitinoprotein fibers appear as fiber reinforced composites. The matrix is an ordered set of crystalline chitin fibrils surrounded by sheaths of ordered proteins (*35, 36*). In arthropod cuticles, for example, the chitinoprotein fibrils are disposed horizontally and parallel in successive planes. However, the direction of the main fibril axis rotates continuously from one plane to the following one. This constitutes the "twisted plywood model" of cuticular structure described by Bouligand (*37, 38*). The pseudohelicoïdal structures of the organic matrix of cuticles are organized as cholesteric liquid crystals. Since this description, many publications confirmed this analogy between the three-dimensional organization of extracellular fibrillar material in several plant (cell walls) and animal phyla (vertebrate bone, fish scales, egg membranes, annelid, Pogonophora and Vestimentifera tubes, and others) and that of molecules in liquid crystals (*35, 39*). In mollusks, the rather less organized chitinoprotein fibrils act as a structural skeleton on which other structural and calcifying proteins are arranged (*40*). In the newly discovered worms Vestimentifera from the deep-sea hydrothermal vents, tubes are made of β-chitin with high crystallinity and proteins without mineralization (like in Pogonophora), but the crystallite has unusually large microfibrils, 50 nm in section and 3400 nm length, and atypical fibrillar structure (*41*). The latter form a liquid crystal-like structure, differing from the cholesteric arrangement, merely a nematic texture with random twist inversion (*42*). Large microfibrils are also observed in fungi, 9 to 27 nm in diameter and up to 1 µm length.

The chitinoprotein fibrils of the organic matrix leave space for minerals but there is no apparent relationship between chitin content and degree of mineralization (except in gastropod operculi, *43*), hardness or flexibility of the structure. These last properties mainly depend on the protein composition and abundance, but principally on the nature of stabilization forces involved (hydrogen bonded or Van der Waals interchain regular interactions, covalent bridging, quinonic sclerotization, disulfur bridges, etc).

Distribution and Production of Chitin in Benthic Biocoenoses and Sediments

Rocky Benthic Substrates. Jeuniaux and collaborators studied the biomass and production of chitin by benthic communities on rocky substrates in the Bay of Calvi (Corsica, Mediterranean Sea; *2*). Chitin biomass in infralittoral communities of photophilous algae (characterized by several species of *Cystoseira* between 5 and 30 m depth), sampled by diving, was principally due to sessile colonies of Bryozoa and Hydrozoa and to agile species of crustaceans on the other hand. The whole benthic biological cover contributed to a mean chitin biomass close to 1 g m^{-2}. In sciaphilous communities, in semi-dark caves or below overhanging rocks, the chitin biomass due to encrusting colonies of Hydrozoa and Bryozoa was significantly lower, but the contribution of large crustaceans, while highly variable, was sometimes very important. The mean chitin biomass value was tentatively estimated at 1.4 g m^{-2} (*44-46*).

The rate of chitin production by infralittoral communities living on rocky shores was estimated by measuring the amount of chitin accumulated after pioneering communities allowed to settle and to grow on naked substrates immersed at different depths (6 to 37 m) in natural conditions. One must assume that, in this case, predation and mortality are negligible. The species composition of the biological cover on the plates was, after a few months, roughly similar to that of neighboring benthic communities. The results (*2*) showed that during the first year, the pioneering communities developed and chitin production was low (ca 0.3 g m^{-2} year^{-1}); it was quicker during the second year (ca 1 g m^{-2} year^{-1}). Chitin production appeared to be lower at greater depths (750 mg m^{-2} around 19-30 m), compared to a shallower experimental sites (1200 to 1400 mg m^{-2} from 5 to 11 m,) (*2, 45*). In a *Posidonia* meadow, the main chitin producers are still Hydrozoa and Bryozoa: chitin production by the epiphytic biological cover increased from 0.5 mg m^{-2} of the leaf surface during spring and summer, to 1.3 mg m^{-2} during autumn and winter. Thus annual chitin production in the meadow of Calvi Bay was estimated at 75 mg m^{-2} year^{-1} (*44*).

We have already noted that crustaceans are the main chitin producers in infralittoral benthic communities growing on rocky substrate (2). However, large decapod species inhabiting crevices (e.g. lobsters, crabs) were not really taken into account in these estimations. The importance of these crustaceans as chitin producers is obvious, owing to the part occupied by crabs and lobsters in fisheries and canning industries. Some authors (47) estimated that approximately 39000 tons of chitin is the amount available yearly as wastes of fisheries. Although, the specific productivity studies of natural populations are scarce, it was possible to calculate the annual production of chitin in a population of lobsters (1.5 g m^{-2} yr^{-1}) based on a natural population of the spiny lobster *Panulirus homarus* inhabiting a small isolated reef of the Natal Coast (South Africa) (2, 48). Obtained value concerning a single species is on the same order as the estimations of chitin production obtained so far for the whole pioneering benthic communities growing on naked substrates.

Sediment. In order to evaluate chitin biomass, Poulicek (49) screened one hundred marine sediments of various origins. The chitin biomass of marine sediments is very variable, from 2 up to 2800 µg g^{-1} of decalcified sediment. Most sediments have a low or very low chitin biomass content (67 % below 100 µg g^{-1} of decalcified sediment). No significant difference related to depth nor climatic influence was observed, except that all sediments richer in chitin (above 500 µg g^{-1} of decalcified sediment) are on the continental shelf (less than 200 m depth). Moreover, the chitin content is higher in coarse, much calcified sediments of organoclastic origin with Bryozoa, and shelly sands and gravels being the richest.

To our knowledge, no data on production of chitin in sediment are available. Nevertheless, considering potential "rain" of chitinous material arising from benthic communities or from plankton and settling suspended matter, the very low chitin content of most sedimentary environments provides an evidence for rapid degradation of this biopolymer in an open water. The latter process occurs via the microbial loop, and/or at the sedimentary interface, or in the surficial layers of sediments.

Distribution and Production of Chitin in Open Water

Planktonic Biocenoses. The biomass and production of chitin by typical oceanic zooplankton was estimated in the bay of Calvi (Corsica) in the Mediterranean Sea (50, 51). Plankton was sampled and analyzed during two annual cycles (50) and the results expressed with respect to a square meter of surface water for a water column of 100 m depth. The zooplankton biomass was dominated by five species of copepods (belonging to the genus *Acartia, Calanus, Clausocalanus, Centropages* and *Oithona*: more than 90 % of total biomass) and by cladocera. The daily production of every species was estimated by the "cumulative growth method" of Winberge (52) adapted for species with several larval instars during the life span. Chitin content (expressed in % of dry weight) was found to be relatively constant during larval development and in adults of the same species. However, variations were noted between species, from 3.1 % in *Clausocalanus* spp. and 8.6 % in *Acartia clausi* to as much as 12.2 % in the dominant cladoceran *Evadne* spp. (2, 3, 51). The mean daily chitin production by planktonic crustaceans (copepods, cladocerans and decapod larvae) was calculated. Seasonal variations are important, with a maximum production of 20 mg m^{-2} day^{-1} during the spring bloom in May, corresponding to a chitin biomass around 400 mg m^{-2}. Mean annual chitin production was estimated at approximately 1.0 g m^{-2} yr^{-1}, for a mean chitin biomass of 26 mg m^{-2} (*2, 3, 51*). The values calculated for chitin production must be considered as minimum ones, as exuviae and peritrophic membranes were not taken into account (although it has been shown that part of the chitin of the cuticle is hydrolyzed and recovered by the molting animal before ecdysis).

Other data on chitin production by plankton are very scarce and merely concern fresh water environments, however, the some estimates (*2*), appear to suggest the biomass and production values of the same order of magnitude as those obtained for marine zooplankton.

Krill. The role played by euphausiids in the production of chitin must be considered carefully in the present review, as krill is regarded as one of the most important components of the marine pelagic food web. Recent estimates of annual production of Antarctic krill varied from 16 to 1350 10^6 tons per year (*53*) or, more likely, from 100 to 500 10^6 tons (wet weight) per year (*54*). Jeuniaux and collaborators estimated chitin production by krill using results of exhaustive studies of populations in the North Atlantic and South Pacific oceans. Following Lindley (*55*), the chitin biomass of krill would be highest in the Norwegian Sea (0.6 mg m^{-3}, principally *Meganyctiphanes norvegicus*), whereas in other parts of the North Sea and North Atlantic Ocean, chitin biomass of krill varies from 0.2 to 0.4 mg m^{-3}. Thus, the chitin production values (without exuviae) range from 0.2 to 0.3 mg m^{-3} yr^{-1} in most ocean areas, to more than 0.9 mg m^{-3} yr^{-1} in the North Sea and near the Atlantic. If the scattered vertical distribution of krill "swarms" is taken into account, chitin biomass and production in a water column of 100 m depth would vary respectively from 10 to 30 mg m^{-2} (biomass) and from 10 to 40 mg m^{-2} yr^{-1} (production). The latter values are lower than those obtained for zooplankton in the Mediterranean Sea, but the calculated values for krill are probably underestimated (i.e. without exuviae being taken into account). The data of Ritz and Hosie (*56*) on a krill community in the South Pacific dominated by *Nyctiphanes australis*, are more reliable as they include the production of exuviae. For a mean annual chitin biomass of 0.4 mg m^{-3}, the total chitin production by exuviae was estimated to be 16 mg m^{-3} yr^{-1}. After extrapolation for a water column of 100 m depth, a chitin biomass of 20 mg m^{-2} and a chitin production of about 800 mg m^{-2} yr^{-1} was estimated (*2, 3*).

The total chitin production (calculated for adults + exuviae) would be 1.1 g m^{-2} yr^{-1} (values overestimated according to *57*). Other calculated values for *Euphausia superba* are in the range of 100 to 500 mg m^{-2} $year^{-1}$ (*3*). These values more nearly approach the 1.0 g m^{-2} yr^{-1} obtained for the Mediterranean zooplankton.

Particulate Organic Matter (POM). Although the general and elementary composition of particulates has been widely reported, their chemistry has not been thoroughly analyzed. The difficulty in collecting sufficient amount of material for such studies is limiting factor. Suspended material in open water is composed of both organic (from less than 1 % to more than 95 %) and inorganic components, with wide variability in composition (*58, 59, 60*). The input of detrital chitin to the open waters is certainly very high with the presence of fragments of planktonic crustacean cuticles, exuviae, and fecal pellets with chitinous peritrophic membranes, diatoms, and other contributors (58, 59) Figures 8-11.

Nevertheless, a thorough study of detrital chitin distribution in the North Sea reveals a very low chitin content in suspended matter measured during several cruises (Table I) and, thus, indirectly indicates that chitin degrades at a high rate (*61*).

Table I Distribution of detrital particulate chitin in the suspended matter of the North Sea (*61, 62*).

Season	Chitin (µg/l) Mean [extr. values] *(% org. matter)*	Total POM (mg/l) Mean [extr. values]	Organic compounds (µg/l) mean [extr. values] *(% total susp. matter)*
Winter	1.7 [0.3 - 4.8]; *1.9*	5.4 [0.5 - 30.0]	90.1 [19.0 - 305.0]; *1.7*
Spring	4.9 [0.9 - 26.1]; *3.0*	6.2 [1.3 - 24.8]	164.9 [27.5 - 669.3]; *2.7*
Summer	3.1 [0.8 - 11.7]; *2.2*	5.1 [0.7 - 19.9]	141.5 [20.6 - 387.1]; *2.8*
Autumn	1.6 [0.2 - 6.6]; *1.7*	2.5 [0.3 - 21.0]	90.8 [18.8 - 226.4]; *3.6*

The concentration of detrital particulate chitin in January was the highest (in µg l^{-1}) in southern and eastern parts of the North Sea, and lower in the northern part, and in the Skagerrak (*61, 62*). The proportion of chitin in suspended matter (in µg mg^{-1}) showed the reverse trend (same tendency in the distribution of the content of the whole organic matter in the suspended material determined by ashing and with the distribution of particulate organic carbon (POC)). In March, the chitin content of suspended particles was much higher than in January, with the highest concentrations in the Skagerrak and along the eastern side of the North Sea (*61, 62*). This indirectly reflects the distribution of primary production, which begins earlier in the southern and the eastern North Sea than in the northern part. Distribution patterns in October were similar to those in January, but not as clear; July displayed intermediate values.

Chitin concentrations are generally very low compared to those of lipids and proteins. Proportions of proteins in the North Sea suspensions (70-85 % of the total organic matter) are always higher than those of other compounds, such as lipids (13-25 %) and chitin (2-3 %) (*61, 62*).

Other data on chitin distribution in suspended particles are scarce. Chitin concentration estimated near the mouth of the Delaware Bay (estuarine conditions) ranged from 4 to 21 µg l^{-1} (*63*), very similar to the data obtained for the North Sea samples close to the mouth of Schelde river (*61*). The decrease in chitin concentrations from 21 to 4 µg l^{-1} reflects the seasonal decline in primary production in Delaware Bay. Concentration of chitin in both, the Delaware Bay and the North Sea are also similar to those of the subarctic Pacific (Gulf of Alaska, 50°N, 150°W; 4 to 10 µg l^{-1}; *63*), and to north occidental Mediterranean Sea (< 5 µg l^{-1}; *58, 59*).

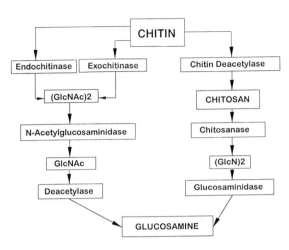

Figure 1. Alternative pathways of chitin biodegradation in marine environments. Note: GlcNAc: acetyl glucosamine; $(GlcNAc)_2$: chitobiose; $(GlcN)_2$: dimer of glucosamine.

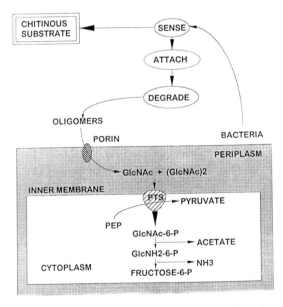

Figure 2. Chitin biodegradation processes by *Vibrio* (adapted and transformed from *127*). Note: GlcNAc: acetyl glucosamine; $(GlcNAc)_2$: chitobiose; GlcNAc-6-P: acetyl glucosamine-6-phosphate; $GlcNH_2$-6-P: glucosamine-6-phosphate; PEP: phosphoenol pyruvate; PTS: phosphoenolpyruvate phosphotransferase system.

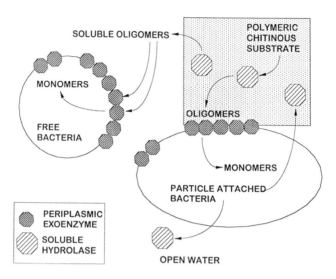

Figure 3. Conceptual model of cooperation between free and attached bacteria for organic polymer biodegradation (adapted from *60*).

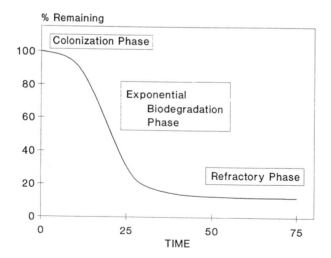

Figure 4. Chitin biodegradation kinetic curve according to our conceptual model.

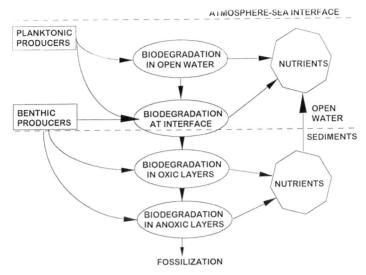

Figure 5.   Box model displaying main interrelations between different compartments in marine environment

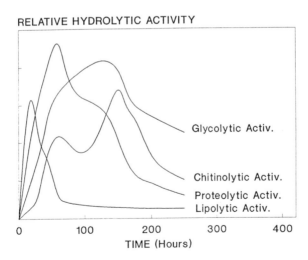

Figure 6.   Succession of hydrolytic activities on experimental detritic material suspended in open water.

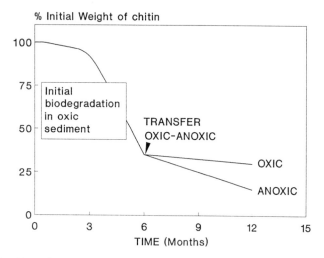

Figure 7. Transfer experiment between oxic and anoxic layers during chitin biodegradation

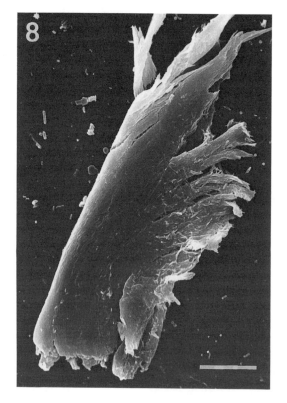

Figure 8. "Fresh" cuticular piece of zooplanktonic organism in the particulate organic matter (POM) collected at 75 m depth in Mediterranean Sea. Scanning Electron Microscopy (SEM) scale is 5 μm.

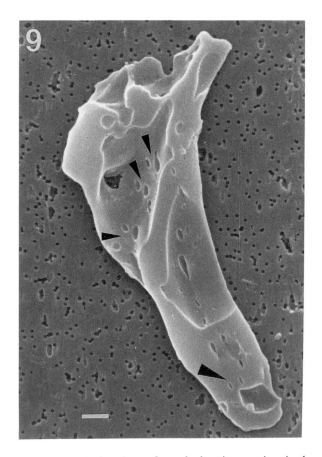

Figure 9. Weathered cuticular piece of zooplanktonic organism in the particulate organic matter (POM) collected at 675 m depth in Mediterranean Sea. SEM scale is 1 μm. Observe "smooth" shape and bacterial pitting of the surface (arrows).

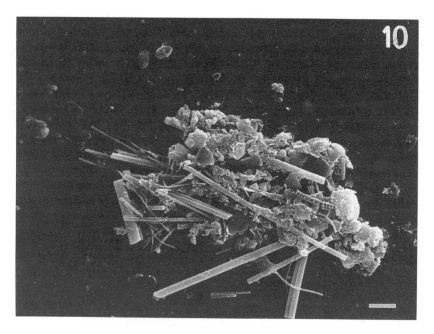

Figure 10. Partly desagregated copepod fecal pellet without peritrophic membrane collected at 250 m depth in Mediterranean Sea. SEM scale is 10 μm.

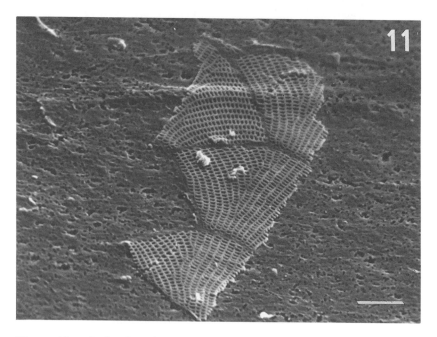

Figure 11. Isolated peritrophic membrane collected at 675 m depth in Mediterranean Sea. SEM scale is 1 μm.

Assuming refractory character of chitin, its concentration would be expected to be highest in the Delaware Bay estuary and the eutrophic North Sea and lower in subarctic Pacific or mesopelagic samples of the Mediterranean Sea due to the higher production level in the former. However, heterotrophic bacteria (60, 61), readily use chitin like other compounds such as dissolved free amino acids or hexoses. The concentration of these compounds in the highly productive Delaware Bay and North Sea are similar to that found in the less productive subarctic Pacific (*60, 61, 64-66*).

Vertical fluxes of chitin associated with settling suspended matter were also measured in sediment trap experiments (*63*). The large decrease in chitin flux in the subarctic Pacific, from 11.2 mg m^{-2} day^{-1} at 10 m to 1.5 mg m^{-2} day^{-1} at 70 m supports rapid degradation of chitin. Chitin flux remained almost constant at about 1.3 to 1.8 mg m^{-2} day^{-1} between 70 m and 500 m depth. The C flux decreased similarly with depth. The vertical flux of chitin in the Gulf of Alaska was 0.5 to 1.1 % of the total C and N flux (*63*). Muller et al. (*67*) observed that the percent of total C flux contributed by glucosamine was greater in deep (3.6 %) than in the surface waters (1.9 %).

In Table II, we use data of Honjo and coworkers (*68*) to calculate chitin fluxes in the deep sea (results from PARFLUX experiment with five sediment trap arrays deployed at four deep ocean stations: Söhm abyssal plain (Sargasso Sea), Demerara abyssal plain (tropical Atlantic), East Hawaii abyssal plain, and north central Panama Basin). We used mean proportions of chitin in particulate organic matter estimated in different settings (*58-61, 63*) to estimate the flux ranges. The total mass flux in these experiments was relatively constant below a depth of 1000 m, except in the Panama Basin, where it increases linearly (Table II). Biogenic materials are very abundant and accounted for 60 to 90 % of the material collected in sediment traps (*68*). There is a general trend of increase proportion of lithogenic material with depth: the main biogenic particles are fecal pellets, crustacean cuticle exuviae and fragments, amorphous aggregates including free cells, fragments, pigmented granules ("olive cells" of authors) and "waxy" particles. This composition is very similar to that found in deep Mediterranean samples (*58, 59*).

Calculated chitin fluxes are much higher (6-280 µg m^{-2} day^{-1}) than in other settings (except Hawaii) and appear frequently to stay at high levels, even in very deep stations. Nevertheless, all investigations (*58-63,67, 68*) indicate that the chitin flux associated with settling particulate matter is relatively small (< 0.3 mg m^{-2} day^{-1}). The material associated with the chitin (e.g. protein and carbohydrates in fecal pellets) may still be a large fraction of the total C and N flux. Chitinolytic marine bacteria associated with settling POM, in superficial down to mesopelagic layers (*58*, 59), could degrade and utilize the portions of the chitin fibrils that are readily susceptible to hydrolysis by chitinase (*13, 60, 69, 70*). The relatively undegraded chitin that sinks lower to the ocean depths (below the most productive, mixed area, *58-59*) may be heavily cross-linked with sclerotized proteins which limits further degradation by bacterial exohydrolytic activity. This may explain the large turnover times (> 2000 days for the detritic chitin pool below the mixed layer, *63*).

Table II Flux of suspended material in four deep stations as recorded in
sediment traps experiments (PARFLUX data recalculated according to *68*)

Localization	Depth m	Mass flux mg/m²/day	Biogenic material % total flux	Organic material % tot. flux	Chitin (*) µg/m²/day
Sargasso Sea	976	19.6	90.1	27.2	27-51
(Söhm abyss.	3694	18.4	84.0	10.8	10-19
plain)	5206	13.0	-	29.6	19-36
Demerara	389	69.4	91.9	19.7	68-128
abyssal	988	49.2	83.0	17.4	43-81
plain	3755	46.4	76.2	10.3	24-45
	5068	47.0	69.3	10.5	25-47
East Hawaii	378	11.4	-	59.5	34-64
abyssal	978	7.5	-	16.2	6-11
plain	2778	17.1	92.6	14.0	12-23
	4280	16.8	91.7	10.7	9-17
	5582	11.1	84.8	13.5	7-13
North	667	114.1	81.1	22.5	128-242
Panama	1268	104.5	80.1	19.2	100-189
Basin	2265	125.1	73.3	17.0	106-200
	2869	158.0	68.7	17.0	134-253
	3769	179.3	58.2	13.6	122-231
	3791	179.6	58.9	16.5	148-280

* calculated from data in (*68*)

The sinking rates of individual chitinous particles are too low to contribute
significant amounts of organic matter to the deep-sea sediments (Figures 8, 9).
Smayda (*71*) reported sinking rates ranging from < 0.1 m day[-1] up to > 500 m day[-1] for
very large and heavy particles, Lännergren (*72*) recorded a range of 0-9 m day[-1], and
Bienfang (*73*) found from 0.3-1.7 m day[-1]. A more important route is the sinking of
fecal pellets (*74-83*). Tumer (*84*) determined sinking rates for fecal pellets of the
copepod *Pontella* to be 15-150 m day [-1], and Fowler & Small (*85*) estimated rate of
240 m day[-1] for euphausiid fecal pellets. The sinking rate of individual fecal pellets
was not constant due to their small volume and density, the factors very susceptible to
variations in water column micro-structure. Even assuming laboratory conditions and
the fact that the plankton in experimental enclosures were fed on unialgal diets rather
than a natural mix of phytoplankton cells, values obtained give an indication of the
likely rates in the sea. Knauer *et al.* (*74*) estimated the flux of fecal pellets in the
northeastern Pacific to be about 1000 fecal pellets m[-2] day[-1], whereas Wiebe *et al.* (*75*)
measured a flux of 160 fecal pellets m[-2] day[-1] off the Bahamas. Furthermore, the
confirmatory laboratory studies found sinking rates of 50-940 m day[-1] with a mean of
159 m day[-1]. Some fecal pellets therefore sink quickly enough to supply chitin to
deep-sea sediments. But even whilst sinking, organic matter is likely to be broken

down by bacterial activity (*81-83, 86, 87*) (Figure 12). Thus, the amount of organic matter reaching the sediments will be reduced.

Honjo & Roman (*83*) studied the rate of breakdown of copepod fecal pellets and found that degradation of the peritrophic membrane of the pellet took 3h at 20°C, however, at temperature close to that of the deep ocean (i.e. 5°C), peritrophic membranes remained intact for up to 20 days. After degradation of the membrane, the pellet lost its integrity and broke up into small amorphous aggregates (Figures 10, 11). Therefore, if fecal pellets "survive" bacterial degradation whilst sinking through the warm surface waters, there is a good chance that they could survive to sink further down. Iturriaga (*88*), studying the bacterial alterations occur in sedimenting organic matter, found that at 20°C dead phytoplankton decomposed at a rate of 35 % day^{-1} whereas zooplanktonic remains decomposed at 18 % day^{-1}, and at 5°C the rates were lower (about 3 % day^{-1} for phytoplankton and 8 % day^{-1} for zooplankton). Comparable decomposition rates were recorded by Poulicek and coworkers (*69, 70, 90, 91*) and Harding (*89*) who found that killed copepods decomposed within 3 days at 20°C and within 11 days at 4°C. Harding (*89*) commented that copepods, which die in the surface waters, are unlikely to be recognizable as copepods if caught by a plankton net within one or two days of death.

Conclusion. The results considered above allow us to propose a mass balance of chitin production and accumulation in the oceans. Table III summarizes the biomasses and fluxes of chitin.

Assuming chitin production in a pelagic systems to be 2.3 10^9 tons per year , and chitin mean biomass in POM as 5 μg l^{-1}, then on a global basis, calculated residence time of chitin in the open ocean is 103 days. This value is very close to the 119 days estimated for turnover in cold surface waters (approx. 12°C, surface to depth of 50 m), based on the chitin flux and concentrations of suspended chitin (*63*). It is also similar to the turnover time of 140 days at 15°C calculated by Seki (*92*, estimation based on chitin degradation and probable number of chitinolytic bacteria in the water column). Taking into consideration the mean ocean depth (3729 to 3824 meters according to *93*), and the time needed for fine particles to settle through the water column, this can explain the low chitin fluxes at intermediate and abyssal depths (6 to 280 μg m^{-2} day^{-1}), even if shallower fluxes are much more greater (1 to 11 mg m^{-2} day^{-1}). The very effective biodegradation pathways of chitin and chitinoproteic complexes in the open water environment can only explain the latter phenomenon. It also explains the generally low chitin content in the sediments, even where fast sinking fecal pellets, sheltering relatively undegraded chitinous particles, and skeletal pieces of macrofauna (molluscan shells, crustacean cuticles, etc.) are embedded in the sedimentary layers.

Biochemical Aspects of Chitin Biodegradation

Chitin Biodegradation Enzymes. Chitin is degraded by at least two different enzymes (*15, 16*), i) the endoenzyme chitinase (more precisely defined as poly-β-1,4-(2-acetamido-2-deoxy)-D-glucoside glycanohydrolases, E.N. 3.2.1.14), which splits

178

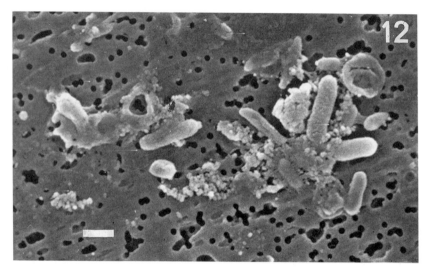

Figure 12. Bacterial clusters associated with small organic particles in the POM collected at 675 m depth in Mediterranean Sea. SEM scale is 1 μm.

**Table III Chitin production and accumulation in a global perspective
(compiled from *2, 3, 17, 44, 45, 49, 51, 61-63*)**

PRODUCTION	Extent	Chitin production	
	10^6 km^2	g m^{-2} year^{-1}	tons 10^3 year^{-1}
Pelagic systems	360		
Zooplankton		1.0 *	360 000
Krill		5.3 **	1 938 000
Other shrimps		?	?
Benthic systems			
Rocky substrates	1.4		
Epifauna		1.0	1 500
Large Decapods		1.5	2 100
Reefs	1.5	?	?
Continental slope	30	?	?
Sediments	320	?	?
Total	360		>>2 301 600
ACCUMULATION	Extent	Chitin biomass	
	10^6 km^2	µg g^{-1}	µg l^{-1}
Pelagic system			
MOP	360		1 - 26
Benthic systems			
Sediments	320	2 - 2 800	

* for a water column, 100 meters depth; ** for a water column 500 meters depth

chitin into oligomers, and ii) *N*-acetyl-β-D-glucosaminidase (2-acetamido-2-deoxy-β-D-glucoside acetamido-deoxygluco hydrolase E.N.3.2.1.30), which degrades these products to monomers (*4, 15, 94*). Both enzymes are in most cases soluble and generally secreted. The exochitinase activity, splitting the glycosidic bond two residues distant from the non reducing end of each chitin chain, was described by Ohtakara (*95, 96*), and separated from endochitinase activity by DEAE-cellulose column chromatography (*97-99*). Endochitinase cleaves bonds randomly along the chitin strand to form loose ends that can be further hydrolyzed by exochitinase (*100*).

Chitinases comprise families 18 and 19 of glycosyl hydrolases (based on their amino acid sequences, *99*). The family 19 is homogeneous and contains only plant enzymes whereas family 18 is more diverse, containing chitinases from plants, fungi, bacteria and viruses. Family 18 includes eukaryotic chitobiases (i.e. *N*-acetyl-β-D-glucosaminidase), but prokaryotic enzymes constitute family 20 together with hexosaminidase (*98, 99*). The isoelectric point and optimum pH of the reactions is usually in the range of about 5.0-5.5 for both enzymes, except in the digestive system of reptiles (*15, 16*). Their apparent molecular masses are between 40 and 120 kDa for endochitinases, 60 kDa for exochitinases, and around 100 kDa for *N*-acetyl-β-D-glucosaminidase. Numerous chitinases have been purified, the corresponding gene isolated, cloned and the primary and tertiary structures of the enzymes determined (*111, 112*).

Chitinases are widely distributed hydrolases synthesized by bacteria, fungi, higher plants and digestive glands of animals whose diet includes chitin (*15*). They allow many plants and animals to perform vital functions such as digestion, growth and defense against parasites and chemical attack. The widespread distribution of endo- and exochitinases and *N*-acetyl-β-D-glucosaminidase in bacteria, molds and Protozoa suggests a primitive nature of their function. In animals, diploblastic Metazoa are characterized by both ecto- and endodermic secretion of these enzymes. The ectodermic secretion of *N*-acetyl-β-D-glucosaminidase is maintained whereas most Protostomia (except nematodes and, of course, arthropods, generally loses that of endochitinase where it functions in molting (15, 101). The endodermic secretion of endochitinase is preserved in most invertebrates (except for groups with very specialized food containing no chitin). In Deuterostomia, and especially in vertebrates, a marked tendency towards a loss of this function is clearly observed (regressive evolution by so called "enzymapheresis") (*102*). In fungi, there are three roles attributed to chitinases (*103*) involving [1] the gross autolysis associated with the release of spores, [2] a nutritional role as in the case of saprophytes (enabling fungi to use insect or other fungal debris as a food source or, for pathogen species, to penetrate their arthropod or fungal host), [3] a morphogenetic role in the growth and differentiation of hyphae. In plants, chitinase is ubiquitous, found in seeds, leaves, stems and roots (*104-108*), but present in various quantities. *In vitro* studies have shown the ability of plant chitinase to attack and to degrade fungal cell (*104*). Moreover, the enzyme products found may act as elicitors of phytoalexin formation and thereby contribute to plant defense (*109, 110*).

Endo-β-*N*-acetylmuramidases, more commonly known as lysozymes (mucopeptide *N*-acetyl muramoylhydrolase, E.N. 3.2.1.17), have been isolated from a great variety of sources, animals, plants and microorganisms (*4*). Endo-β-*N*-acetylmuramidase hydrolyses the glycan of the bacterial cell walls to oligosaccharides of *N*-acetylglucosaminyl-*N*-acetylmuramic acid type, while endo-β-*N*-acetyl glucosaminidases give saccharides with *N*-acetylglucosamine at the reducing end. However, the ability of lysozyme to degrade chitin has been reported in several studies (*4, 113*).

Chitosan Biodegradation Enzymes. Chitosan is a polymer composed of β-1,4-linked glucosamine residues. It is usually obtained through artificial deacetylation of chitin with concentrated NaOH solution. The occurrence of "natural" chitosan was evidenced by the discovery that it is a major component of cell wall of mucoraceous fungi, formed by concerted action of chitin synthetase and chitin deacetylase (*114-116*).

A new class of enzymes, chitosanases (E.N. 3.2.1.99) active in hydrolyzing chitosan, was proposed by Monaghan (*117*). This kind of enzyme is probably widely dispersed (activity found in 25 fungal and 15 bacterial strains out of 200, *118*). The molecular weight of chitosanase was estimated to be about 31 kDa, and its isoelectric point at pH 8.3. Optimum pH is 5.6, but the enzyme is stable from pH 4.5 to 7.5. It is inactivated more quickly at alkaline than at acidic pH. In *Streptomyces*, extracellular chitosanase synthesis is induced by glucosamine. The purified enzyme hydrolyzes chitosan, but not chitin or carboxymethylcellulose (*119*). Chitin deacetylase (i.e. chitin amidohydrolase) has been purified and characterized (*120, 121*), and cDNA isolated, characterized and sequenced (*122*). The enzyme (EN.3.5.1.41) is an acidic glycoprotein (30 to 67 % carbohydrate content) of 75 to 150 kDa with very narrow specificity for β-1,4-linked *N*-acetyl-D-glucosamine homopolymers (at least three to four residues) in *Mucor*, but less stringent in *Colletotrichum* (*123*). The effectiveness of the latter enzyme in deacetylating various chitin substrates is ranked as follows: carboxymethyl chitin>glycol chitin>amorphous chitin>crystalline chitin. Optimum pH for the reaction varies, from around 4.5 to 5.5 in *Mucor* to 8.5 in *Colletotrichum* (*123*). Three factors can affect the deacetylation process of chitin and chitin derivatives, i) the properties and mode of action of specific strains of deacetylase, ii) the structural properties of chitin (depending on the nature of stabilization forces, and thus the origin of the chitinous material), and iii) the mode of interaction between the enzyme and the chitin molecule, partly under environmental control (*123*). Chitin alteration is a result of weathering processes, under biological mediation or not, which can modify the conformation of the polymer and, as a result, influence the interaction between chitin and the enzyme and the subsequent deacetylation reaction.

Chitin Biodegradation by Microorganisms

Chitin mineralization is primarily a microbial process, and chitin can act as the sole source of carbon and nitrogen for many microbes (bacteria and fungi). Its degradation is usually assumed to be via the "traditional" pathway, *i.e.* via chitinolysis to *N*-acetylglucosamine. However, the alternative pathways can be defined (*124, 125*), and, namely the "chitosan pathway" is of ecological importance (Figure 1).

According to the traditional models, chitin is depolymerized by either endo- or exochitinases, generally secreted as soluble, diffusive hydrolases. The action of "endo" hydrolases results in the formation of a variety of oligosaccharides of varying chain length. The endochitinases can rarely act on chains with less than three acetyl glucosamine residues, and the rate of hydrolysis is proportional to the degree of polymerization. Consequently, the result of endocleavage is a mixture of short chain oligosaccharides of which the disaccharides (chitobiose) are dominant. The exochitinase commonly acts after endochitinase has exposed a significant number of chain ends. Its action may result in the direct formation of disaccharides. The oligosaccharides formed are hydrolyzed by an acetylglucosaminidase, freeing acetyl glucosamine monomers that can be directly assimilated by the cells.

The "chitosan pathway" involves a partial or total deacetylation step, where chitin is transformed into chitosan. Chitosan is further hydrolyzed by chitosanase to oligomers of glucosamine. The hydrolysis of these oligomers by glucosaminidase frees glucosamine residues as direct substrates for cells.

Bacterial Processes. Chitin turnover is essential for recycling carbon and nitrogen in marine ecosystems. By 1937, ZoBell & Rittenberg (*126*) had shown that marine chitinovorous bacteria are ubiquitous, very abundant, diverse, widely distributed and able to live in extreme environments such as deep water.

However, how can we understand the process of chitin mineralization by such bacteria? This multi-step process can be subdivided into stages, each of them complex and highly regulated. The bacteria must be able to sense the presence of an adequate substrate, *i.e.* chitin, and/or come into contact with it by active directional swimming or by random collision. Then, they attach to the polymer and degrade it to oligosaccharides, and further into N-acetylglucosamine (or glucosamine via the chitosan pathway), either extracellularly (in the periplasm) or intracellularly. These steps were detailed in the case of *Vibrio furnissii* (*127-129*). Vibrionacea are among the commonest marine bacteria, rod shaped Gram negative, flagellated and mobile, facultative anaerobes.

A key step in the chitin biodegradation process is the adhesion of marine bacteria to chitin-containing particulates. *Vibrio* species were therefore surveyed for their ability to bind to immobilized carbohydrates. One strain of *V. furnissii* adhered to glycosides of three sugars, *N*-acetylglucosamine (the preferred ligand), D-mannose, and D-glucose (*127*). A single lectin disposed at the tips of fimbriae is responsible for binding to the three sugars. Sugar specific lectins are widely distributed among bacteria and the process of lectin-mediated adhesion to insoluble carbohydrates is considered to play a very important role in marine fouling and biodegradation processes. Schrempf (*130*) described a lectin (CHB1) secreted by several streptomycete species and specific for α-chitin, but not binding to β- or γ-chitin nor chitosan. *Vibrio* cells adhering to the chitin analogue divided at the same rate as cells in liquid culture, but the population gradually shifted to a large fraction of free swimming cells, a process that may be necessary for colonization. Metabolic energy is required for cell adhesion to the glycosides. Both the initiation and maintenance of lectin-mediated adhesion requires continuous protein synthesis, and so lectin activity is a major priority of these cells (*127*). The adhesion/de-adhesion apparatus is apparently used as a nutrient sensorium (to continuously monitor the nutrient status of the environment, *131*). In an incomplete medium (or presumably when the environment is unfavorable), cells de-adhere, presumably to migrate to a more favorable environment. This adhesion/de-adhesion behavior of *V. furnissii* (*127*) probably catalyzes the first step in colonizing chitin.

A second step was shown to be chemotaxis to chitin hydrolysis product (*128*). *V. furnissii* swarms toward chitin oligomers $[GlcNAc]_n$, n=1-6, at initial concentrations as low as 10 μmol. Two (or more) independently induced receptors (perhaps the most potent reported for bacteria) with overlapping specificities, recognize $[GlcNAc]_n$, with n = 2-4 (*128*). Expression of the receptor(s) for $[GlcNAc]_5$ and $[GlcNAc]_6$ apparently requires special induction conditions (*128*). The chemotactic response was greatly affected by growth and conditions and the presence of nutrients in the environment. Chemotaxis to the sugars increased 2- to 3-fold when the cells were starved. Nutrients, especially compounds that feed into or are part of the Krebs cycle, were potent inhibitors of taxis to the sugars (*128*). Since most of the catabolites are inhibitory, chemotactic behavior is displayed mainly when the Krebs cycle is operating at the low rate, and is inhibited when the cycle functions at high rate. When the environment fulfills the metabolic needs of the bacteria, then the chemotaxis system is either inhibited or not induced. Thus, the adhesion/de-adhesion system and chemotactic behavior both optimize chitin resource utilization by the bacteria (Figure 2).

It is not clear whether and when microbial cells release extracellular (i.e. soluble) enzymes during growth, but it is clear that polymers like chitin, potential substrates use for metabolism, must be hydrolyzed to yield oligosaccharides before they can act as substrates for periplasmic or cytosolic disaccharases (*132*). The importance of extracellular hydrolysis is evident in the observation that often 80 to 100 % of seawater isolated strains can utilize *N*-acetylglucosamine whereas less than 20 % can hydrolyze chitin. This cooperative chitin degradation is often attributed to different pools of bacteria, free (i.e. open water bacteria) and particles bound (*60*) (Figure 3).

Some steps in the catabolism of the oligosaccharides are known in *V. furnissii* (*129*). Acetyl glucosamine (GlcNAc) and its oligomers ([GlcNAc]$_2$ and [GlcNAc]$_3$) are very rapidly consumed by intact cells. Tetramer ([GlcNAc]$_4$) is utilized somewhat more slowly. During these processes, there is virtually no release of hydrolysis products by the cells (*129*). The oligosaccharides resulting from random hydrolysis of chitin by extracellular (secreted) exo- or endochitinases enter the periplasmic space (via specific porins ?) and are hydrolyzed by a unique membrane-bound endoenzyme (new enzyme called by the authors chitodextrinase, *129*) and an exoenzyme (*N*-acetyl-β-glucosaminidase). Chitodextrinase cleaves soluble oligomers (but not chitin) to di- and trisaccharides, while the periplasmic *N*-acetyl-β-glucosaminidase hydrolyzes the terminal acetyl glucosamine from the oligomers. The end products in the periplasm, GlcNAc and [GlcNAc]$_2$ (possibly [GlcNAc]$_3$), are catabolized as following several pathways (*129*) (Figure 2):

(a) Disaccharide pathway, where a [GlcNAc]$_2$ permease is apparently expressed by *Vibrio furnissii*. Translocated [GlcNAc]$_2$ is rapidly hydrolyzed by a soluble cytosolic *N*-acetyl-β-glucosaminidase, and the GlcNAc is phosphorylated by an ATP-dependent, constitutive kinase to GlcNAc-6-P.

(b) Monosaccharide pathway, where a periplasmic GlcNAc is taken up by the phosphoenolpyruvate phosphotransferase system, yielding GlcNAc-6-P, the common intermediate for both pathways.

Finally, GlcNAc-6-P generated by both pathways is deacetylated and deaminated, producing fructose-6-P, acetate and NH$_3$ (Figure 2).

Chitobiose, [GlcNAc]$_2$ is probably the "true" inducer of the chitin degradative enzymes and, depending on its concentration in the environment, differentially induces the periplasmic and cytosolic *N*-acetyl-β-glucosaminidase (*129*). The disaccharide pathway appears to be the most important when the cells are confronted with low concentrations of the oligomers, but when there is a greater supply of [GlcNAc]$_2$, both monosaccharide and disaccharide pathways are used.

Chitin Susceptibility to Biodegradation Processes. Highly crystalline β-chitin occurs in the absence of other polymer only in diatoms (*132*) and this kind of chitin can be hydrolyzed directly. In most cases, however, chitin is associated with other macromolecules that control its reactivity. Chitin and chitosan are associated with other polysaccharides in the fungal cell wall, while chitin is associated with proteins in animals. Therefore, where chitin occurs naturally in association with proteins or other polysaccharides, it might be expected to have properties slightly different from those of isolated chitin, particularly its susceptibility to biodegradation.

The composition and the architecture of the fungal wall are known in several groups of fungi. Chitin is probably a universal component of fungal cell walls (*27*). The basic pattern of organization of the hyphal walls of most fungi includes an inner layer composed of chitin microfibrils in a β-glucan matrix. The latter matrix also contains proteins and an outer layer of α-glucan (glucan in the chytrids; chitosan and polyglucuronic acid in the zygomycetes and mannoproteins and glucans in the ascomycetes and basidiomycetes) (*133,134*). A layer of a β-glucan mucilage may also be present on the outer surface, and treatment with β-glucanase is essential before chitinase can effectively attack the walls. Thus, β-glucan and protein may form the matrix of a fibrillar chitinous layer, which forms the inner compact surface of the wall, but is looser and uneven on the outer side (*27*). What is quite specific to fungi is that chitin is continuously subjected to various chemical modifications such as deacetylation or covalent linking to glucans and/or proteins (*133*). The involvement of protein in the cell walls has not been clarified, but protein molecules may link chitin to carbohydrates. The glucan-chitin complex is highly resistant because the glucan chains are linked to chitin through their reducing ends via amino acids, particularly lysine (*135, 136*). Amino acid analysis shows significant amounts of threonine, serine, aspartic acid and glutamic acid, which may contribute, to the linkages between carbohydrates and protein moieties in the cell walls. Also noteworthy is the high content of alanine and glycine, since both amino acids are involved in the cross-linking of peptidoglycans in bacterial cell walls. Both chitin and β-glucan chains are connected by hydrogen bonds (*27, 133, 135*). This results in a cross-linked network of β-glucan-chitin complex, which confers rigidity and chemical stability to the cell wall.

There is an experimental evidence of covalent bonding between chitin and proteins in animals, especially in arthropods (*4, 137, 138*). The covalent proteoglycan is, however, accompanied by other proteins, specifically or non-specifically associated with the former by weaker forces and forming a chitinoproteic complex. *N*-acetylglucosamine and chitin can react with α-amino acids, peptides and cuticular proteins to give a glycoprotein complex (*139*). Chitin, whether in its α- or β-form, is covalently linked to arthropodins or sclerotins in cuticles or other types of proteins in mollusk shells, to form more or less stable glycoproteins. The latter process occurs probably through aspartyl, glutamyl, seryl, glycyl and histidyl residues (usually carboxyl to the amino group of glucosamine but other types of bond may also occur) (*140-143*). In the internal shell of cephalopods, the greater part of the protein moiety can be removed, while the remaining chitin is bound to a protein rich in aspartic acid (*4*). Enzymatic studies and acidic hydrolysis of cuticles have also led to the conclusion that there is a stable linkage between proteins and chitin through a non-aromatic amino acid (*144, 145*).

Structural proteins from animal exoskeletons are generally more or less stabilized through different kinds of bonding, such as hydrogen, hydrophobic, and ionic bonds, Van der Waals interchain regular interactions, disulfur bridges and sclerotization. The last two stabilization forces have an important consequences on the biodegradation processes.

According to Peter *et al.* (*138*), very little is known about the molecular mechanism of sclerotization and, despite many efforts to analyze the chemical details of that process, and a number of excellent reviews (*145-146*), this important process is still a matter of controversy. The sclerotization process in arthropods and mollusks begins in the cuticle or the periostracum by an enzymatic oxidation of a diphenolic substrate (for example hydroquinones, catechol, *N*-acetyl dopamine, DOPA, *146, 147*). This oxidation yields the corresponding o-quinone and/or p-quinonemethide (reactive intermediate). According to the results of a model reactions and analyses of cuticular extracts, crosslinking of cuticle proteins then results from Michael type conjugate addition and Schiff's base formation with free peptidic amino groups. Chitin may also be involved in the o-quinone or p-quinonemethide mediated formation of crosslinks, either via unacetylated amino groups or hydroxy oxygen (*138*). Moreover, non-covalent interactions between oxidation products of the sclerotization agents and polypeptides and/or chitin may contribute to the stability of sclerotized chitinoproteic complexes (*136, 138*). The resulting chitin-tanned protein complex gains considerable stability as well as hardness, rigidity and resistance to enzymatic hydrolysis.

Disulfur bridges (as in keratin) are generally of a little importance in the stabilization of invertebrate skeletal proteins except in the case of *Halocynthia* tunical cuticle, and in *Vestimentifera*. These deep-water "worms" build dwelling tubes that can reach lengths of one meter in *Riftia*, and some tens of centimeter in *Tevnia*. The tubes are made of a chitinoproteic complex (*148-150*) of a very high chemical stability, interpreted by authors as the result of strong self-interaction of proteins via numerous sulfur bridges.

The chitin-protein complexes may be complemented by the deposition of other substances, such as waxes and lipoproteins, giving the complex some impermeability. Animals for a number of different functions have extensively exploited such chitinous structures. The chitin-protein framework also provides a structural organic framework around which mineral may be deposited (calcification and silicification) (*40, 151*).

In our opinion, one of the most important controls on the rate of environmental chitin degradation is the level of chitin accessibility by hydrolytic activities. Chitinous structures almost always consist of a covalently-linked complex of chitin and proteins or glucans. In addition, the proteins themselves can form stable structural bonds (disulfide bonds, sclerotization of chitinoproteic complexes) rendering chitin even less accessible. This explains why some chitinous structures are much more resistant to degradation than others in the same experimental setting.

Morphological, Ecological and Biogeochemical Aspects of Chitin Biodegradation

Pathways of Biodegradation. The process of chitin biodegradation can be conceptually divided into three successive steps (Figure 4).

The first step is the colonization process during which microorganisms settle on the detrital material, multiply and begin degradation. The amount of material biodegraded is almost negligible. The duration of this phase varies from a few hours (in the case of zooplanktonic remains in open water) to some months (mineralized skeletons within sediments such as mollusk shells, echinoid plates, etc.).

The second step involves the most active biodegradation. Many microorganisms adapted to the substrate, weather and mineralize the organic components of the chitinous skeleton. The duration of this phase varies from a few days (zooplanktonic remains in open water or at the water-sediment interface) to several months (mineralized skeletons within the sediments). The rate of chitin (or other compounds) degradation during this phase is generally a negative exponential, of first or second order.

The last step involves the slowing down of the process during which all labile compounds have been mineralized and only refractory molecules or complexes remain. Only very specialized microorganisms can manage to degrade such compounds. This phase last until the material is totally broken down (from some weeks up to several years).

In addition to the above model of degradation, we propose a non linear parametric transition functions that allows direct comparison between several parameters such as different settings, seasons, etc. The logistic curve based upon parameters relating to actual biological phenomena (e.g. microbial growth) is preferred for such comparison where % of chitin remaining in the material (Y) is plotted against time (X) according to the following equation:

$$\text{Chitin remaining} = a + \frac{b}{1 + \exp\left[-(x-c)/d \right]}$$

where a: transition height=maximum amount of degradable material (before refractory phase begins); b: transition center=when maximum degradation rate is supposed to be observed; c: transition width calculated as 2.197 c (=duration of the exponential degradation phase, after colonization steps and before refractory phase); d: shape factor.

Chitin biodegradation processes are quite different in the main compartments of the marine environment (Figure 5).

Open Water Conditions. Planktonic communities are especially rich in chitin producers due to the abundance of crustaceans. Through mechanical and biodegradation processes (including ingestion, egestion, digestion processes, coprorhexia, etc.) (*69, 70, 77, 90*), this chitinous material is partly incorporated into suspended matter in the oceans.

Poulicek and co-workers designed an *in situ* experimental device to study zooplankton biodegradation in the open water (*70, 90, 91, 152-154*). Dead organisms are first degraded from inside by autolytic processes (*69, 70, 90, 91, 152-154*). These processes induce the lysis of most muscles and organs during about 50 hr *post mortem* (at 13 to 22°C in an oligotrophic setting, Figures 13, 14). This autolysis results from the activity of the hydrolytic enzymes of the digestive tract and other organs and of the symbiotic microorganisms in the intestine (*69, 70, 153-154*). The latter stage occurs 6-20 h *post mortem* and leads to the opening of the cuticle generally along the articular membrane between two sclerites (where the cuticle is less or not hardened by quinone tanning, Figure 13) (*70, 89, 154*). The heterotrophic microorganisms can invade the corpse through that gap and further degrade it (Figure 14). At 50-100 h *post mortem*, the chitinoproteic layers of the cuticle are deeply altered (Figures 16, 17), so that pieces of appendages (legs, antenna, etc.) are lost (*69, 155*). These fragments can form macroscopic aggregates often known as "marine snow", falling relatively fast through the water column (*69, 90*) (Figure 15). The formed aggregate material is degraded during sedimentation by a rich heterotrophic microorganism community (Figures 14, 15) (*70, 89, 153, 155-157*).

A wide variety of hydrolases have been detected during biodegradation. The importance of each enzymatic activity varies with time (Figure 6) (*69, 70, 154*). Lipolytic activities, which are mainly autolytic, occur first, culminating after 12-40 h. Proteolytic and glycolytic activities develop next, and culminate 80-120 h *post mortem*. This prepares the cuticles for hydrolysis by chitinolytic activities, that reach a maximum between 60 and 70 h (first squad) and 120 and 180 h *post mortem* (second more functional squad). All mentioned activities are due mainly to the development of successive populations of heterotrophic microorganisms (Figures 14, 15, 17) regulated by ciliates and flagellates (Figure 14). Autolytic processes are negligible after 50 h. The number of active bacterial strains on the experimental material (measured as plate count or microbial ATP estimations) increases according to a logarithmic curve during the first 100-150 h *post mortem* then decreases slowly, reflecting the progressive hydrolysis of metabolizable substrates (*70, 153, 154*).

The hydrolytic activities described above lead to a rapid decrease of the organic content of the animal remains (*69, 70, 154*). The overall decrease of the organic content follows the negative exponential kinetic curve. The fate of chitin is essentially similar. After a short delay during which proteins and lipids gradually disappear, the chitin content of the material reduces quickly (maximum variation between 50 and 120 h *post mortem*). After about 150 h, approximately 90% of the initial chitin have disappeared but the chitin remaining at the beginning of the refractory phase is only very slowly mineralized and is likely to settle on the sediment-water interface.

Biodegradation Processes at the Sediment-Water Interface. Material of planktonic origin settling onto sediment has generally already undergone extensive biodegradation. However, since most of decomposing microorganisms remains

188

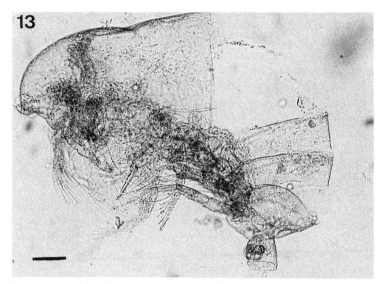

Figure 13. Empty and broken copepod "ghost" after 72 hours incubation *in situ* (Calvi Bay experimental setting at 37 m depth, temperature 16°C). Light Microscopy (LM) scale is 100 μm.

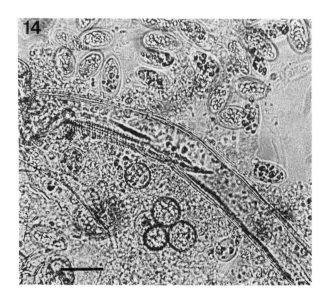

Figure 14. Detail of Figure 13 showing muscles and cuticular remains embedded in a rich sheath of bacteria. Note abundant Ciliates. LM scale is 10 μm.

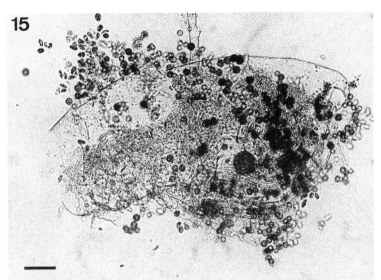

Figure 15. Cuticular remains aggregates with bacteria and ciliates constituting "marine snow" after 120 hours incubation in situ (Calvi Bay experimental setting at 37 m depth, temperature 16°C). LM scale is 100 μm.

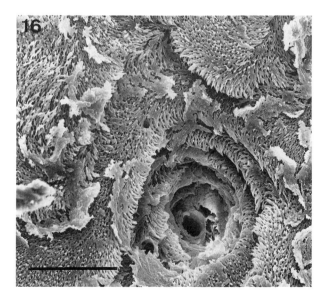

Figure 16. Bored and exfoliating crustacean cuticular surface after 72 hours incubation in situ (Calvi Bay experimental setting at 37 m depth, temperature 16°C). SEM scale is 10 μm.

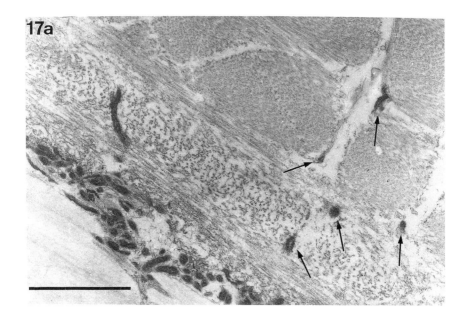

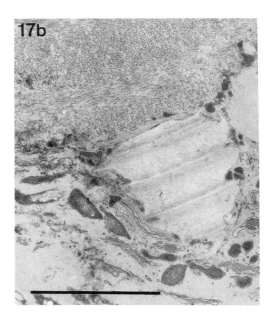

Figure 17. Bacterial weathering of the calcified layers of a crustacean cuticle (*Carcinus maenas*), Calvi Bay experimental setting at 37 m depth, 72 h immersion. Transmission Electron Microscopy (TEM) scale is 5 µm. a) Bacterial mat biodegradation from the internal side (membranous layer); pore canals are used as preferential routes for deeper invasion (arrows); b) Detail of bacterial weathering of the membranous layer.

attached to the settling material (Figure 21), the matter deposited on the sediment surface continues to degrade (69, 70, 153). The apparent decrease of the process is mainly due to the prior loss of most labile compounds. The material now consists mainly of more or less refractory molecules, and chitin remains (generally less than 5 to 10 % of the initial content). In shallow water, mechanical disruption and abrasion occur, due to hydrodynamic forces resuspending material. Weakening of the remaining cuticle leads to complete fragmentation and pulverization of the planktonic material (153). Such mechanical disruption does not occur deeper in the oceans and chitin incorporation into the oxic sedimentary layers should be possible. Whatever the input, it has been shown that the number of chitinoclastic bacterial strains is a logarithmic function of the chitin content of the sediment at the interface (49). This means that a slight increase of the chitin content readily available should induce an immediate growth of chitinoclastic bacteria resulting in the rapid degradation of this input. In such conditions, the bacteria behave as typical opportunistic organisms.

Biodegradation in Oxic Marine Sediments. Since accumulation of particulate chitin in open water as a result of downward flux appears unlikely, the chitin content of sediments is related to their composition. Furthermore, the organoclastic sediments typically harbor much more chitin than terrigenous, volcanic or authigenic ones. Mollusk shells and bryozoans are the main contributors to sedimentary chitin (49, 159).

In the oxic layers of marine sediments, biodegradation of the chitinoproteic matrices of animal skeletons occurs mainly through the activity of microborers (for literature review see 160-168). A well-defined sequence of microorganisms settle onto detrital fragments and some of them are able to bore holes of 1-150 μm in diameter (Figures 18-20). Although, the cyanobacteria are abundant, bacteria distributed within the organic matrices and fungi are the main biodegraders (69, 161, 167) (Figures 18-20). Marine borers, of calcified tissues, are involved in constructive and destructive processes. Some of these 'borers' pit and corrode carbonate surfaces while others bore into them, changing the pH of seawater in some micro-environments and possibly by secreting chelating metabolites (Figures 16-18, 20). The mechanism by which endoliths bore remains unknown, though it is thought to be by chemical rather than by mechanical means. These processes increase the porosity and weaken the surface layers of inhabited substrata, particularly when the chitinoproteic matrices inside the skeletons are deeply altered. Moreover, microphytes provide food for herbivores (e.g. gastropods, Polyplacophora) promoting this way, carbonate removal by abrasion (169).

Endolithic microphytes are also involved in producing "constructive" and "destructive" micrite envelopes around carbonate particles thus protecting from further biodegradation. These microphytes bore into substrates at 0.5 to 50 μm day^{-1}. Filament densities within detrital substrata may range from 50,000 mm^{-3} to 500,000 cm^{-2}. In experiments, typical densities of boreholes are of the order of 150,000 to 400,000 cm^{-2} in mollusk shells and echinoid plates after one to four years. There is no significant difference between experimental settings (Mediterranean Sea, English Channel, Papua New Guinea or the Caribbean Virgin Islands). Most genera of

192

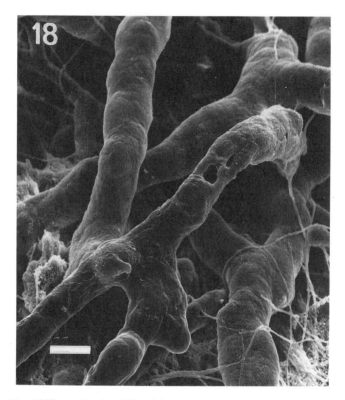

Figure 18. Different kinds of fungi boring into a mollusk shell after 6 months incubation in situ at water-sediment interface (Virgin Islands experimental setting at 15 m depth, temperature 26°C). SEM scale is 10 μm.

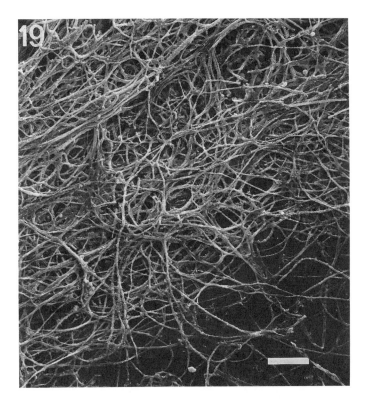

Figure 19. Filamentous bacteria associated with chitinoproteic organic matrices in a mollusk shell after 9 months incubation in situ at water-sediment interface (Calvi Bay experimental setting at 37 m depth). SEM scale is 10 μm.

194

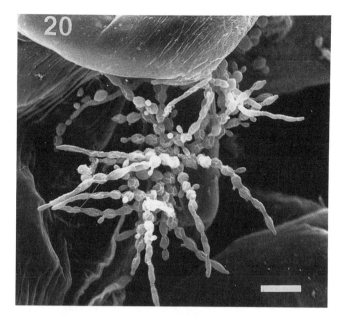

Figure 20. Undetermined organism of fungal affinity boring into a detritic mollusk shell at water-sediment interface (Indian Ocean, 4870 m depth). SEM scale is 10 µm.

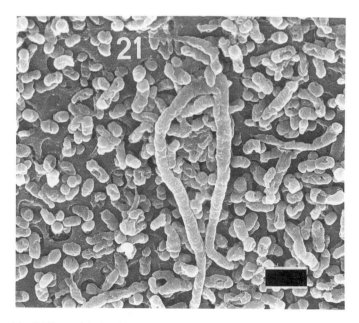

Figure 21. Different kinds of bacteria attached to a planktonic crustacean cuticular remain collected in a sediment trap disposed in nepheloid layer (Mediterranean Sea, 560 m depth). SEM scale is 5 µm.

endolithic microborers have a worldwide distribution, although there is some endemism at the species level.

Endolithic algae, adapted to life with very little light, occurred at surprisingly great depths within oceans, with *Plectonema*, a cyanobacteria, being found down to 370 meters on the continental slope near Florida (*169*). Fungi, which are not light-limited, have been reported from 4780 meters (Figure 20) (*160, 163, 166, 167*).

Most of the vacated boreholes (Figure 16) (only 10-15% is occupied by the borer) are used by bacteria to reach the organic veils inside the skeleton (*161, 163, 164, 167*). With the exception of cyanobacteria, most microorganisms isolated from weathered skeletons are able to secrete extracellular hydrolases, like proteolytic and chitinolytic enzymes, when cultivated *in vitro* (*161, 167*). These hydrolytic activities also occurred in detrital skeletons (*161, 167, 168*). Estimations of enzymatic activity inside weathering skeletons show a great diversity of hydrolases and high levels of hydrolytic activity (*163*). Chitinolytic activity is important in mollusk shells and crab cuticles, and can explain a quick decrease of chitin content in experimental samples. More than 90 % of the initial chitin content in crustacean cuticles disappears in less than two months (Figure 17a, 17b, 23, 24), whereas the same percentage loss takes more than one year in mollusk shells (*162, 163*) (Figure 22a, b). The kinetics of chitin loss are the same in different experimental settings: once the microborers settle on the experimental material and invade the cortical layers (low hydrolytic activity), there is a fast decrease in chitin content, synchronous with the exponential growth of endolith populations. After this rapid biodegradation, the rate reduction corresponds to lowering of the concentration of the readily available growth substrates (refractory phase). The remaining organic compounds are mainly refractory or structurally bonded (tanned proteins of the chitinoproteic complex), thus being much less accessible.

Biodegradation in Anoxic Marine Sediments. Although anoxic superficial sediments cover only 10% of the sea floor, they underlie most of the shallow highly productive areas of the world oceans. These sediments accumulate more than 90% of the total organic matter buried in sediments annually (*69*). Hence processes of anaerobic decomposition are very important in determining to what extent some chitinous compounds "survive" rapid mineralization in the surface biotic layers and are buried long term. Most aerobic microorganisms can oxidize a wide range of substrates to CO_2 whereas individual anaerobic microorganisms metabolize (often incompletely) a rather restricted range of molecules. Although, the microorganisms inhabiting anaerobic microenvironments are much less diversified, the anaerobic communities can be very efficient in decomposing relatively refractory molecules (*164-168, 170*). Thus, aquatic anaerobic communities (164-168) can decompose most of the organic polymers. Anaerobic biodegradation of skeletal substrates is mainly due to bacteria, diatoms and fungi (*165, 170*). Despite the fact that few forms of decomposer are common to both aerobic and anaerobic conditions, the biodegradation patterns are very similar in both cases. The same colonization curve is observed, culminating in the same densities of borers (*163, 167*). The results of estimates of enzymatic activity show that, in the case of mollusk shells, there is no significant

196

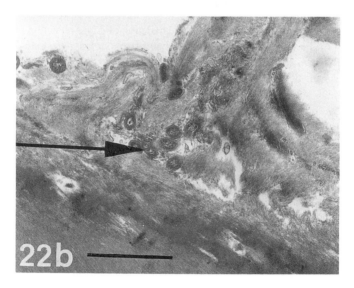

Figure 22. Prismatic layer of *Pinna nobilis* shell weathered for 12 months in Calvi Bay experimental setting in oxic sediment at 37 m depth. a) The organic matrices between crystallites almost disappeared. Area marked with a star enlarged in Figure 22b. SEM scale is 10 μm. b) Bacteria within organic matrix remains. TEM scale is 1 μm.

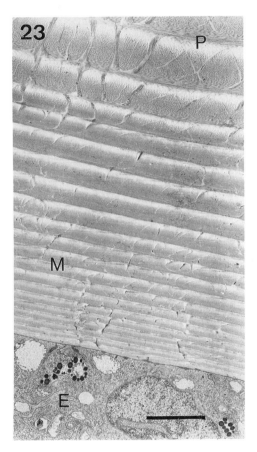

Figure 23. Fresh cuticle of the crab *Carcinus maenas*. TEM scale 1 μm. Note: E, epithelium; M, membranous uncalcified layer; P, principal calcified layer.

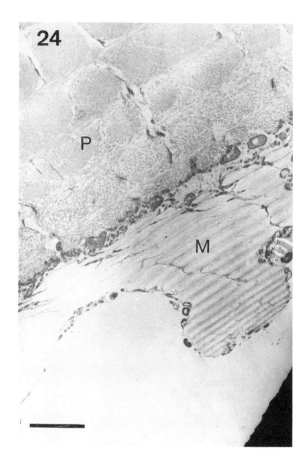

Figure 24. Bacteria weathering the cuticular layers of *Carcinus maenas* after 12 days in Calvi Bay experimental setting at water-sediment interface, 37 m depth: membranous layer (M) almost completely disappeared, begin of principal layer (P) alteration. TEM scale is 10 μm.

difference between hydrolytic levels (*166, 168*). However, in other samples, where anaerobic hydrolytic processes appear lower, the kinetics of biodegradation appears essentially similar for chitin, with comparable rate constants (*166-168*). Slowing down of the process after initial rapid biodegradation is also observed, but the composition of the refractory compounds may be somewhat different in the two experimental conditions, as shown by the results of transfer experiments (*69*) (Figure 7). Skeletal material (mother of pearl) was laid down *in situ* in oxic sedimentary layers for six months, after which, part of the experimental material was transferred to anoxic sedimentary conditions (other samples remained in the same environment for a further six months period). After a rapid decrease of the organic content during the first six months (70% degradation), the extraction of the organic compounds slowed down, a normal phenomena at the beginning of the refractory phase (speed constant K_{aer}=0.91). The material that was transferred to anaerobic conditions weathered three times faster than that remaining in aerobic conditions (K_{ana}=3.00). The latter phenomena can be interpreted as a reflection of the adaptation of anaerobic microbiocenoses to the biodegradation of the less labile compounds that remained after aerobic degradation. The cumulative effect of both processes (aerobic and anaerobic weathering) results in optimal recycling of the chitinous compounds of skeletal substrates.

Biodegradation in Deep-Sea Environments. Enormous bulk of mineralized skeletal structures, sheltering chitinoproteic complexes, settles onto deep-sea sediments, sometimes in the form of deep calcareous oozes. Extraction of organic matter from calcareous skeletons, for example Pteropod or other Mollusk shells, occurs at any depth in marine sediments where microborers are ubiquitous (*13, 163, 167, 169, 171*). In the aphotic environment, fungi (Figure 20), colorless cyanobacteria and bacteria are supposed to function mainly heterotrophically, thus using the organic matrix of skeletal remains as a nutrient source (*69, 163, 171*). Most endolithic organisms live in close contact with the organic sheaths around crystallites (*164, 165*). Thus, the extracellularly secreted enzymes are able to hydrolyze the organic compounds of those sheaths, as experimentally shown in shallow depth (*13*). Despite the importance of the deep-sea ecosystem, data on enzymatic degradation under high hydrostatic pressure are scarce (*172, 173*): chitinase is a highly barotolerant enzyme, active and very stable under deep-sea conditions (1000 atm and 2°C), and psychrophilic and barophilic bacteria can secrete this hydrolase under *in situ* conditions.

Preliminary observations *in situ* around deep-hydrothermal vents indicate that the Vestimentifera tube is a very stable structure (*174*). An experimental *in situ* approach showed that after a 6 months of exposure, and in contrast with exoskeleton fragments of the crab *Bythograe*, the tube samples appeared relatively unaltered, and the relative amount of chitin was comparable to that of the control samples (Figures 25, 26). This was confirmed with an experiment at shallow depths in Calvi Bay (*174*).

The results of experiments with the vent crab exoskeleton in both environments are comparable: a rapid and significant decrease (30 % after 12 days) of

Figure 25. Structure of the fresh *Riftia pachyptila* (Vestimentifera) tube. White dots are chitin microfibrils. TEM scale is 1 μm.

201

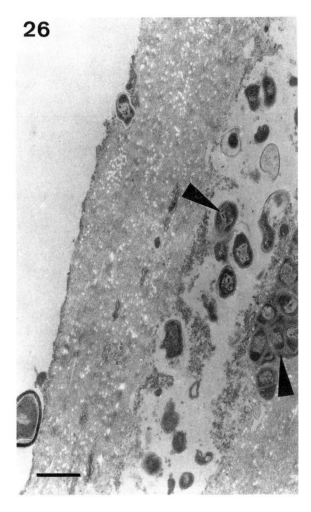

Figure 26. Structure of the most cortical layer of an *in situ* weathered *Riftia pachyptila* tube. Relatively unaltered structure after 12 days in situ incubation around hydrothermal vents (depth 2600 m). Clusters of bacteria embedded in the tube walls (arrows) are also frequent in fresh material (not illustrated). TEM scale is 1 μm.

the total organic content in the vent site, is paralleled in the Calvi site where quite 80 % of this fraction is dissolved after 20 days (*174*). The latter observation agrees with experiments on sclerites of the shore crab *Carcinus maenas* in Calvi Bay (*162, 175*) (Figures 23, 24). While the loss of organic material in the vent crab results mainly from a decrease of the chitin content (32 % and 80 % after 12 and 20 days respectively), the alkalo-soluble protein content degrades more slowly. In contrast to the vent crab, the *Riftia* tube exhibits a radically different "biodegradation" profile. The whole organic content of the samples is stable over a period of about 3 weeks, either in the vent site or in shallow environment (*174*). After 6 months, there is a loss of about 20 % of material mainly the protein fraction. There was almost no variation in the chitin content of the *Riftia* tubes even after 6 months of vent exposure. The comparison of the sites suggests that resistance to degradation is independent of environmental conditions and is a function of the structural and/or chemical parameters that determine chitin accessibility to biodegraders.

Conclusions: Biogeochemical Implications of Chitin Biodegradation

Chitin produced in the marine environment is unlikely to accumulate to a large extent or to be preserved in sediments. From open water to anoxic sedimentary layers, the efficiency of chitinoclastic microorganisms is very high. The microbial degradation of chitin in the marine environment is a highly complex and structured process involving a variety of microorganisms and microbial consortia. The specialized features of some of them (mode of life of microborers in skeletal remains in sediments, adaptation of anaerobic decomposers to more or less refractory compounds, etc.) lead to an optimization of recycling. This explains why chitin does not accumulate to a large extent in most marine sediments. For the general marine "economy" to maintain the steady state of geochemical cycles, chitin mineralization processes are of great ecological significance as it contributes to both carbon and nitrogen cycles.

The biodegradation and breakdown of chitin in calcified skeletons is accompanied by an increased dissolution rate of a biogenic carbonates, thus has repercussions for chemical equilibria (the organic sheaths around $CaCO_3$ crystallites partly protect them from dissolution in thermodynamically undersaturated media, i.e. deep and/or cold environments). After extraction of the organic matter, the calcified skeletons, less "protected" from the inside, are more rapidly dissolved particularly where microborers increase the surface accessible to dissolution.

Considering the biomass of the chitin and chitinoproteic complexes produced annually in the oceans, and their ability to complex transition elements without interference of alkali nor alkali-earth ions (*4, 13, 14*), the amount of heavy metals ions associated with them should be taken into account when computing global geochemical cycles. While humic and fulvic substances form relatively stable complexes with metal ions, chitin-metal complexes are more unstable, particularly as chitinase is activated by some of the metals bound on chitin (*13, 14*). The latter process makes chitin a potentially important agent for the transport of metal in the marine environment, while humic substances should merely act as sinks for heavy metals. In open water, metals bound on chitinoproteic complexes of suspended matter are released or chelated from/on the chitin through biological weathering (deacetylation and hydrolysis), either in the water column or in the upper layers of the sediments. Decomposition of these complexes in areas of high biological productivity, together with other vector mechanisms (clays, metal oxides and hydroxides, etc.), may be a source of localized metal rich interstitial waters and sediments in the marine environment.

Resistant biopolymers are known to be significant contributors to kerogen (*176-180*). Recently, new investigations have shown that not only plant remains, but also animals yield highly resistant macromolecules (*176-182*). However, the organic material found in most fossil invertebrates (especially Paleozoic and Mesozoic), unlike plant remains, have no clear source in the living animal, as the cuticle chemistry appears altered during diagenesis (*176-184*). The preservation potential of chitin in the fossil record is lower than that of other macromolecules (like lignin for example; *176*). Even when chitin is cross-linked, as in thick sclerotized cuticles (*178-180*), or associated with proteins stabilized by disulfur bridges as in Vestimentifera (*174*), it will only be preserved in very special diagenetic environments (bituminous shales or strongly reducing bottom conditions). The primary control on chitin preservation in the geological record is not only time, but rather the nature of depositional environments inhibiting diagenetic alteration (*179*). Chitin is rarely detected in substantial amounts even in modern sediments (*49*).

References

1. Gooday, G.W. *Adv. Microbial Ecol.* **1990**, *11*, 387-430.
2. Jeuniaux, Ch.; Voss-Foucart, M.F. *Bioch. Syst. Ecol.* **1991**, *19*, 347-356.
3. Goffinet, G. In *Chitin in Life Sciences*; Giraud-Guille M.M. Ed. ; 1st Summerschool European Chitin Society ; Jacques André Publisher, Lyon (France), 1996, pp 53-65.
4. Muzzarelli, R.A.A. *Chitin;* Pergamon Press: Oxford, **1977**, 309 pp.
5. Peter, M.G. *J. Macromol. Sci.-Pure Appl. Chem.* **1995**, *A32*, 629-640.
6. Domard, A. In *Chitin in Life Sciences;* Giraud-Guille M.M. Ed.; 1st Summerschool European Chitin Society; Jacques André Publisher, Lyon (France), 1996, pp 98-109
7. *Chitin in Nature and Technology;* Muzzarelli, R.A.A.; Jeuniaux, Ch.; Gooday, G.W., Eds ; Plenum Press:New York, 1985.
8. *Chitin and Chitosan;* Skjak-Bræk, G.; Anthonsen, Th.; Sandford, P., Eds.; Elsevier Applied Sciences:London, UK, 1989.
9. *Chitin World;* Karnicki, Z.S.; Wojtasz-Pajak, A.; Brzeski, M.M.; Bykowski, P.J., Eds.; Wirtschaftsverlag NW: Bremerhaven, Germany, 1994.
10. *Advances in Chitin Science;* Domard, A.; Jeuniaux, Ch.; Muzzarelli, R.A.A.; Roberts G., Eds.; Jacques André Publisher: Lyon, France, 1996, Vol.1.
11. Schaefer, J.; Kramer, K.J.; Garbow, J.R.; Jacob, G.S.; Stejskal, E.O.; Hopkins, T.L.; Speirs, R. *Science* **1987**, *235*, 1200-1204.

12. Kramer, K.J.; Hopkins, T.L.; Schaefer, J. *Insect Biochem. Molec. Biol.* **1995**, *25*, 1067-1080.
13. Poulicek, M.; Machiroux, R.; Toussaint, Cl. In *Chitin in Nature and Technology;* Muzzarelli, R.A.A.; Jeuniaux, Ch.; Gooday, G.W., Eds.; Plenum Press: New York, 1985, pp 523-530.
14. Poulicek, M. In *Proc. Metal cycling in the Environment;* SCOPE Belgium ed.: Brussel, 1985, pp 153-165.
15. Jeuniaux, Ch. *Chitine et chitinolyse. Un chapitre de la biologie moléculaire;* Masson Ed: Paris, 1963.
16. Jeuniaux, Ch. In *Comprehensive Biochemistry;* Florkin M.; Stotz E.H., Eds; Elsevier: Amsterdam, NL, 1971, 26C, pp 595-632.
17. Jeuniaux, Ch. *Bull. Soc. Zool. France* **1982**, *107*, 363-386.
18. Herth, W. *J. Ultrastruct. Res.* **1979**, *68*, 6-15.
19. Herth, W. *J. Ultrastruct. Res.* **1979**, *68*, 16-27.
20. Smucker, R.A. *Biochem. Syst. Ecol.* **1991**, *19*, 357-369.
21. Herth, W.; Kuppel, A.; Schnepf, E. *J. Cell Biol.* **1977**, *73*, 311-321.
22. Herth, W.; Schnpef, E. In *Cellulose and other natural polymer systems;* Brown M., Ed.; Plenum Publ. Corp.: New York, 1982, pp 185-206.
23. Herth, W.; Mulisch, M.; Zugenmaier, P. In *Chitin in Nature and Technology;* Muzzarelli, R.A.A.; Jeuniaux, Ch.; Gooday, G.W., Eds.; Plenum Press: New York, 1985, pp 107-120.
24. Mulisch, M. *Europ. J. Protistol.* **1993**, *29*, 1-18.
25. Mulisch, M. In *Chitin in Life Sciences;* Giraud-Guille M.M., Ed.; 1st Summerschool European Chitin Society; Jacques André Publisher: Lyon, France, 1996, pp 30-40.
26. Ruiz-Herrera, J. *Fungal Cell Wall: Structure, Synthesis and Assembly;* CRC Press: Boca Raton, FL, USA, 1992.
27. Gooday, G.W. In *Chitin in Life Sciences;* Giraud-Guille M.M., Ed.; 1st Summerschool European Chitin Society; Jacques André Publisher: Lyon, France, 1996, pp 20-29.
28. Peters, W. *Experentia* **1966**, *22*, 820-821.
29. Wagner, G.P.; Lo, J.; Laine, R.; Almeder, M. *Experentia* **1993**, *49*, 317-319.
30. Jeuniaux, Ch.; Compère, Ph.; Toussaint, Cl.; Decloux, N.; Voss-Foucart, M.F. In *Advances in Chitin Science;* Domard, A.; Jeuniaux, Ch.; Muzzarelli, R.A.A.; Roberts G., Eds.; Jacques André Publisher: Lyon, France, 1996, Vol.1, pp 18-25.
31. Rudall, K.M.; Kenchington, W. *Biol. Rev.* **1973**, *48*, 597-636.
32. Blackwell, J.; Weih, M.A. *J. Mol. Biol.* **1980**, *137*, 49-60.
33. Hadley, N.F. *Sientific American* **1986**, *235*, 98-106.
34. Giraud-Guille, M.M.; Bouligand, Y. In *Chitin in Nature and Technology;* Muzzarelli, R.A.A.; Jeuniaux, Ch.; Gooday, G.W., Eds.; Plenum Press: New York, 1985, pp 29-35.
35. Giraud-Guille, M.M. In *Chitin in Life Sciences;* Giraud-Guille M.M., Ed.; 1st Summerschool European Chitin Society; Jacques André Publisher: Lyon, France, 1996, pp 1-10.

36. Giraud-Guille, M.M.; Chanzy, H.; Vuong, R. *J. Struct. Biol.* **1990**, *103*, 232-240.
37. Bouligand, Y. *C. R. Acad. Sci. (Paris)* **1965**, *261*, 3665-3668.
38. Bouligand, Y. *Tissue and Cell* **1972**, *4*, 189-217.
39. Giraud-Guille, M.M. *Int. Rev. Cytol.* **1996**, *166*, 59-101.
40. Poulicek, M.; Voss-Foucart, M.F.; Jeuniaux, Ch. In *Chitin in Nature and Technology;* Muzzarelli, R.A.A.; Jeuniaux, Ch.; Gooday, G.W., Eds; Plenum Press: New York, 1985, pp 7-12.
41. Gaill, F.; Chanzy, H.; Vuong, R. *Biol. Cell.* **1989**, *66*, 8a.
42. Gaill, F.; Schillito, B.; Chanzy, H.; Goffinet, G.; Da Conceciao, M.; Vuong, R. In *Advances in Chitin and Chitosan;* Brine, C.J.; Sandford, P.A.; Zikakis, J.P., Eds.; Elsevier Applied Publ.: London, New York, 1992, pp 225-231.
43. Poulicek, M. *Biochem. Syst. Ecol.* **1983**, *11*, 47-54.
44. Jeuniaux, Ch.; Bussers, J.C.; Voss-Foucart, M.F.; Poulicek, M. In *Chitin in Nature and Technology;* Muzzarelli, R.A.A.; Jeuniaux, Ch.; Gooday, G.W., Eds; Plenum Press: New York, 1985, pp 515-522.
45. Jeuniaux, Ch.; Voss-Foucart, M.F.; Gervasi, E.; Bussers, J.C.; Poulicek, M. *Bull. Soc. R. Sci. Liège* **1988**, *57*, 287-299.
46. Jeuniaux, Ch.; Voss-Foucart, M.F.; Poulicek, M.; Bussers, J.C. In *Chitin and Chitosan;* Skjak-Bræk, G.; Anthonsen, Th.; Sandford, P., Eds.; Elsevier Applied Sciences: London, UK, 1989, pp 3-11.
47. Allan, G.G.; Fox, J.R.; Kong, N. In *Proc. 1st Interntl. Conf. Chitin/Chitosan;* Muzzarelli, R.A.A.; Pariser, E.R., Eds.; MIT Sea Grant Report, 1978, pp 68-78.
48. Berry, P.F.; Smale, M.J. *Mar. Ecol. Progr. Ser.* **1980**, *2*, 337-343.
49. Poulicek, M.; Jeuniaux, Ch. In *Chitin and Chitosan* ; Skjak-Bræk, G.; Anthonsen, Th.; Sandford, P., Eds.; Elsevier Applied Sciences: London, UK, 1989, pp 151-160.
50. Dauby, P. *Prog. Belgian Oceanog. Res.* **1985**, *1*, 442-450.
51. Gervasi, E.; Jeuniaux, Ch.; Dauby, P. In *Aspects récents de la Biologie des Crustacés;* IFREMER ed: Brest, France, 1988, pp 33-38.
52. Winberge, G.G. *Methods for estimation of production of aquatic animals;* Academic Press: London, 1971, 175 p.
53. Everson, I. *The living resources of the Southern Ocean;* Ocean Fisheries Survey Program; U.N. Development Progr., F.A.O. ed.: Roma, 1977.
54. Ross, R.M.; Quetin, L.B. *Comp. Biochem. Physiol.* **1988**, *90B*, 499-505.
55. Lindley, J.A. *Mar. Biol.* **1982**, *71*, 7-10.
56. Ritz, D.A.; Hosie, G.W. *Mar. Biol.* **1982**, *68*, 103-108.
57. Nicol, S.; Hosie, G.W. *Biochem. Syst. Ecol.* **1993**, *21*, 181-184..
58. Bianchi M. *European Microbiology of Particulate Systems Final Report;* MTP MAST II ECC Reports, CEE, Brussel , 1996, Vol 1.
59. Bianchi M. *European Microbiology of Particulate Systems Final Report;* MTP MAST II ECC Reports, CEE., Brussel, 1996, Vol 2.

60. Poulicek, M. *Biodégradation des molécules organiques naturelles et artificielles en milieu marin;* Cours d'Océanographie Européens; Oceanis (20)4/5, Inst. Océanog.: Paris, 1994.
61. Poulicek, M.; Jeuniaux, Ch. *Bull. Soc. R. Sci. Liège* **1994**, *63*, 89-165.
62. Lardinois, D.; Eisma, D.; Chen, S. *Neth. J. Sea Res.* **1995**, *33*, 147-161.
63. Montgomery, M.T.; Welschmeyer, N.A.; Kirchman, D.L. *Mar. Ecol. Prog. Ser.* **1990**, *64*, 301-308.
64. Coffin, R.B. *Limnol.Oceanog.* **1989**, *34*, 531-542.
65. Kirchman, D.L.; Hoch, M.P. *Mar. Ecol. Prog. Ser.* **1988**, *45*, 169-178.
66. Kirchman, D.L.; Keil, R.G.; Wheeler, P.A. *Deep Sea Res.* **1989**, *36*, 1763-1776.
67. Muller, P.J.; Suess, E.; Ungerer, C.A. *Deep Sea Res.* **1986**, *33*, 819-838.
68. Honjo, S.; Manganini, S.J.; Cole, J.J. *Deep Sea Res.* **1982**, *29*, 609-625.
69. Poulicek, M.; Jeuniaux, Ch. *Bioch. Syst. Ecol.* **1991**, *19*, 385-394.
70. Poulicek, M.; Nisin, O.; Loizeau, V.; Bourge, I.; Toussait, Cl.; Voss-Foucart, M.F. *Bull. Soc. R. Sci. Liège* **1992**, *61*, 129-142.
71. Smayda, T.J. *Oceanogr. Mar. Biol. Ann. Rev.*, **1970**, *8*, 353-414.
72. Lânnergren, C. *Mar. Biol.* **1979**, *54*, 1-10.
73. Biengfang, P.K. *Mar. Biol.* **1980**, *61*, 69-77.
74. Knauer, G.A.; Martin, J.H. ; Bruland, K.W. *Deep Sea Res.* **1979**, *26A*, 97-108.
75. Wiebe, P.H.; Boyd, S.H.; Winget, C. *J. Mar. Res.* **1980**, *34*, 341-354.
76. Martens, P. ; Krause, M. *Helgol. Meeres.* **1990**, *44*, 9-19.
77. Lampitt, R.S.; Noji, T.; von Bodungen, B. *Mar. Biol.* **1990**, *104*, 15-23.
78. Andrews, C.C.; Karl, D.M.; Small, L.F., Fowler, S.W. *Nature* **1984**, *307*, 539-541.
79. Bathmann, U.; Noji, T.T.; Voss, M.; Peinert, R. *Mar. Ecol. Progr. Ser.* **1987**, 38, 45-51.
80. Dunbar, R.B.; Berger, W. *Bull. geol. Soc. Am.* **1981**, *92*, 212-218.
81. Emerson, C.W.; Roff, J.C. *Mar. Ecol. Progr. Ser.* **1987**, *35*, 251-257.
82. Gowing, M.M.; Silver, M.W. *Mar.Biol.* **1983**, *73*; 7-16.
83. Honjo, S.; Roman, M.R. *J. Mar. Res.* **1978**, *36*, 45-57.
84. Turner, J.T. *Mar.Biol.* **1977**, *40*, 249-259.
85. Fowler, S.W.; Small, L.F. *Limnol. Oceanog.* **1972**, *17*, 293-296.
86. Peduzzi, P.; Herndl, G.J. *Mar.Biol.* **1986**, *92*, 417-424.
87. Turner, J.T. *Trans. Am. microsc. Soc.* **1979**, *98*, 131-135.
88. Iturriaga, R. *Mar.Biol.* **1979**, *55*, 157-169.
89. Harding, G.C.H. *Limnol. Oceanog.* **1973**, *18*, 670-673.
90. Dave, D.; Poulicek, M. *Bull. Soc. R. Sci. Liège* **1988**, *57*, 301-312.
91. Poulicek, M. *Bull. Soc. R. Sci. Liège* **1992**, *61*, 113-128.
92. Seki, H. *J. Oceanog. Soc. Jap.* **1965**, *21*, 261-269.
93. Kennisch, M.J. *Practical Handbook of Marine Science;* CRC Press: Boca Raton FL, USA, 1990.
94. Tews, I.; Wilson, K.S.; Vorgias, C.E. In *Advances in Chitin Science;* Domard, A.; Jeuniaux, Ch.; Muzzarelli, R.A.A.; Roberts G., Eds.; Jacques André Publisher: Lyon, France, 1996, Vol.1., pp 26-33.

95. Ohtakara, A. *Agr. Biol. Chem.* **1963**, *27*, 454-460.
96. Ohtakara, A. *Agr. Biol. Chem.* **1964**, *28*, 745-751.
97. Bade, M.L.; Hickey, K. In *Chitin and Chitosan;* Skjak-Bræk, G.; Anthonsen, Th.; Sandford, P., Eds.; Elsevier Applied Sciences: London, UK, 1989, pp 179-183.
98. Perralis, A.; Ouzounis, Chr.; Wilson, K.S.; Vorgias, C.E. In *Advances in Chitin Science;* Domard, A.; Jeuniaux, Ch.; Muzzarelli, R.A.A.; Roberts G., Eds.; Jacques André Publisher: Lyon, France, 1996, Vol.1, pp 34-41.
99. Henrissat, B.; Bairoch, A. *Biochem. J.* **1993**, *293*, 781-788.
100. Berger, L.R.; Reynolds, D.M. *Biochem. Biophys. Acta* **1958**, *29*, 522-534.
101. Jeuniaux, Ch.; Compère, Ph.; Goffinet, G. *Boll. Zool.* **1986**, *53*, 183-196.
102. Florkin, M. *Aspects moléculaires de l'adaptation et de la phylogénie;* Masson Ed.: Paris, 1966.
103. Gooday, G.W.; Humphreys, A.M.; Mac Intosh, W.H. In *Chitin in Nature and Technology;* Muzzarelli, R.A.A.; Jeuniaux, Ch.; Gooday, G.W., Eds.; Plenum Press: New York, 1985, pp 83-91.
104. Boller, T. In *Chitin in Nature and Technology;* Muzzarelli, R.A.A.; Jeuniaux, Ch.; Gooday, G.W., Eds.; Plenum Press: New York, 1985, pp 223-230.
105. Powning, R.F.; Irzykiewicz, H. *Comp. Biochem. Physiol.* **1965**, *14*, 127-134.
106. Boller, T.; Gehri, A.; Mauch, F.; Vögeli, U. *Planta* **1983**, *157*, 22-31.
107. Molano, J.; Polacheck, I.; Duran, A.; Cabib, E. *J. Biol. Chem.* **1979**, *254*, 4901-4911.
108. Boller, T.; Vögeli, U. *Plant Physiol.* **1984**, *74*, 442-456.
109. Darvill, A.G.; Albersheim, P. *Ann. Rev. Plant Physiol.* **1984**, *35*, 243-287.
110. Nitzsche, W. *Theor. Appl. Gen.* **1983**, *65*, 171-179.
111. Perrakis, A.; Tews, I.; Dauter, Z.; Wilson, K.S.; Vorgias, C.E. In *Chitin World;* Karnicki, Z.S.; Wojtasz-Pajak, A.; Brzeski, M.M.; Bykowski, P.J., Eds.; WirtschaftsVerlag NW: Bremerhaven, Germany, 1994, pp 408-415.
112. Tews, I.; Vincentelli, R.; Perrakis, A.; Dauter, Z.; Wilson, K.S.; Vorgias, C.E. In *Chitin World;* Karnicki, Z.S.; Wojtasz-Pajak, A.; Brzeski, M.M.; Bykowski, P.J., Eds.; Wirtschafts Verlag NW: Bremerhaven, Germany, 1994, pp 416-423.
113. Oishi, K.; Ishikawa, F.; Nomoto, M. In *Chitin and Chitosan;* Skjak-Bræk, G.; Anthonsen, Th.; Sandford, P., Eds.; Elsevier Applied Sciences: London, UK, 1989, pp 185-195.
114. Araki, Y.; Ito, E. *Eur. J. Biochem.* **1975**, *55*, 71-78.
115. Calvo-Mendez, C.; Ruiz-Herrera, J. *Exp. Mycol.* **1987**, *11*, 128-140.
116. Davis, L.L.; Bartnicki-Garcia, S. *Biochemistry* **1984**, *23*, 1065-1073.
117. Monaghan, R.L.; Eveleigh, D.E.; Tewari, R.P.; Reese, E.T. *Nature New Technol.* **1973**, *245*, 78-80.
118. Monaghan, R.L. *Discovery, distribution and utilization of chitosanases;* Univ. Microfilms: Ann Arbor, Vol. 724, 1975, pp 75-24.
119. Price, J.S.; Storck, R. *J. Bacteriol.* **1975**, *124*, 1574-1585.
120. Kafetzopoulos, D.; Martinou, A.; Bouriotis, V. *Proc. Natl. Acad. Sci. USA*, **1993**, *90*, 2564-2568.

121. Martinou, A.; Kafetzopoulos, D.; Bouriotis, V. *J. Chromatgr.* **1993**, 644, 35-41.
122. Kafetzopoulos, D.; Thireos, G.; Vournakis, J.; Bouriotis, V. *Proc. Natl. Acad. Sci. USA*, **1993**, *90*, 8005-8008.
123. Tsigos, I.; Martinou, A.; Varum, D.; Kafetzopoulos, D.; Christodoulidou, A.; Tzanodaskalaki, M.; Bouriotis, V. In *Chitin World;* Karnicki, Z.S.; Wojtasz-Pajak, A.; Brzeski, M.M.; Bykowski, P.J., Eds.; Wirtschafts Verlag NW: Bremerhaven, Germany, 1994, pp 98-107.
124. Davis, B.; Eveleigh, D.E. In *Chitin, chitosan and related enzymes;* Ziokakis, J.P., Ed.; Academic Press: Orlando, FL, USA, 1984.
125. Gooday, G.W.; Prosser, J.I.; Hillman, K.; Cross, M.G. *Biochem. Syst. Ecol.* **1991**, *19*, 395-400.
126. ZoBell, C.E.; Rittenberg, S.C. *J. Bacteriol.* **1937**, *35*, 275-287.
127. Yu, Ch.; Lee, A.M.; Bassler, B.L.; Roseman, S. *J. Biol. Chem.* **1991**, 266, 24260-24267.
128. Bassler, B.L.; Gibbons, P.J.; Yu, Ch.; Roseman, S. *J. Biol. Chem.*, **1991**, *266*, 24268-24275.
129. Bassler, B.L.; Yu, Ch.; Lee, A.M.; Roseman, S. *J. Biol. Chem.* **1991**, 266, 24276-24286.
130. Schrempf, H. In *Advances in Chitin Science;* Domard, A.; Jeuniaux, Ch.; Muzzarelli, R.A.A.; Roberts G., Eds.; Jacques André Publisher: Lyon, France, 1996, Vol.1, pp 123-128.
131. Yu, Ch.; Lee, A.M.; Roseman, S. *Biochem. Biophys. Res. Commun.* **1987**, *149*, 86-120.
132. Smucker, R.A.; Kim, C.K. In *Microbial enzymes in aquatic environments*, Chrost R.J., Ed.; Springer Verlag: New York, 1991, pp 249-269.
133. Gooday, G.W. In *Chitin and Chitosan;* Skjak-Bræk, G.; Anthonsen,Th.; Sandford, P., Eds.; Elsevier Applied Sciences: London, UK, 1989, pp 13-22.
134. Wessels, J.G.H. *Int. Rev. Cytol.* **1986**, *104*, 37-79.
135. Sietsma, J.H.; Vermeulen, C.A.; Wessels, J.G.H. In *Chitin in Nature and Technology;* Muzzarelli, R.A.A.; Jeuniaux, Ch.; Gooday, G.W., Eds.; Plenum Press: New York, 1985, pp 63-69.
136. Sietsma, J.H.; Wessels, J.G.H. *J. Gen. Microbiol.* **1979**, *114*, 99-106.
137. Hunt, S.; Nixon, M. *Comp. Biochem. Physiol.* **1981**, *68B*, 535-546.
138. Peter, M.G.; Kegel, G.; Keller, R. In *Chitin in Nature and Technology;* Muzzarelli, R.A.A.; Jeuniaux, Ch.; Gooday, G.W., Eds.; Plenum Press: New York, 1985, pp 21-28.
139. Hackman, R.H. In *Biology of the Integument;* Bereiter-Hahn, J. et al., Eds.; Springer Verlag: Berlin, 1984, 583.
140. Jeuniaux, Ch.; Compère, Ph.; Goffinet, G. *Boll. Zool.* **1986**, *53*, 183-196.
141. Brine, Ch.J.; Austin, P.R. *Comp. Biochem. Physiol.* **1981**, *70B*, 173-178.
142. Brine, Ch.J.; Austin, P.R. In *Chitin and Chitosan: Proc. IId Intern. Conf. Chitin Chitosan* ; Hirano, S.; Tokura, S., Eds.; Sapporo, Japan, 1982, 105-110.
143. Attwood, M.M.; Zola, H. *Comp. Biochem. Physiol.* **1967**, *20*, 993-998.
144. Lipke, H.; Geoghegan, T. *Biochem. J.* **1971**, *125*, 703-716.

145. Lipke, H.; Sugumaran, M.; Henzel, W. *Adv. Insect Physiol.* **1983**, *17*, 1-34.
146. Brunet, P.C.J. *Insect Biochem.* **1980**, *10*, 467-500.
147. Waite, J.H.; Andersen, S.O. *Biol. Bull.* **1980**, *158*, 164-173.
148. Gailll, F.; Voss-Foucart, M.F.; Gerday, Ch.; Compère, Ph.; Goffinet, G. In *Adv. Chitin Chitosan;* Brine, C.J.; Sandford, P.A.; Zikakis, J.P., Eds.; Elsevier Appl. Sci.: London, 1992, pp 232-236.
149. Gaill, F.; Hunt, S. *Mar. Ecol. Prog. Ser.* **1986**, *3*, 267-274.
150. Gaill, F.; Hunt, S. *Rev. Aquat. Sci.* **1991**, *4*, 107-137.
151. Giraud-Guille, M.M.; Bouligand, Y. In *Chitin World;* Karnicki, Z.S.; Wojtasz-Pajak, A.; Brzeski, M.M.; Bykowski, P.J., Eds.; Wirtschafts Verlag NW: Bremerhaven, Germany, 1994, pp 136-144.
152. Loizeau, V. *Essai de paramétrisation des flux de polychlorobiphényles à travers l'écosystème pélagique de la baie de Calvi (Corse);* Mém. Maîtrise Sci. Techn.: Corte University, 1989, (unpublished).
153. Bourge, I. *Etude des diverses modalités de la dégradation du zooplancton à l'interface sédimentaire;* Mém. Licence Sci. Zool.: Liège University, 1990, (unpublished).
154. Nisin, O. *Les premières étapes de la dégradation du zooplancton en pleine eau en baie de Calvi. Approches microscopique, bactériologique et biochimique,* Mém. Licence Sci. Zool.: Liège University, 1991, (unpublished).
155. Alldredge, A.L.; Silver, M.W. *Prog. Oceanog.* **1988**, *20*, 41-82.
157. Herndl, G.J.; Peduzzi, P. *Mar. Ecol.* **1988**, *9*, 79-90.
158. Kirchner, M. *Hegol. Meeresunters.* **1995**, *49*, 210-212.
159. Poulicek, M.; Poppe, G. *Bull. Soc. Roy. Sci. Liège* **1981**, *11-12*, 519-542.
160. Poulicek, M. *J. Molluscan Study* **1983**, *12A*, 136-141.
161. Poulicek, M.; Jaspar-Versali, M.F. *Bull. Soc. Roy. Sci. Liège* **1984**, *53*, 114-126.
162. Poulicek,M.; Goffinet, G.; Voss-Foucart, M.F.; Jaspar-Versali, M.F.; Bussers, J.C.; Toussaint, Cl. In *Chitin in Nature and Technology;* Muzzarelli, R.A.A.; Jeuniaux, Ch.; Gooday, G.W., Eds.; Plenum Press: New York, 1985, pp 547-550.
163. Poulicek, M.; Goffinet, G.; Jeuniaux, Ch.; Simon, A.;Voss-Foucart, M.F., In *Recherches Océanographiques en Mer Méditerranée;* I.R.M.A. ed., 1988, pp 107-124.
164. Simon, A.; Mulders, M.; Poulicek, M. *Annls Soc. r. zool. Belg.* **1989**, *119*, 169-173.
165. Simon, A.; Poulicek, M. *Cah. Biol. Mar.* **1990**, *31*, 95-105.
166. Simon, A.; Poulicek, M ; Machiroux, R.; Thorez, J. *Cah. Biol. Mar.* **1990**, *31*, 365-384.
167. Poulicek, M.; Jaspar-Versali, M.F. *Malacological Reviews, Proc . IXth Malac. Congress (Edinburgh),*1992, pp 263-274.
168. Simon, A.; Poulicek, M ; Velimirov, Br.; MacKenzie, F.T. *Biogeochemistry* **1994**, *25*, 167-195.
169. Lukas, K.J. In *SEM/1979/II;* SEM Inc, AMF O'Hare ed., IL, USA, 1979, pp 441-455.

170. May, J.A.; Perkins, R.D. *J. Sed. Petrol.* **1979**, *49*, 357-378.

171. Zeff, M.L.; Perkins, R.D. *Sedimentology* **1979**, *26*, 175-201.

172. Kim, J.; ZoBell, Cl.E. *J. Oceanog. Soc. Jap.* **1972**, *28*, 131-137.

173. Helmke, E.; Weyland, H. *Mar. Biol.* **1986**, *91*, 1-7.

174. Gaill, F.; Voss-Foucart, M.F.; Shillito, B.; Goffinet, G. In *Advances in Chitin Science;* Domard, A.; Jeuniaux, Ch.; Muzzarelli, R.A.A.; Roberts G., Eds.; Jacques André Publisher: Lyon, France, 1996, Vol.1, pp 143-148.

175. Voss-Foucart, M.F.; Bussers, J.C.; Goffinet, G.; Poulicek, M.; Toussaint, Cl.; Jeuniaux, Ch. *Ann. Soc. Roy. Zool. Belg.* **1984**, *114*, 145-146.

176. Tegelaar, E.W.; de Leeuw, J.W.; Derenne, S.; Largeau, C. *Geochim. Cosmochim. Acta* **1989**, *53*, 3103-3106.

177. van Bergen, P.F.; Collinson, M.E.; Briggs, D.E.G.; deLeeuw, J.W.; Scott, A.C.; Evershed, R.P.; Finch, P. *Acta Bot. Neerl.* **1995**, *44*, 319-342.

178. Stankiewicz, B.A.; Briggs, D.E.G.; Evershed, R.P. *Energy Fuels* **1997**, *11*, 515-521.

179. Stankiewicz, B.A.; Briggs, D.E.G.; Evershed, R.P.; Flannery, M.B.; Wuttke, M. *Science* **1997**, *276*, 1151-1153.

180. Stankiewicz, B.A.; Briggs, D.E.G.; Evershed, R.P.; Duncan, I.J. *Geochim. Cosmochim. Acta* **1997**, *61*, 2247-2252.

181. Baas, M.; Briggs, D.E.G.; van Heemst, J.D.H.; Kear, A.J.; de Leeuw, J.W., *Geochim. Cosmochim. Acta* **1995**, *59*, 945-951.

182. Briggs, D.E.G.; Kear, A.J.; Baas, M.; de Leeuw, J.W.; Rigby, S. *Lethaia* **1995**, *28*, 15-23.

183. Brumioul, D.; Voss-Foucart, M.F. *Comp. Biochem. Physiol.* **1977**, *57B*, 171-175.

184. Voss-Foucart, M.F.; Jeuniaux, Ch. *J. Paleontol.* **1972**, *46*, 769-770.

Chapter 12

The Fate of Chitin in Quaternary and Tertiary Strata

B. Artur Stankiewicz[1,2,5], Derek E. G. Briggs[1], Richard P. Evershed[2], Randall F. Miller[3], and Anja Bierstedt[1,4]

[1]Biogeochemistry Research Centre, Department of Geology, University of Bristol, Bristol BS8 1RJ, United Kingdom
[2]Organic Geochemistry Unit, School of Chemistry, University of Bristol, Bristol BS8 1TS, United Kingdom
[3]Steinhammer Palaeontology Laboratory, New Brunswick Museum, Saint John, New Brunswick E2K 1E5, Canada
[4]Department of Chemistry, Fritz-Foerster-Bau, Dresden Technische Univeristät, Dresden, D-01069, Germany

Chitin is one of the most abundant biopolymers on earth. It occurs in a range of organisms but is particularly important as a constituent of arthropod cuticles. Experiments have demonstrated that chitin is more resistant to degradation than protein, but it is rarely preserved in the fossil record. The chitin content of beetle cuticles from 11 Quaternary deposits in Canada, UK and USA was estimated using pyrolysis-GC/MS and quantitative colorimetric assay. The proportion preserved ranged from 3 to 37 dry weight %. Analyses of insects and fresh-water crustaceans from several European Tertiary biotas revealed that the chitin biopolymer is preserved at levels varying from 2 to 38%. Chitin can survive, even for millions of years, in non-marine clastic sediments that provide favourable environmental conditions, but it is much more susceptible to degradation in marine settings. The 25 Ma lacustrine deposit of Enspel, Germany, preserves the oldest reliable evidence of chitin reported to date. The differences in the proportion of chitin preserved in fossils reflect the environment of deposition more than their age. Chitin is more likely to be found in fossils preserved in terrestrial than in marine strata.

The fate of various macromolecules that make up biological tissues has been the subject of geochemical studies for many years (*1-4*). This research has focused on biopolymers which contribute to coal and kerogen. Foremost among these are

[5]Current address: Shell E&P Technology Company, 3737 Bellaire Boulevard, Houston, TX 77025

211

major components of plants (i.e. cellulose and lignin) and algae (i.e. aliphatic polymers), which are relatively resistant to bio- and chemical degradation during diagenesis (2). Interest in less stable macromolecules (e.g. chitin, proteins) has been limited due to their faster degradation in the biosphere and rare survival into the geosphere.

Chitin is one of these less decay-resistant biopolymers (2). Although the annual production in the biosphere is estimated to be second only to cellulose (5), chitin has received surprisingly little attention among biogeochemists. Chemically it is a polysaccharide, a major component of arthropod exoskeletons where it is cross-linked with proteins (see earlier chapter in this volume). Although chitin is relatively resistant to chemical and physical degradation, it is rapidly broken down in most modern marine and terrestrial ecological systems (6-9, earlier chapter in this volume) by chitinoclastic microorganisms through enzymatic hydrolysis (10). Many studies have investigated the degradation of chitin under both oxic and anoxic conditions (6-9). However, the role of degraders such as fungi, and the influence of physical factors such as pH and temperature of the environment, rate of burial, and organic productivity are still poorly understood.

Several recent studies have investigated the preservation of chitin in fossils ranging in age from Silurian (ca 420 Ma) to Holocene (11-17). These studies showed that under favourable environmental conditions chitin can be preserved in sedimentary rocks as old as 25 Ma (16). Chitin has also been studied using stable isotope ratios as a source of paleoenvironmental information (12, 18-20). However, a lack of sufficient material for such analyses is often a limiting factor.

The structure of arthropod cuticles appears to be a very important factor in retarding chitin decomposition and promoting its preservation. Thick, heavily sclerotized cuticles, such as those of some beetles, may preserve chitin even in Tertiary strata (16). Beetle elytra are abundant in Quaternary deposits (21-24) where they provide an important indicator of climatic change (25, 26). However, few geochemical studies have investigated these fossils (12, 15).

An important limiting factor in studies of chitin in fossil specimens is the small quantity of cuticle normally available and the recalcitrant nature of the fossil material. Flash pyrolysis, in combination with gas chromatography and mass spectrometry (py-GC/MS), has proved to be a very efficient tool for elucidating the molecular composition of many modern and fossil materials (3, 4, 27). This method was used to demonstrate the preservation of the oldest fossil chitin in beetles from Enspel, Germany (16). Colorimetric assay (28) has allowed the quantitative estimation of chitin in fossils, particularly Quaternary specimens (17). The application of Fourier-transform infrared spectroscopy to the recognition of chitin in fossils is also being investigated (Mastalerz M., Indiana University, Geological Survey, in preparation).

Here we present the results of our analyses of the fate of chitin in Quaternary sediments, using both py-GC/MS and colorimetric assay techniques, and briefly review the preservation of chitin in Tertiary sediments. Our goal is to summarize possible factors controlling the preservation of this very important biopolymer in the geosphere.

Sample description

The Quaternary beetle elytra analyzed (Table 1) represent a range of environmental settings and ages (*15, 23, 24, 29-33*). The youngest sample is estimated to be 1600 years old (Thornog Bog) and the oldest *ca* 80,000 years old (Moose Point).

The Tertiary fossils came from several European localities (*16, 17*, McCobb *et al.* in review). The youngest are insects and the crayfish *Astacus* sp. from the Pliocene lake Willershausen in Germany (ca 3 Ma). A beetle from the Upper Miocene (8 Ma) of St. Bauzile, France, is preserved in diatomite. The curculionid beetles from the Oligocene (25 Ma) of Enspel, Germany, are preserved in the volcanoclastic sediments of a maar-lake (*16*). The insects of the Late Eocene (36 Ma) Bembridge Marls of the Isle of Wight occur in brackish lagoonal and estuarine sediments (McCobb *et al.* in review).

Experimental

Quaternary beetles were separated from the sediments by kerosene flotation and sorted using ethanol. The cuticle of the Tertiary specimens was separated mechanically from the rock matrix. Prior to chemical analysis all specimens were extracted ultrasonically with CH_2Cl_2 and methanol to remove extractable constituents and any contaminants introduced by handling.

Colorimetric assay was performed on most of the specimens to establish the amount of chitin preserved in the cuticle. The assay is based on KOH hydrolysis, which deacetylates chitin to the glucosamine polymer chitosan, and further conversion of the latter to 2,5-anhydrohexoses using $KHSO_4$ and $NaNO_2$ (Sigma-Aldrich Chemical Co). The color reaction was developed using freshly prepared 3-methyl-2-benzothiazolone hydrochloride (MBTH; Sigma-Aldrich Chemical Co.) and $FeCl_3$ (Sigma-Aldrich Chemical Co.). The absorbance was recorded at 650 nm against reference blank solutions (prepared by following exactly the same protocol but without introducing a sample) using a Cecil CE 2292 Series 2 Digital Ultraviolet Spectrophotometer (see *28, 17* for details of the procedure).

Pyrolysis-gas chromatography/mass spectrometry (Py-GC/MS) was performed on 0.05 and 0.5 mg of solvent-extracted cuticle using a CDS (Chemical Data System, Oxford, Pennsylvania) 1000 pyroprobe coupled to a Carlo Erba (Milan, Italy) 4130 GC, interfaced with a Finnigan (San Jose, California, USA) 4500 MS. Samples were placed in quartz holders and pyrolyzed in a flow of helium for 10 seconds in a platinum coil at 610°C. Compounds were separated using a Chrompack (Middleburg, The Netherlands) CP Sil-5 CB column (50 m x 0.32 mm i.d., film thickness 0.4 mm). The GC oven was operated as follows: isothermal for 5 min at 35°C; temperature programmed at 4°C min^{-1} to 300°C and then isothermal for 10 min. The MS was operated in full scan mode (35-650 daltons, 1 scans/sec., 70 eV electron energy). The identification of pyrolysis products derived from chitin and proteins is described in detail elsewhere (*14-16, 27*).

Table I. Samples of Quaternary beetles analysed in the course of this study and the results of the colorimetric assay.

Sample name	Age[†]	Lithology	Locality	Chitin [%]
Beetle (*Tenebrio molitor*)	Modern		UK (Bristol)	39.47
Beetle	Ho (*ca* 1.6 Ka)	Peat	N. Ireland (Thornog Bog)	36.54
Beetle (Scarabaeidae)	Ple (25-28 Ka)	Silty clay	UK (Woolpack Farm)	31.46
Beetle	Ple (25-28 Ka)	Silty clay	UK (Welland Bank)	26.00
Beetle (*Olophrum boreale*)	Ple (<10.8 Ka)	Sand (peaty)	Canada, NS (West Mabou)	32.54
Beetle (*O. boreale, O. rotundicolle*)	Ple (<10.8 Ka)	Silty clay	Canada, NS (Benacadie)	21.97
Beetle (*O. rotundicolle*)	Ple (*ca* 11.1 Ka)	Peat	Canada, NS (Lantz)	10.52
Beetle (*O. rotundicolle*)	Ple (*ca* 11.5 Ka)	Peat	Canada, NS (Amaguadees)	3.39
Beetle (*O. rotundicolle*)	Ple (*ca* 11.8 Ka)	Silty peat	Canada, NS (Hirtles)	22.63
Beetle (*O. rotundicolle*)	Ple (33-40 Ka)	Sandy peat	Canada, BC (Lynn Canyon)	23.03
Beetle (*O. rotundicolle*)	Ple (*ca* 80 Ka)	Silty peat	Canada, NS (Moose Point)	33.15
Beetle (Tenebrionidae)	Ple (20 Ka)	Asphalt	USA (La Brea, CA)	22.62

[†] key to abbreviations: Ho - Holocene, Ple - Pleistocene; NS - Nova Scotia, BC - British Columbia

The cuticles were mounted on aluminum stubs using silver dag, and coated in gold. Their morphology was examined using a Cambridge Stereoscan 250 Mk3 scanning electron microscope at 7-12 kV.

Results

Quaternary. The beetles from Thornog bog in Northern Ireland are estimated to be 1600 years old and were collected from an active peat deposit. All of the beetle samples from Canada belong to the genus *Olophrum* and were collected from either late glacial or interglacial sediments deposited as detritus in shallow, standing or slow-moving water, or from peats where beetles were likely deposited *in situ* (Table 1). Late-glacial silts and clays were deposited in depressions on the newly deglaciated landscape. Some ponds eventually filled in to become peat bogs (*30*). The peat layers were overlain by mineral-rich sediments, sometimes with a high organic content. Overlying sediments, consisting of clean sands, mudflows or till, are interpreted as proglacial or glacial sediments resulting from ice advance during the Younger Dryas (*30, 34*). Older sites were deposited on developed landscapes in later stages of the last interglacial or during an interstadial, but in similar settings. The Pleistocene specimens from the UK were preserved in similar settings to the Canadian sites (i.e. shallow depressions within channel deposits), but represent the last interglacial (Ipswichian). The beetle specimens from La Brea, California, are preserved in tar pits (*ca* 20 Ka) (*15*).

All beetle specimens yielded abundant pyrolysis products characteristic of chitin such as acetamide, acetamidofurans, acetamidopyrones or oxazoline derivatives (Figures 1, 2). The total ion chromatograms closely resemble the trace obtained from commercially purified chitin, confirming the presence of chitinous moieties. However, even though the specimens all yielded components derived from chitin, the absolute amounts vary between samples, as revealed by the colorimetric assay (Table I). The lowest amount of chitin was detected in the specimens from the 11.5 Ka peat of Amaguadees (3.4 dry wt. %) and the highest values were obtained for the 1.6 Ka beetle (36.5%) and in specimens in the 80 Ka silty peat from Moose Point (33.1%) (Table I).

All the beetle elytra investigated showed high quality structural preservation including details of the external morphology and almost intact procuticle (Figure 4).

Tertiary. The Tertiary fossils show greater variability of chitin preservation than the Quaternary. The youngest samples are insect and crayfish remains from the Pliocene of Willershausen, Germany. Pyrolysis of these cuticles yielded products derived from chitin demonstrating its survival in all the specimens, but the absolute content varied from 2% in an orthopteran and 5% in *Astacus* to up to 37.9% in a weevil (*17*). The beetle from the Miocene deposits of St. Bauzile, France yielded a pyrolyzate dominated by chitin markers and the chitin content was 12.7 dry wt. % (Figure 3). The oldest reliable record of chitin to date is from the 24.7 Ma Oligocene of Enspel, Germany, where chitinous moiety was detected in several specimens of curculionid beetle (Figure 4). In contrast insect cuticles from the

216

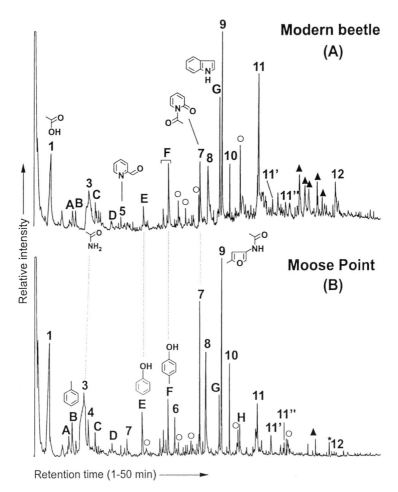

Figure 1. Total ion chromatograms (pyrolysis at 610°C for 10 sec) of specimens of A) modern beetle elytra (*Tenebrio molitor*), B) Pleistocene beetle elytra (*Olophrum rotundicolle*) from Moose Point, Nova Scotia, Canada. Chitin markers: 1 = acetic acid, 2 = pyridine, 3 = acetamide, 4 = 2-methylpyridine, 5 = 2-pyridinecarboxaldehyde, 6 = N-acetyl-N-ethenylacetamide, 7 = acetylpyridone, 8 = 3-acetamidofuran, 9 = 3-acetamido-5-methylfuran, 10 = 3-acetamido-4-pyrone, 11-11" = oxazoline derivatives, 12 = 1,6-anhydro-2-acetamido-2-deoxyglucose, O = other chitin markers. Amino acid markers: A = pyrrole, B = toluene, C = methylpyrroles, D = ethylpyrrole, E = phenol, F = 3- and 4-methylphenol, G = indole, H = methylindole, ▲ = 2,5-diketopiperazines; * - contaminants.

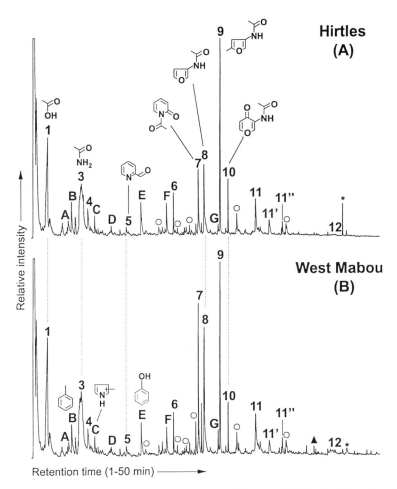

Figure 2. Total ion chromatograms (pyrolysis at 610°C for 10 sec) of specimens of A) Pleistocene beetle elytra (*Olophrum rotundicolle*) from Hirtles, Nova Scotia, Canada, B) Pleistocene beetle elytra (*O. boreale*) from West Mabou, Nova Scotia, Canada. Key to the peaks as in Figure 1.

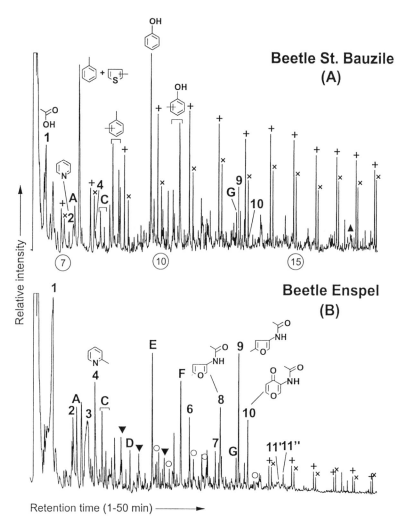

Figure 3. Total ion chromatograms (pyrolysis at 610°C for 10 sec) of specimens of A) Upper Miocene (*ca* 6 Ma) beetle elytra from St. Bauzile, France, B) Oligocene (24.7 Ma) beetle elytra from Enspel, Germany. Key to the peaks as in Figure 1; ▼ = indicate products derived from non-chitinous polysaccharides (e.g. mucopolysaccharides); + =*n*-alk-1-enes, x = *n*-alkanes, which carbon number is indicated in circles under chromatogram.

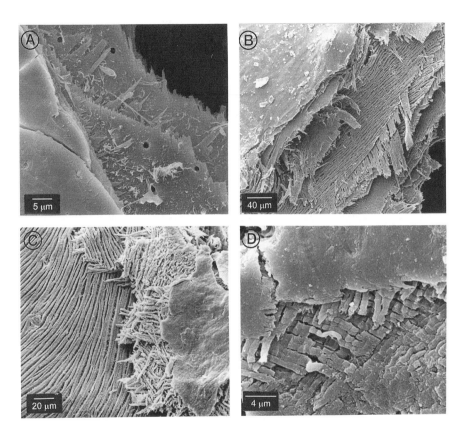

Figure 4. Scanning electron photomicrographs of fossil beetle cuticles revealing overlapping layers of the cuticle (exo- and endocuticle). A) beetle (*Olophrum rotundicolle*) from Pleistocene (*ca* 80 Ka) Moose Point, Canada, B) beetle (Tenebrionidae) from Pleistocene (*ca* 20 Ka) tar pits of La Brea, California, USA, C) weevil from Pliocene (3 Ma) lake Willershausen, Germany, D) beetle (Curculionidae) from Oligocene (25 Ma) maar-lake Enspel, Germany. Note the obvious fibrous structures in specimens B-D.

Eocene (*ca* 36 Ma) Bembridge Marls of the Isle of Wight, UK, did not reveal pyrolysis products derived from chitin but only aliphatic components (McCobb *et al.* in review). Likewise crustacean cuticles from marine deposits in Japan ranging in age from Pliocene to Eocene do not preserve detectable chitin components (*17*).

The Tertiary fossils frequently revealed morphological details comparable to the cuticles of modern species, including the preservation of chitinous fibers in the endocuticle (Figure 4).

Discussion

The results obtained from this and several other studies (*12, 15, 16, 20*) clearly indicate that the chitinous moiety is preserved in terrestrial deposits of wide-ranging age and sedimentary setting. Although there are numerous studies showing degradation of this polymer in marine and terrestrial settings (*6-10, 35*), our investigations indicate that chitin can survive to a degree dependent on several factors such as the sedimentology or primary productivity of the environment.

Quaternary. The striking contrast in the chitin content of samples from Amaguadees, which is only 11.5 Ka but yielded just 3.4%, and Moose Point, where chitin from 80 Ka beetles survived in an almost unaltered state (33.15% vs *ca* 40% in modern beetles), indicates that the primary control on preservation is environmental, not age. Miller et al. (*12*) likewise noted higher amounts of chitin in interglacial 5a and 5e beetle specimens (80-120 Ka) than in late glacial ones (*ca* 11 Ka). In the modern environment, the beetle *Olophrum*, recovered from Canadian sites, lives in clumps of *Carex* or moss at the edge of lakes or small streams and in bogs. It is also found in moist litter of willow and alder (*36*). *Olophrum* fossils recovered from thick, matted peat deposits may have lived and died in the bog, whereas specimens recovered from sandy or silty, organic-rich layers were likely washed into the deposits from their habitat along the margin of open water.

Our study shows that the lowest amounts of chitin occur in beetles from peat horizons, whereas higher values characterize specimens preserved in layers with an increased input of clastic sediment. Thus the strongly acidic environment of peat bogs may enhance chitin degradation (acidic hydrolysis) even though it also reduces the activity of microorganisms (*37*), thus slowing the degradation of various macromolecules (e.g., lignin or cellulose). Bogs vary greatly in acidity and additional study of the moss flora, stratigraphy and morphology of the fossil deposits may provide information about the pH. Although all the specimens investigated came from organic-rich lithologies in a peat bog setting, the addition of clastics such as sand or silt may have shifted the pH towards neutral values. Acid hydrolysis of (1'4)-β-glycosidic bonds linking hexosamine monomers within the chitin polymer may be more common where a lower pH prevails in the environment of deposition. However, the possibility that the differences in the chitin content of the fossils may reflect variation in the length of exposure before burial, thus enhanced aerobic degradation (litter stage), can not be ruled out. The proportion of chitin in beetle cuticles from some Pleistocene sites (e.g. Hirtles, Lynn Canyon,

Benacadie) is as high as that in specimens from the La Brea tar pits of California. Thus, while the preservational properties of asphalts are well documented (*15*), conditions similarly favourable for preservation may occur in Pleistocene clastic sediments.

Tertiary. The preservation of chitin in Tertiary deposits is much rarer than in younger strata. Although it appears that Pliocene deposits may contain arthropod cuticles that are well preserved chemically, only one example of the survival of chitin in sedimentary deposits older than Miocene has been documented (*16*). Thus, it is apparent that the preservation of chitin is limited to special environmental conditions and may also depend on the type of organism in question. Studies of the cuticles of 46 Tertiary crustaceans from Europe (*11*) and 16 from Japan and elsewhere (*17*) did not reveal the presence of chitin. In contrast, most of the cuticles of Tertiary beetles investigated yielded evidence of chitinous moieties. However, those of other insects (e.g. flies) from the same sedimentary settings failed to do so. In some cases the latter showed the presence of an aliphatic polymer characteristic of many Paleozoic and Mesozoic arthropod exoskeletons rather than N-containing chitin or proteins (*14*).

Crustacean cuticles, present mainly in the marine environment, appear to degrade more rapidly than the heavily sclerotized tissues of terrestrial insects. In laboratory conditions crustacean cuticles showed rapid degradation of both chitin and proteins within 8 weeks (*13, 38*). Although chitin is more resistant to degradation than coexisting proteins, it decays rapidly once the biomineralized components of the cuticle have dissolved (*38*). In contrast, similar experiments conducted on insects (crickets, beetles, cockroaches), both in the laboratory (oxic conditions) and in a freshwater pond (Bristol, UK), did not reveal any significant change in chitin content even after 25-35 weeks of decay (Duncan, I. J., University of Bristol, unpublished data; Stankiewicz *et al.*, in preparation). These and other investigations (*35*), and the rapid degradation of chitin in the oceans (*10*), strongly suggest that chitin is more likely to be found in fossils preserved in terrestrial than in marine strata.

Chemical vs morphological preservation. All Quaternary cuticles investigated here showed exceptional structural preservation of the cuticle on both a macro- and microscopic scale. In general, the well exposed meshwork of chitinous fibers observed in beetle cuticle from the tar pits of La Brea, California (*14*), contrasted with the disintegration of cuticle in specimens from glacial sediments (Figure 4) even though both contained a similar content of chitin. Therefore, it is difficult at the present time to directly relate the presence of well exposed chitinous fibers (revealed by degradation of the proteinaceous matrix) and the chitin content in the cuticle. Several Tertiary specimens (e.g. Willershausen, Enspel) that preserve chitin, were also characterized by fibers clearly visible under the SEM (Figure 4). Although a close relationship between changes in cuticle morphology and chemistry has been demonstrated in laboratory decay experiments (*38*), the relationship between chemical and morphological decay has yet to be fully documented.

222

Conclusions

The fate of chitin in the bio- and geosphere depends on factors that control the decomposition of biological tissues in the natural environment, such as the intensity of animal, fungal and bacterial degradation (both aerobic and anaerobic). Anoxia, temperature, sediment type (*35*), rate of burial, and primary productivity in the environment of deposition are all known to influence the preservation of fossil molecules, but the exact role of individual factors in the survival of chitin remains unknown. Our study shows that chemical factors such as the pH of the environment may be significant. More important is the structure and composition of the cuticle, and therefore the type of arthropod incorporated into the fossil record. It is clear that the preservation of chitin does not require "unusual fossilization conditions, such as...amber or...oil seeps" (*11*), at least in the terrestrial realm. The chitin content of beetles from glacial sediments may be higher than that in specimens from the tar pits of La Brea. Time seems to be a less important element than previously thought. Chitin which survives initial degradation may be transformed subsequently to a composition that we presently do not recognize as chitin-derived.

Acknowledgments

This research was funded by a NERC research grant to DEGB and RPE through the Ancient Biomolecules Initiative (GST/02/1027). NERC also provided financial support for mass spectrometry facilities (F14/6/13). Late-glacial studies in eastern Canada were supported by National Geographic Society Research Grant 4588-91. This investigation would not have been possible without the generosity of many individuals who donated samples: L. McCobb - Quaternary beetles; B. Riou - for assistance in field sampling at St. Bauzile; D. Meischner - Pliocene insects and crayfish from Willershausen; M. Wuttke - insects from Enspel; J. Harris and C. McNassor - insects from La Brea; E. Jarzembowski - insects from the Bembridge Marls.

References

1. Nip, M.; Tegelaar, E.; de Leeuw, J. W.; Schenck, P. A.; Holloway, P. J. W. *Org. Geochem.* **1986**, *10*, 769-778.
2. Tegelaar, E. W.; de Leeuw, J. W.; Derenne, S.; Largeau, C. *Geochim. Cosmochim. Acta* **1989**, *53*, 3103-3106.
3. de Leeuw, J. W.; van Bergen, P. F.; van Aarssen, B. G. K.; Gatellier, J.-P. L. A.; Sinninghe Damsté, J. S.; Collinson, M. E. *Phil. Trans. R. Soc. Lond. B* **1991**, *333*, 329-337.
4. van Bergen, P. F.; Collinson, M. E.; Briggs, D. E. G.; de Leeuw, J. W.; Scott, A. C.; Evershed, R. P.; Finch P. *Acta Bot. Neerl.* **1995**, *44*, 319-342.
5. Gooday, G. W. In *Advances in Microbial Ecology*; Marshall, K. C., Ed.; Plenum Press; New York and London, 1990, Vol. 11; pp 387-430.

6. Poulicek, M.; Goffinet, G.; Jeuniaux, C.; Simon, A.; Voss-Foucart, M. F. *Bull. Soc. R. Liège* **1988**, *57*, 313-330.

7. Poulicek, M.; Jeuniaux, C. In *Chitin and Chitosan*; Skjak-Braek, G.; Anthonsen, T.; Sanford, P., Ed.; Elsevier: London, 1989; pp 151-160.

8. Warnes, C. E.; Randles, C. I. *Ohio J. Sci.* **1977**, *77*, 224-230.

9. Warnes, C. E.; Randles, C. I. *Ohio J. Sci.* **1980**, *80*, 250-255.

10. Poulicek, M.; Gail, F.; Goffinet, G. This volume.

11. Schimmelmann, A.; Krause, R. G. F.; DeNiro M. J. *Org. Geochem.* **1988**, *12*, 1-5.

12. Miller, R. F.; Voss-Foucart, M-F.; Toussaint, C.; Jeuniaux, C. *Palaeogeogr. Palaeoclimat. Palaeoecol.* **1993**, *103,* 133-140.

13. Baas, M.; Briggs, D. E. G.; van Heemst, J. D. H.; Kear, A.; de Leeuw, J. W. *Geochim. Cosmochim. Acta* **1995**, *59*, 945-951.

14. Stankiewicz, B. A.; Briggs, D. E. G.; Evershed, R. P. *Energy Fuels* **1997**, *11*, 515-521.

15. Stankiewicz, B. A.; Briggs, D. E. G.; Evershed, R. P.; Duncan, I. J. *Geochim. Cosmochim. Acta* **1997**, *61*, 2247-2252.

16. Stankiewicz, B. A.; Briggs, D. E. G.; Evershed, R. P.; Flannery, M. B.; Wuttke, M. *Science* **1997**, *276*, 1541-1543.

17. Bierstedt, A.; Stankiewicz, B. A.; Briggs, D. E. G.; Evershed, R. P. *Analyst* **1998**, *123,* 139-146.

18. Miller, R. F. *Atlantic Geol.* **1994**, *30,* 65-66.

19. Schimmelmann, A.; Miller, R. F.; Leavitt, S. W. In *Climate Change in Continental Isotopic Records*; Swart, P. K.; Lohmann, K. C.; McKenzie, J.; Savin, S., Ed.; American Geophysical Union: Geophysical Monograph 78, 1993, pp 367-374.

20. Miller, R. F.; Fritz, P.; Morgan, A. V. *Palaeogeogr. Palaeoclimat. Palaeoecol.* **1988**, *66*, 277-288.

21. Hoganson, J. W.; Ashworth, A. C. *Quater. Res.* **1992**, *37*, 101-116.

22. Cong, S.; Ashworth, A. C.; Schwert D. P. *Quater. Res.* **1996**, *45*, 216-225.

23. Miller, R. F. *Atlantic Geol.* **1995**, *31*, 95-101.

24. Miller, R. F. *Can. J. .Earth Sci.* **1996**, *33*, 33-41.

25. Morgan, A. V.; Morgan A. *Geosci. Can.* **1980**, *7*, 22-29.

26. Elias, S. A. *Quaternary Insects and Their Environments.* Smithsonian Institution Press: Washington, D.C., 1994.

27. Stankiewicz, B. A.; van Bergen, P. F.; Duncan, I. J.; Carter, J.; Briggs, D. E. G.; Evershed, R. P. *Rapid Comm. Mass Spectrom.* **1996**, *10*, 1747-1757.

28. Hackman, R. H.; Goldberg M. *Anal. Biochem.* **1981**, *110*, 277-280.

29. Miller, R. F. *Can. J. .Earth Sci.* **1997**, *34*, 247-259.

30. Miller, R. F. *Quater. Proc.* **1998**, *5*, in press.

31. Miller, R. F. (in press). *Atlantic Geol.*

32. Hebda, R.; Hicock, S. R.; Miller, R. F.; Armstrong, J. E. *Geologists Association of Canada Annual Meeting* **1983**, Abstracts 8/31.

33. Stea, R. R.; Mott, R. J. *Friends of the Pleistocene 53rd Annual Reunion* **1990**, p.73.

34.   Stea, R. R.; Mott, R. J. *Boreas* **1989**, *18*, 169-187.
35.   Allison, P. A. *PALAIOS* **1991**, *5*, 432-440.
36.   Campbell, J. M. *Can. Entomolog.* **1983**, *115*, 577-622.
37.   Logan, G. A.; Collins, M. J.; Eglinton, G. In *Taphonomy: Releasing the Data Locked in the Fossil Record*; Allison, P. A.; Briggs, D. E. G., Ed.; Plenum Press: New York, 1991, pp 1-24.
38.   Stankiewicz, B. A.; Mastalerz, M.; Hof, C.; Bierstedt, A.; Flannery, M. B.; Briggs, D. E. G.; Evershed, R. P. *Org. Geochem.* **1998**, *28,* 67-76.

SOURCES OF ORGANIC NITROGEN IN SEDIMENTARY ORGANIC MATTER

Chapter 13

Chitin: 'Forgotten' Source of Nitrogen

From Modern Chitin to Thermally Mature Kerogen: Lessons from Nitrogen Isotope Ratios

A. Schimmelmann[1], R. P. Wintsch[1], M. D. Lewan[2], and M. J. DeNiro[3]

[1]Department of Geological Sciences, Indiana University, Bloomington, IN 47405–1403
[2]U.S. Geological Survey, Box 25046, MS 977, Lakewood, CO 80225
[3]Department of Geological Sciences, University of California, Santa Barbara, CA 93106

Chitinous biomass represents a major pool of organic nitrogen in living biota and is likely to have contributed some of the fossil organic nitrogen in kerogen. We review the nitrogen isotope biogeochemistry of chitin and present preliminary results suggesting interaction between kerogen and ammonium during thermal maturation. Modern arthropod chitin may shift its nitrogen isotope ratio by a few per mil depending on the chemical method of chitin preparation, mostly because N-containing non-amino-sugar components in chemically complex chitin cannot be removed quantitatively. Acid hydrolysis of chemically complex chitin and subsequent ion-chromatographic purification of the "deacetylated chitin-monomer" D-glucosamine (in hydrochloride form) provides a chemically well-defined, pure amino-sugar substrate for reproducible, high-precision determination of $\delta^{15}N$ values in chitin. $\delta^{15}N$ values of chitin exhibited a variability of about one per mil within an individual's exoskeleton. The nitrogen isotope ratio differed between old and new exoskeletons by up to 4 per mil. A strong dietary influence on the $\delta^{15}N$ value of chitin is indicated by the observation of increasing $\delta^{15}N$ values of chitin from marine crustaceans with increasing trophic level. Partial biodegradation of exoskeletons does not significantly influence $\delta^{15}N$ values of remaining, chemically preserved amino sugar in chitin. Diagenesis and increasing thermal maturity of sedimentary organic matter, including chitin-derived nitrogen-rich moieties, result in humic compounds much different from chitin and may significantly change bulk $\delta^{15}N$ values. Hydrous pyrolysis of immature source rocks at 330°C in contact with ^{15}N-

enriched NH_4Cl, under conditions of artificial oil generation, demonstrates the abiogenic incorporation of inorganic nitrogen into carbon-bound nitrogen in kerogen. Not all organic nitrogen in natural, thermally mature kerogen is therefore necessarily derived from original organic matter, but may partly result from reaction with ammonium-containing pore waters.

The widespread and quantitatively enormous occurrence of chitin not only in fauna, but also in fungi and microbiota (1) makes chitin an important source of organic nitrogen in the food chain. This ecological significance, together with chitin's availability from paleontological and archaeological sources make it a valuable substrate for ecological, environmental and physiological studies (see articles in 2, 3, 4, to name a few). This article offers a synopsis of several nitrogen isotope studies on chitin focusing on sources of natural isotopic variability, including: diet, exoskeleton heterogeneity, molting, populations, environments, chitin degradation (5-8). It is documented that an isotopically conservative substrate is maintained as long as the amino sugar building blocks of chitin remain chemically intact and can be used as a pure substrate for isotopic characterization.

Although chitin is efficiently recycled by most heterotrophs, notably excluding humans for our lack of the chitin-depolymerizing enzyme chitinase, a small amount of chitin-derived organic nitrogen is likely to remain preserved in the geological record in the form of sedimentary organic nitrogen. Enzymatic and thermal chemical decomposition of chitin and the subsequent partial incorporation of chitin biomass into macromolecular humic substances (such as humic and fulvic acids, protokerogen, and kerogen) may entail changes in the isotopic composition of the overall organic nitrogen pool. We present experimental evidence from hydrous pyrolysis of kerogen that $\delta^{15}N$ values of organic matter can shift dramatically if sufficiently high temperatures allow chemical dehydration reactions with ammonia, for example in thermally elevated geologic settings in the presence of NH_3-bearing pore waters. In effect, such dehydration reactions are a non-biogenic source of organic nitrogen and may isotopically overprint biogenic N-isotopic signals in some organic geochemicals with a history of higher-temperature diagenesis, catagenesis, and metamorphism.

Choice of a Substrate for Isotopic Measurements in Chitin

Chitin consists mostly of polycondensated N-acetyl-glucosamine units, but is partly deacetylated, contains various amounts of covalently bonded proteinaceous, carbohydrate, and lipidic components, and in the case of crustacean chitin may be closely linked with biologically deposited minerals (1; articles in this volume). The biochemical pathways leading to the biosynthesis of amino sugars for chitin discriminate strongly against ^{15}N, making amino sugars up to 12 per mil more depleted in ^{15}N than proteinaceous muscle biomass (6). Chemical purification procedures for chitin must therefore aim to decrease the non-amino-sugar components and to arrive at a substrate where all nitrogen is in the form of amino sugars. Such a substrate should yield reliable and reproducible $\delta^{15}N$ values representative of only the amino sugar biochemical nitrogen pool. In practice, the complete elimination of N-containing proteinaceous components from macromolecular chitin is an elusive goal (5, and references therein; Figure 1). Additional experiments using various conditions for deproteinization in alkaline media, both employing heterogeneous suspension and homogeneous solution of chitin, resulted in unsatisfactory isotopic shifts of up to 2.9 per mil and poor reproducibility (5). The analytical problem was solved by a sequence of preparative steps resulting in pure, crystalline D-glucosamine hydrochloride from chitin. In brief, mineral components were first removed in dilute hydrochloric acid, then bulk proteinaceous material was hydrolyzed in hot aqueous 1N NaOH, followed by hydrolysis of remaining chitin in hot hydrochloric acid. The resulting D-glucosamine hydrochloride was purified from neutral and amino acid contaminants by ion-exchange chromatography, then freeze-dried and used as a crystalline substrate for highly reproducible determination of $\delta^{15}N$ values (5). All following $\delta^{15}N$ values in this article relating to chitin were determined using D-glucosamine hydrochloride from chitin.

Factors of Natural $\delta^{15}N$-Variability in Modern Chitin

Variability due to Diet. Populations of five different species of arthropods were raised on controlled diets as described elsewhere (9-11) and the $\delta^{15}N$ values of the diets compared against the corresponding isotopic compositions of chitin (Figure 2). Differences in $\delta^{15}N$ values due to metabolic fractionation effects range from -17.5 per mil to -3.3 per mil, averaging -9.5 ± 3.6 per mil, and compare well with similar differences between chitin and muscle biomass in individuals (6). The $\delta^{15}N$ value of chitin from pupal cases of the dypterid fly *Calliphora vicina* is about 6 per mil more negative than that of the exoskeletons of the adult flies that emerged from the cases, suggesting ongoing metabolic fractionation during stages of maturation.

Variability with Changing Trophic Level. A comparison among several species representing different trophic levels at four geographic locations yields a general trend towards more positive $\delta^{15}N$ values in chitin with increasing trophic level (Figure 3). This result is in agreement with a large number of trophic isotope studies utilizing substrates other than chitin, establishing $\delta^{15}N$ values as ecological indicators for trophic structure (12-16, to name a few). The isotopic enrichment of chitin in ^{15}N along the food chain is balanced by the simultaneous loss of ^{15}N-depleted inorganic nitrogen species via excretion, for example ammonia and urea.

Variability within Individual Exoskeletons and through Molting. The variability of isotope ratios within single exoskeletons must be considered when it is not practical or possible to process an entire individual. Chitin from five body sections of a Northern lobster (*Homarus americanus*) yielded a 1.3 per mil range of $\delta^{15}N$ values, with a mean of +0.5 ± 0.5 per mil (6). A crustacean's exoskeleton ages and will eventually be shed in the process of molting (ecdysis). The old carapace will be partially leached before ecdysis, and the new carapace will gradually accumulate chitin for many days after molting (17, 18), giving an opportunity for metabolic isotope fractionation. A comparison of chitin from old versus new carapaces of

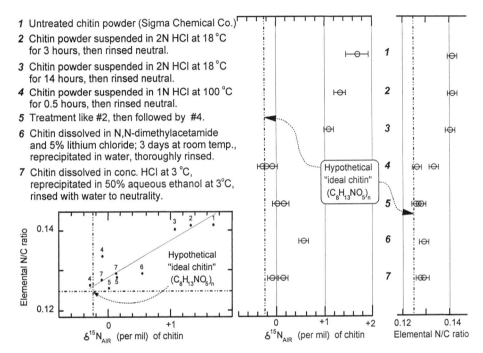

1 Untreated chitin powder (Sigma Chemical Co.)

2 Chitin powder suspended in 2N HCl at 18 °C for 3 hours, then rinsed neutral.

3 Chitin powder suspended in 2N HCl at 18 °C for 14 hours, then rinsed neutral.

4 Chitin powder suspended in 1N HCl at 100 °C for 0.5 hours, then rinsed neutral.

5 Treatment like #2, then followed by #4.

6 Chitin dissolved in N,N-dimethylacetamide and 5% lithium chloride; 3 days at room temp., reprecipitated in water, thoroughly rinsed.

7 Chitin dissolved in conc. HCl at 3 °C, reprecipitated in 50% aqueous ethanol at 3°C, rinsed with water to neutrality.

Figure 1. $\delta^{15}N$ values and elemental N/C ratios of chitin isolates prepared from the same commercial chitin powder (Sigma Chemical Co.). The $\delta^{15}N$ value for "ideal chitin" (hypothetical pure polymer of N-acetylglucosamine) is based on the measured $\delta^{15}N$ value of D-glucosamine hydrochloride prepared from commercial chitin. The cross-plot shows that the chemical removal of nitrogen-rich, ^{15}N-enriched proteinaceous matter from commercial chitin tends to lower the C/N ratio and the $\delta^{15}N$ value in the direction towards the C/N ratio and the $\delta^{15}N$ value of "ideal chitin".

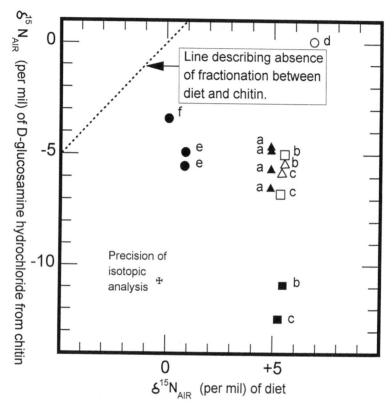

Figure 2. $\delta^{15}N$ value of D-glucosamine hydrochloride from chitin versus $\delta^{15}N$ value of diet. Diets of (a) a mixture of 50% rice bran and 50% corn bran, (b) pork, (c) horsemeat, (d) milkweed seeds, (e) wheat seedlings, and (f) corn seedlings were fed to the following species: *Melanoplus sanguinipes* (grasshopper, ●), *Artemia salina* (brine shrimp, ▲), *Calliphora vicina* (blow fly; pupal cases: ■; organisms: Δ), *Oncopeltus fasciatus* (milkweed bug O, *Musca domestica* (house fly, pupal cases: □). Note that chitin is always isotopically depleted in ^{15}N relative to diet, and that some species fractionate more than others.

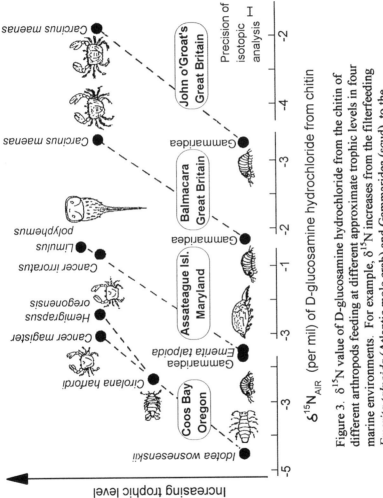

Figure 3. δ^{15}N value of D-glucosamine hydrochloride from the chitin of different arthropods feeding at different approximate trophic levels in four marine environments. For example, δ^{15}N increases from the filterfeeding *Emerita talpoida* (Atlantic mole crab) and Gammaridea (scud), to the omnivore *Cancer irroratus* (Atlantic rock crab), and to the carnivore *Limulus polyphemus* (horseshoe crab).

several individuals from three species of crustaceans showed non-systematic but significant isotopic differences, with a mean of $\Delta\delta^{15}N = -1.0 \pm 1.9$ per mil, and a maximum of 4.8 per mil (6). It is of interest that the largest difference between old and new exoskeletons is 4 per mil and thus similar to the difference between pupal cases and adults of *Calliphora vicina* (Figure 2).

Variability within Crustacean Populations. The chitin from individuals of four populations of different species of crustaceans were studied, along with the influences of age and sex (6). The right claws of ten adult male red swamp crayfish (*Procambarus clarkii*) yielded a mean $\delta^{15}N = +0.6 \pm 0.6$ per mil, with a range of 1.9 per mil. Populations of ghost crabs (*Ocypode ceratophtalmus*), wharf crabs (*Sesarma* sp.), and striped shore crabs (*Pachygrapsus crassipes*), including different ages and sexes, did not show any systematic trends in $\delta^{15}N$ that could be assigned to either age or sex, but encompassed individual ranges between 2 and 3 per mil.

Variability among environments. Environmental and climatic influences on $\delta^{15}N$ in chitin from arthropods were evaluated using a large variety of arthropods sampled in the Hawaiian Islands, the Caribbean, North and South America, and Europe (6). Surprisingly, marine chitins with a mean $\delta^{15}N$ value of -1.1 ± 2.8 per mil (range 13.8 per mil; n = 43) were not clearly distinguishable from terrestrial or freshwater chitins with a mean $\delta^{15}N$ value of -1.7 ± 2.7 per mil (range 11.7 per mil; n = 16), despite the fact that marine invertebrate bulk organic matter is generally enriched in ^{15}N relative to terrestrial and freshwater invertebrate bulk organic matter (19). Subdividing the chitin samples on the basis of various categories of marine and terrestrial environments does not clarify the basis for the lack of separation between marine and terrestrial ^{15}N values (Figure 4). Only severe anthropogenic pollution of a near-coastal site off San Pedro in Southern California caused a significant shift in the chitin of four crustacean species towards more negative $\delta^{15}N$ values (6), in agreement with comparable nitrogen isotopic data on muscle tissue of ridgeback prawn (*Sicyonia* sp.) in the same area (20).

$\delta^{15}N$ values of D-Glucosamine are Conservative through Partial Biodegradation, Roasting, and Fossilization

Following the death of an arthropod its chitinous exoskeleton is typically subject to biodegradation, although articles in this volume report cases of astonishing preservation. Archaeological chitin specimens may have been cooked or roasted. To test for isotopic post-mortem shifts in chitin, aliquots of isotopically characterized modern ground chitin samples from several species were partially biodegraded in marine mud (for ten weeks under anoxic conditions) or in garden soil (8 weeks), whereas other aliquots of chitin were roasted under oxic conditions for three hours at 275°C or heated in an anoxic water vapor atmosphere for three hours at temperatures between 200°C and 325°C (8). Following the recovery of the treated samples, D-glucosamine hydrochloride from the hydrolyzed chitin was used to determine the extent of degradation (via yield in comparison with the original chitin load) and the accompanying nitrogen stable isotope shift.

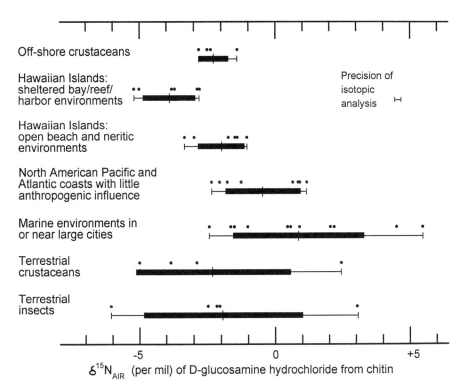

Figure 4. Nitrogen isotopic variability in arthropod chitin from different environments. Each datapoint represents a mean $\delta^{15}N$ value from all chitins sampled in one location. Within each environmental category (group of locations) the $\delta^{15}N$ range, mean, and standard deviation is indicated by lines and bars.

Microbial degradation in mud and soil had accounted for up to 75% loss of chitin biomass during the short time of the experiments, which is in excellent agreement with results from a multitude of chitin biodegradation experiments in marine oxic, estuarine, limnic, and lotic suburban waters, as well as in the rhizosphere (references in *8, 21,* articles in this book). Heating and roasting also reduced the amount of recoverable D-glucosamine hydrochloride. More importantly, in 17 experiments the isotopic shift $\Delta\delta^{15}N$ never exceeded 0.9 per mil, with the exception of one biodegradation experiment in garden soil where a shift of 2.2 per mil was observed.

It was concluded from our data that partial degradation of chitin may produce only small isotopic shifts and does not alter the gross isotopic character of chitin (*8*). The study also demonstrated the use of scanning electron microscopy to ascertain that no fungal neo-formed chitin contaminates partially biodegraded, fossil, or archaeological chitin samples. A variety of chitin samples from Peruvian archaeological excavations were measured isotopically and found to be isotopically similar to modern chitin in comparable environments (*8*). An overview on aspects of paleoecology based on chitin, including applications of stable isotopes, is given by Miller (*22*).

The key for the conservative behavior of $\delta^{15}N$ in chitin is the preservation of intact D-glucosamine units. Isotopic exchange or reaction with chemical species of nitrogen percolating through a burial assemblage, which would alter the original isotopic record, would require thermal or pH conditions that exceed, by far, those usually encountered in shallow depositional environments. Under such extreme conditions there would be no chemical preservation of chitin.

Isotopic Changes in Kerogen during Hydrous Pyrolysis

From Chitin to Mature Kerogen. Biodegradation of arthropod chitin in marine, estuarine, and freshwater environments proceeds quickly, and after a short time the bulk of amino sugar biomass is lost to fungi, bacteria, and scavengers (21, 23-25). Some soil environments seem to have a somewhat better preservation potential, especially for beetle chitin (*22, 26*), which may be due to the relatively high degree of sclerotization of insect chitin (*1*). Very organic-rich environments may help preserve otherwise labile organic components by encapsulation in refractory organic matter (*27*). It was suggested that in some morphologically preserved marine arthropod exoskeletons the original chitinous tissue may have been substituted by more resistant organic matter from other sources (*28*). Early diagenesis of remaining chitinous biomass may cause chemical stabilization via cross-linking within a kerogen matrix and thus offer protection from further enzymatic chitinolytic attack. The high sensitivity of pyrolysis-GCMS has made it possible to unambiguously detect chemically preserved chitinous "molecular fossils" (*29, 30*, articles in this volume) whereas much less sensitive traditional chemical and enzymatic approaches often contradicted each other (*31* and references therein; articles in this volume). At present it is not possible to isotopically characterize nitrogen in humic compounds in a way that would discriminate between potential precursory pools of nitrogen (such as chitin), whereas $\delta^{15}N$ values can be determined for operationally defined bulk substrates such as total organic matter, kerogen, humic acids, and fulvic acids.

Not all Organic Nitrogen in Thermally Mature Organic Matter is Necessarily of Biogenic Origin. Nitrogen cycling between inorganic and organic nitrogen in low-temperature, biologically active environments is known from chemical and isotope studies on plant and soil biomass decomposition (for example, *32-34*, articles in this volume). First experimental evidence for the abiogenic incorporation of ammonia nitrogen into phenolic moieties of humic substances, at temperatures above 100 °C, dates back almost eighty years (*35*). In contrast, the relationship between organic and inorganic nitrogen at higher temperatures and pressures, such as in geological environments where oil and thermogenic gas are generated, has traditionally been considered to be a unidirectional process of continuing loss of organic nitrogen leading to the formation of ammonia (*36*) and elemental nitrogen (*37-40*). The remaining organic nitrogen in kerogen and coal is thus subject to isotopic and C/N mass-balance shifts along thermal maturation (*36, 41, 42*).

Ammonia is dissolved in pore fluids of saturated sediments (*43-45*), in fluid inclusions (*46*) and may be present in clays and feldspars (*47-51*). Sedimentary ammonia is potentially mobile, especially at higher temperatures and with the flow of formation waters and hydrocarbons (*52*). We devised experiments with ^{15}N-enriched ammonia and organic-rich rocks to evaluate chemical interaction between inorganic and organic nitrogen in the process of thermal maturation and oil generation. Shock (*53*) argued on thermodynamic grounds that, at temperatures up to 350°C, ammonia may react with organic matter via dehydration reactions, for example

$$R\text{-}COOH + NH_3 \rightarrow R\text{-}CONH_2 + H_2O,$$

even under "wet" conditions that are typical for basinal sedimentary environments. Amides may participate in subsequent organic reactions, for example resulting in crosslinking of kerogen. Carboxylic acids are generated in significant amounts from kerogens at higher temperatures (*54-56*) and carboxyl groups are present in kerogens.

Hydrous Pyrolysis and Determination of ^{15}N Abundance in Kerogen. The term "hydrous pyrolysis" describes the heating of rocks submerged under water to temperatures below the critical temperature of water (*57*). At temperatures above ca. 270°C many organic-rich source rocks expel oil or wax that rises and floats on the water. We utilized thermally immature, organic-rich rocks and performed anoxic hydrous pyrolysis in stainless steel reactors at 330 ±1 °C for various lengths of time in the presence of either water or aqueous solutions of ammonium chloride with different ^{15}N-abundances. The inner surfaces of the reactors had been passivated via "carburization" (*57*) to avoid catalytic artifacts from metal surfaces (*58-60*). Rock chips were chosen to encompass different types of kerogen, namely Mahogany Shale from the Green River Fm. (type-I kerogen), Clegg Creek Member of the New Albany Shale (type-II kerogen, *61*, p. 16), and Fairfield lignite (WX-3) from the Wilcox Fm., Texas (type-III kerogen). Experiments with ammonium chloride utilized a ratio of

5.5 g of rock chips to 11 mL of aqueous 0.093 molar ammonium chloride solution under nitrogen. Three isotopically different stock solutions of 1g of NH_4Cl in 200ml of distilled water were used, with ^{15}N abundances of approx. 0.3660 atom % (natural abundance, $\delta^{15}N$ = -0.8 per mil), ca.10, and ca. 50 atom %, respectively (ammonium chloride with $\geq$99 atom % ^{15}N was purchased from Isotech Inc.).

After cooling the reactors to room temperature the supernatant oil or wax was removed and the rock chips were washed and dried. Total nitrogen in rock would be an unsuitable substrate for analysis because clays will have adsorbed ammonium and ammonium ions substituting for metal cations in their crystal lattice (49, 50). Even kerogen itself may initially contain adsorbed ammonium and ammonium ionically linked to carboxyl groups, which would not fit the definition of organic nitrogen as being covalently linked to organic carbon. We therefore followed a rigorous series of extractions to prepare kerogen isolates free of NH_3 or NH_4^+. First, we pulverized air-dried rock chips in a ball mill and extracted the powder with chloroform in a Soxhlet apparatus, to remove bitumen and to improve the wetability. After driving off the chloroform, we disintegrated carbonates by stirring the rock powder in excess 2N hydrochloric acid at room temperature overnight, followed by extensive washing with HCl-acidified water. This mild treatment at low pH can be expected to remove most, if not all, of the adsorbed ammonium and NH_4^+ in kerogen and minerals, except intra-crystalline NH_4^+ in siliceous mineral phases. In comparison, even a pH-neutral solution of KCl would be effective in releasing adsorbed ammmonium (62). Although the subsequent extractions represent harsh conditions, we argue that they are prudent and, because of the much reduced concentration of NH_4^+ in solution, are unlikely to have introduced N-isotopic artifacts for organic nitrogen. As a next step, concentrated hydrofluoric acid was used to digest siliceous minerals, but it is also known to produce neo-formed fluoride phases that might incorporate small amounts of NH_4^+. After extensive washing we therefore used hot concentrated hydrochloric acid to dissolve fluoride phases. More washing, a heavy liquid separation (aqueous $ZnBr_2$ solution, $\rho \approx$ 2.4g/ml, to further reduce the presence of mineral phases such as fluorite), washing with HCl-acidified water, freeze-drying, washing with methylene chloride, and final drying resulted in very dark kerogen isolates. Their low ash content was confirmed by elemental analysis (SEM EDS) that listed in type II kerogen the following relative abundances: Si (1.7 %), S (47.8 %), Cl (10.5 %), Ti (4.5 %), Fe (35.5 %), with Al, K and Ca below detection limits. Lignite (type III kerogen) yielded only S (23.7 %) and Cl (68.3 %) as quantifiable signals, with all other elements below detection limits. The virtual absence of K^+, which is chemically similar to ammonium NH_4^+, indirectly confirms that our rigorous acid extraction of kerogens was effective in eliminating cations like NH_4^+.

The off-line combustion of aliquots of kerogen in quartz ampules, the purification and quantification of resulting elemental nitrogen and carbon dioxide on a vacuum line fitted with a Toepler pump and manometers, and the mass-spectrometric isotopic measurements followed the same routines as described elsewhere for chitin and D-glucosamine hydrochloride (6). The unexpectedly large ^{15}N-abundances in kerogens from ^{15}N-enrichment experiments are reported in atom % ^{15}N (Figure 5), whereas some experiments at natural ^{15}N-abundance level are reported as $\delta^{15}N$ values.

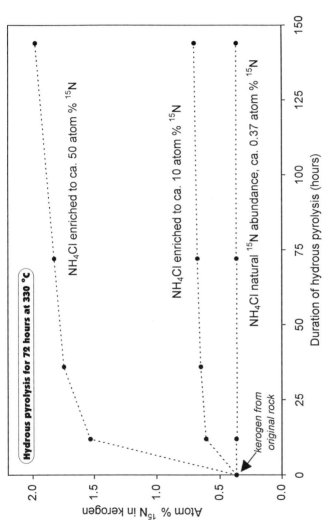

Figure 5. Abiogenic formation of organic nitrogen is suggested by $\delta^{15}N$ values of type II kerogen from thermally immature source rock of the Clegg Creek Member of the New Albany Shale (61, p. 16) following hydrous pyrolysis. Aliquots of kerogen were heated to 330°C for various lengths of time in closed stainless steel reactors, together with 0.093 molar solutions of ammonium chloride with three different levels of ^{15}N-abundance.

Isotopic Evidence for Abiogenic Formation of Organic Nitrogen. The original kerogen type II from immature source rock, without hydrous pyrolysis, has a $\delta^{15}N_{kerogen}$ value of +0.1 per mil. Using water without any addition of ammonium chloride in hydrous pyrolysis experiments caused no significant shift in $\delta^{15}N_{kerogen}$ value. However, using a 0.093 molar solution of ammonium chloride with $\delta^{15}N =$ -0.8 per mil in hydrous pyrolyses lasting 12, 36, 72, and 144 hours resulted in $\delta^{15}N_{kerogen}$ values clustering around + 4.2 ± 0.3 per mil. This demonstrates that the mere presence of regular ammonium chloride during hydrous pyrolysis can significantly shift the $\delta^{15}N_{kerogen}$ value by several per mil.

Unexpectedly dramatic isotopic shifts in kerogen occurred when ammonium chloride enriched to either ca. 10 or ca. 50 atom % ^{15}N was used. The ^{15}N-abundance in type II kerogen increases with time and with increasing concentration of ^{15}N in the ammonium chloride used (Figure 5). It further appears from the slope of connecting lines between data points that even after 144 hours the chemical reactions were still proceeding. The ^{15}N-abundances in different types of kerogen, after 72 hours hydrous pyrolysis in 0.093 molar ammonium chloride solution with ca. 50 atom % ^{15}N, rise from type II (1.8 atom % ^{15}N), to type I (2.3 atom % ^{15}N), to type III (9.7 atom % ^{15}N). Considering that all original kerogens started with natural ^{15}N-abundances close to 0.37 atom % ^{15}N, these increases can be used in mass-balance calculations to estimate how many percent extraneous nitrogen would need to be added to the pre-existing organic nitrogen pool to account for the ^{15}N-enrichment. In the same II, I, III-sequence as above, these estimates amount to 3.0, 4.0, and 23.1 percent, respectively.

Independent evidence for the amounts of nitrogen added to kerogen is available from the elemental composition of kerogen. The elemental C/N ratio of kerogen should decrease if inorganic nitrogen from ammonia is added to the organic nitrogen pool via organic dehydration reactions. However, the elemental C/N ratio of kerogen also decreases through artificial generation of bitumen and oil, because the elimination of hydrocarbons with high C/N ratios leaves carbon-depleted kerogen behind. The latter effect can only be discounted in a close comparison between C/N ratios of pairs of kerogens that underwent identical hydrous pyrolysis conditions, except for the presence of ammonia (Table I). The available pairs of C/N data support the hypothesis of organic dehydration reactions adding to the organic nitrogen pool, again in the sequence II, I and III from least to most addition of nitrogen. Mass balance calculations based on C/N ratios suggest additions of nitrogen to the pre-existing organic nitrogen amounting to 3.3 % (type II, 36h), 4.2 % (II, 144h), 13.3 % (I, 72h), and 22.3 % (III, 72h). The similarity to the above estimates based on isotopic mass-balance calculations, at least for types II and I, is striking.

Our preliminary results do not rule out the possibility of isotopic exchange reactions between organic nitrogen and ammonium nitrogen under conditions of hydrous pyrolysis, which would result in isotopic shifts but entail no changes in the C/N ratio. This will in the future be tested by subjecting some of our ^{15}N-enriched kerogens for a second time to hydrous pyrolysis conditions in the presence of ammonium with natural ^{15}N-abundance.

Shock's (53) hypothesis of organic dehydration reactions under "wet" conditions leading to formation of organic nitrogen is in agreement with our preliminary results. We are not aware of any proposed nitrogen isotopic exchange mechanism that would explain the observed isotopic shifts, and we are confident that ^{15}N-enriched ammonia ions had been quantitatively eliminated from the kerogens by rigorous acid extraction and washing. The results of our in-progress work raise intriguing questions about the possibility that bulk organic $\delta^{15}N$ values in some thermally mature organic matter may have been altered through maturation, at least in environments where ammonia is documented (47, 49, 52, 63-65). The likelihood of altering paleoenvironmental nitrogen isotopic signatures would be diminished if ammonia permeability would be minimized, for example if the movement of ammonia-containing pore fluids would be blocked. The impregnation of pore spaces in source rocks by bitumen disrupting the free capillary flow of hydrous phases could be a factor in the preservation of nitrogen isotope ratios. Future work also needs to consider possible ^{15}N-incorporation from ammonia into kerogen at temperatures lower than typical hydrous pyrolysis temperatures, before the generation of considerable amounts of bitumen and the expulsion of oil.

Table I. C/N Elemental Ratios in Kerogens of Type I, II and III before and after Hydrous Pyrolysis of Rocks at 330°C, in Absence and Presence of Ammonium Chloride in the Hydrous Phase

	Duration of Hydrous Pyrolysis (Hours)	C/N of Kerogen from Original Rock	C/N of Kerogen after Hydrous Pyrolysis without NH_4Cl	C/N of Kerogen after Hydrous Pyrolysis with NH_4Cl
Type I:	0	24.2 ± 0.5 (n=3)	n.a.	n.a.
	72	n.a.	21.3 ± 0.4 (n=3)	18.8 ± 0.3 (n=4)
Type II:	0	37.6 ± 1.4 (n=3)	n.a.	n.a.
	36	n.a.	34.0 ± 0.7 (n=2)	32.9 ± 0.8 (n=6)
	144	n.a.	32.3 ± 1.9 (n=3)	31.0 ± 1.0 (n=6)
Type III:	0	60.8 ± 2.1 (n=2)	n.a.	n.a.
	72	n.a.	58.2 (n=1)	47.6 ± 2.0 (n=3)

n.a. = not applicable

Conclusion

Neither biological nor thermal partial degradation of chitin produced significant changes of $\delta^{15}N$ values as measured in extracted D-glucosamine hydrochloride. Thus, archaeological and fossil chitins (for which electron microscopy indicates the absence of fungal chitin) are useful for paleoenvironmental and paleoclimatic studies, and isotopic ratios of ancient chitins can be interpreted in light of the corresponding values for contemporary chitins from well-characterized environments. Caution is advised when interpreting $\delta^{15}N$ values of bulk organic material that has undergone thermal

240

stress in the presence of ammonia, because isotopically different inorganic nitrogen may have been chemically added to, or exchanged with, original biogenic nitrogen via dehydration reactions, in spite of "wet" conditions.

Acknowledgments

We thank James Lyons (UCSD) for SEM EDS measurements and the editors and an anonymous reviewer for their efforts to improve the clarity of our manuscript. Previous work on chitin was supported by the Lerner-Gray Fund for Marine Research of the American Museum of Natural History, by the Geological Society of America grant 3216-83, by National Science Foundation grant ATM 79-24591, by the Petroleum Research Fund administered by the American Chemical Society, grant 15582-AC2, and by NATO Collaborative Research Programme grant 477/84. Recent work on kerogen has been supported by the Petroleum Research Fund administered by the American Chemical Society, grant ACS-PRF# 31203-AC, and by NATO Collaborative Research Programme grant 971381.

Literature Cited

1. Muzzarelli, R. A. A. *Chitin*; Pergamon Press: Oxford, 1977.
2. *Proceedings of the First International Conference on Chitin/Chitosan*; Muzzarelli, R. A. A.; Pariser, E. R., Eds.; MIT Sea Grant Program; Massachusetts Institute of Technology: Cambridge, Massachusetts, 1978.
3. *Chitin and Chitosan. Proceedings of the Second International Conference on Chitin and Chitosan*; Hirano, S.; Tokura, S., Eds.; Japanese Society of Chitin and Chitosan, Tottori University: Tottori, Japan, 1982.
4. *Chitin in Nature and Technology*; Muzzarelli, R. A. A.; Jeuniaux, C.; Gooday, G. W., Eds.; Plenum Press: New York, 1985.
5. Schimmelmann, A.; DeNiro, M. J. In *Chitin in Nature and Technology*; Muzzarelli, R. A. A.; Jeuniaux, C.; Gooday, G. W., Eds.; Plenum Publishing Corporation: New York, NY, 1986; pp 357-364.
6. Schimmelmann, A.; DeNiro, M. J. *Contrib. Mar. Sci.* **1986**, *29*, pp 113-130.
7. Schimmelmann, A.; DeNiro, M. J. *Geochim. Cosmochim. Acta* **1986**, *50*, pp 1485-1496.
8. Schimmelmann, A.; DeNiro, M. J.; Poulicek, M.; Voss-Foucart, M.-F.; Goffinet, G.; Jeuniaux, C. *J. Archaeol. Sci.* **1986**, *13*, pp 553-566.
9. DeNiro, M. J.; Epstein, S. *Geochim. Cosmochim. Acta* **1978**, *42*, 495-506.
10. DeNiro, M. J.; Epstein, S. *Geochim. Cosmochim. Acta* **1981**, *45*, 341-351.
11. Schimmelmann, A. *Stable isotopic studies on chitin;* Ph.D. Dissertation; University of California at Los Angeles: Los Angeles, CA, 1985.
12. Schoeninger, M.; DeNiro, M. J. *Geochim. Cosmochim. Acta* **1984**, *48*, 625-639.
13. Minagawa, M.; Wada, E. *Geochim. Cosmochim. Acta* **1984**, *48*, 1135-1140.
14. Hobson, K. A.; Welch, H. E. *Mar. Ecol. Prog. Ser.* **1992**, *84*, 9-18.
15. Rau, G. H.; Ainley, D. G.; Bengtson, J. L.; Torres, J. J.; Hopkins, T. L. *Mar Ecol. Prog. Ser.* **1992**, *84*, 1-8.

16. Keough, J. R.; Sierszen, M. E.; Hagley, C. A. *Limnol. Oceanogr.* **1996**, *41*, 136-146.
17. Vigh, D. A.; Dendinger, J. E. *Comp. Biochem. Physiol.* **1982**, *72A*, 365-369.
18. Kulkarni, K. M. *Comp. Physiol. Ecol.* **1983**, *8*, 202-204.
19. France, R. L. *Mar. Ecol. Prog. Ser.* **1994**, *115*, 205-207.
20. Rau, G. H.; Sweeney, R. E.; Kaplan, I. R.; Mearns, A. J.; Young, D. R. *Estuarine Coastal Shelf Sci.* **1981**, *13*, 701-707.
21. Poulicek, M.; Jeuniaux, C. *Biochem. Syst. Ecol.* **1991**, *19*, 385-394.
22. Miller, R. F. *Biochem. Syst. Ecol.* **1991**, *19*, 401-411.
23. Hillman, K.; Gooday, G. W.; Prosser, J. I. *Estuarine Coastal Shelf Sci.* **1989**, *29*, 601-612.
24. Simon, A.; Poulicek, M.; Velimirov, B.; Mackenzie, F. T. *Biogeochem.* **1994**, *25*, 167-195.
25. Boetius, A.; Lochte, K. *Mar. Ecol. Prog. Ser.* **1994**, *104*, 299-307.
26. Miller, R. F.; Voss-Foucart, M.-F.; Toussaint, C.; Jeuniaux, C. *Palaeogeogr. Palaeoclimat. Palaeoecol.* **1993**, *103*, 133-140.
27. Knicker, H.; Hatcher, P. G. *Naturwissenschaften* **1997**, *84*, 231-234.
28. Baas, M.; Briggs, D. E. G.; van Heemst, J. D. H.; Kear, A. J.; de Leeuw, J. W. *Geochim. Cosmochim. Acta* **1995**, *59*, 945-951.
29. Stankiewicz, B. A.; van Bergen, P. F.; Duncan, I. J.; Carter, J. F.; Briggs, D. E. G.; Evershed, R. P.; *Rapid Comm. Mass Spectrom.* **1996**, *10*, 1747-1757.
30. Stankiewicz, B. A.; Briggs, D. E. G.; Evershed, R. P.; Duncan, I. J. *Geochim. Cosmochim. Acta* **1997**, *61*, 2247-2252.
31. Schimmelmann, A.; Krause, R. F.; DeNiro, M. J. *Org. Geochem.* **1988**, *12*, pp 1-5.
32. Flaig, W. *Chem. Geol.* **1968**, *3*, 161-187.
33. Downs, M. R.; Nadelhoffer, K. J.; Melillo, J. M.; Aber, J. D. *Oecologia* **1996**, *105*, 141-150.
34. Carnol, M.; Ineson, P.; Anderson, J. M.; Beese, F.; Berg, M. P.; Bolger, T.; Coûteaux, M.-M.; Cudlin, P.; Dolan, S.; Raubuch, M.; Verhoef, H. A. *Biogeochem.* **1997**, *38*, 255-280.
35. Eller, W. *Brennstoff-Chemie* **1921**, *2*, 129-133.
36. Rohrback, B. G.; Peters, K. E.; Sweeney, R. E. ; Kaplan. I. R. In *Advances in Organic Geochemistry 1981*; Bjorøy, M. et al., Eds.; Wiley: Chichester, 1983; pp 819-823.
37. Boudou, J. P.; Espitalié, J. *Chem. Geol.* **1995**, *126*, 319-333.
38. Krooss, B. M.; Littke, R.; Müller, B.; Frielingsdorf, J.; Schwochau, K.; Idiz, E. F. *Chem. Geol.* **1995**, *126*, 291-318.
39. Littke, R.; Krooss, B; Idiz, E.; Frielingsdorf, J. *Am. Assoc. Petrol. Geol. Bull.* **1995**, *79*, 410-430.
40. Zhiheng Wu; Yasuo Ohtsuka. *Energy Fuels* **1996**, *10*, 1280-1281.
41. Drechsler, M.; Stiehl, G. *Chemie Erde* **1977**, *36*, 126-138.
42. Stiehl, G.; Lehmann, M. *Geochim. Cosmochim. Acta* **1980**, *44*, 1737-1746.
43. Berelson, W. M.; Hammond, D. E.; O'Neill, D.; Xu, X-M.; Chin, C.; Zukin, J. *Geochim. Cosmochim. Acta* **1990**, *54*, 3001-3012.

44. Canfield, D. E.; Jørgensen, B. B.; Fossing, H.; Glud, R.; Gundersen, J.; Ramsing, N. B.; Thamdrup, B.; Hansen, J. W.; Nielsen, L. P.; Hall, P. O. J. *Mar. Geol.* **1993**, *113*, 27-40.

45. Zakutin, V. P.; Chugunova, N. N. *Trans. (Dokl.) Russ. Acad. Sci. Earth Sci. Sect.* **1993**, *328*, 200-205.

46. Pironon, J.; Pagel, M.; Walgenwitz, F.; Barrès, O. *Org. Geochem.* **1995**, *8*, 739-750.

47. Compton, J. S.; Williams, L. B.; Ferrell, R. E, Jr. *Geochim. Cosmochim. Acta* **1992**, *56*, 1979-1991.

48. Ramseyer, K.; Diamond, L. W.; Boles, J. R. *J. Sed. Petrol.* **1993**, *63*, 1092-1099.

49. Daniels, E. J.; Aronson, J. L.; Altaner, S. P.; Clauer, N. *Geol. Soc. Am. Bull.* **1994**, *106*, 760-766.

50. Ward, C. R.; Christie, P. J. *Int. J. Coal Geol.* **1994**, *25*, 287-309.

51. Williams, L. B.; Ferrell, R. E. Jr.; Hutcheon, I.; Bakel, A. J.; Walsh, M. M.; Krouse, H. R. *Geochim. Cosmochim. Acta* **1995**, *59*, 765-779.

52. Williams, L. B.; Wilcoxon, B. R.; Ferrell, R. E.; Sassen, R. *Appl. Geochem.* **1992**, *7*, 123-134.

53. Shock, E. L. *Geochim. Cosmochim. Acta* **1993**, *57*, 3341-3349.

54. Kawamura, K.; Kaplan, I. R. *Geochim. Cosmochim. Acta* **1987**, *51*, 3201-3207.

55. Lewan, M. D.; Fisher, J. B. In *Organic Acids in Geological Processes*; Pittman, E. D.; Lewan, M. D., Eds.; Springer: Heidelberg, 1994; pp 70-114.

56. Barth, T.; Andresen, B.; Iden, K.; Johansen, H. *Org. Geochem.* **1996**, *25*, 427-438.

57. Lewan, M. D. In *Organic Geochemistry*; Engel, M. H.; Macko, S. A., Eds; Plenum Press: New York, NY, 1993; pp 419-442.

58. Dabbagh, H. A.; Shi, B.; Davis, B. H.; Hughes, C. G. *Energy Fuels* **1994**, *8*, 219-226.

59. Michels, R.; Langais, P.; Philp, R. P.; Torkelson, B. E. *Energy Fuels* **1994**, *8*, 741-754.

60. Michels, R.; Langais, P.; Philp, R. P.; Torkelson, B. E. *Energy Fuels* **1995**, *9*, 204-215.

61. Lewan, M. D.; Comer, J. B.; Hamilton-Smith, T.; Hasenmueller, N. R.; Guthrie, J. M.; Hatch, J. R.; Gautier, D. L.; Frankie, W. T. *U. S. Geol. Survey Bull.* **1995**, *2137*, 1-31.

62. Laima, M. C. J. *Biogeochem.* **1994**, *27*, 83-95.

63. Daniels, E. J.; Altaner, S. P. *Am. Min.* **1990**, *75*, 825-839.

64. Lindgreen, H. *Clay Minerals* **1994**, *29*, 527-537.

65. Whelan, J. K.; Seewald, J.; Eglinton, L.; Miknis, F. P. *Proc. ODP, Sci. Results* **1994**, *139*, 485-494.

Chapter 14

Timing and Mechanisms of Changes in Nitrogen Functionality During Biomass Fossilization

S. Derenne[1], H. Knicker[2], C. Largeau[1], and P. Hatcher[3]

[1]Chimie Biorganique et Organique Physique, UMR CNRS 7573, ENSCP, 11 rue Pierre et Marie Curie, 75231 Paris Cedex 05, France
[2]Institut für Physikalische Biochemie und Biophysik, Universität Regensburg, D-93040, Regensburg, Germany
[3]Center for Environmental Chemistry and Geochemistry, The Pennsylvania State University, University Park, PA 16802

Due to the number of problems associated with nitrogen presence in fossil organic matter, the origin of nitrogen-containing compounds and their fate upon fossilization is especially important to understand. Nitrogen in living organisms chiefly occurs as amide groups whereas in fossil fuels, it mostly corresponds to heterocyclic structures. Based on results reported in previous studies and those we recently obtained using solid state ^{15}N NMR, it appears that (i) some amides exhibit a higher resistance than expected and (ii) the above change in nitrogen functionality is completed in immature kerogens and might be an early diagenetic process. The types of pathways which can be invoked to explain amide resistance are discussed along with those which might be implicated in N functionality change.

Nitrogen in fossil organic matter presents a number of major problems associated with the utilization of this fossil organic matter. For example, N-containing compounds can poison petroleum-reforming catalysts and pyrroles are known to make liquid fuel unstable upon storage due to tar formation (*1-3*). Health and environmental hazards are also related to the presence of such compounds in coals and oils via the formation of toxic combustion products like nitrogen oxides (*4-6*). In addition, N-containing constituents of oils are known to affect wettability in reservoirs (*7*). Finally, a number of accumulations of natural gases throughout the world show high contents (up to almost 100 % in Germany) of molecular nitrogen (*8-12*). The occurrence of these N_2-rich natural gases represents a serious exploration risk. Their origin is not fully elucidated although it is often accepted that they result from the thermal degradation of N-containing constituents of sedimentary organic matter.

Consequently, it is especially important to understand the origin and the fate of N-containing compounds in the sedimentary record. It is well known that N in living organisms mainly occurs as amide groups in proteins and peptides. In marine organisms, proteins account for more than 50% of the total biomass and contain 85% of the total organic nitrogen (*13, 14*). Amide groups are usually considered to be highly sensitive to degradation (*15*). Nonetheless, a nitrogen content of 0.5 to 4% is still observed in the organic matter of Recent sediments and low maturity ancient sediments (*e.g. 16-19*). In fossil fuels, N mostly exists associated with heterocyclic structures. Studies of nitrogen compounds in petroleum has led to the identification of a number of products whose structure is based on pyrrole, pyridine, quinoline or carbazole rings (*e.g. 5, 20-23*). Furthermore, the main nitrogen-containing moieties in coals have been shown to be structures of pyrrole and pyridine type (*24-27*).

Tetrapyrroles were among the first products identified in the sedimentary record and they have been extensively studied since then (*28-35*). Such compounds were suggested to be directly related to chlorophylls and haemin. This so-called Treibs hypothesis was recently supported by the identification of intermediates between chlorophyll and sedimentary alkylporphyrins thus leading to propose a scheme of successive transformations (*36*). Porphyrin moieties were also shown to be present in macromolecular fractions. Indeed, dealkylated porphyrins were released upon hydrous pyrolysis of the Messel oil shale (*37*) and incorporation of porphyrins into macromolecular fractions via S bridges was shown to occur (*38-41*). It was thus suggested that tetrapyrrole pigments can significantly contribute to sedimentary nitrogen (*38*). Nevertheless, it was also mentioned that heterocyclic compounds only account for a minor proportion of total nitrogen in living organisms hence a low contribution to organic matter (*42*). Similarly, nucleic acid bases such as purines and pyrimidines were found in some marine sediments but they only represent a very small fraction of the nitrogen (*43, 44*).

Conspicuous differences in the main nitrogen functionality are thus observed between the deposited source organisms (amide groups) and derived fossil fuels (nitrogen-containing heterocycles). Such a change can be explained in two ways: either a transformation of amides into heterocycles or a selective enrichment of the heterocycles along with amide degradation. However, whatever the involved mechanism, the timing of this change is an important question. Indeed, numerous studies have been performed in order to examine whether such changes correspond to early diagenetic processes, or whether they take place during later stages of diagenesis or are associated with the thermal transformations occurring at depth during catagenesis.

The present paper deals with the results previously reported on this topic along with more recent studies, based on solid state ^{15}N NMR. These studies are concerned with resistant biomacromolecules isolated from two types of freshwater microalgae and their fossil counterparts. We also discuss the mechanisms which might be implicated in these N functionality changes associated with diagenesis.

C/N ratio

C/N molar ratios, derived from elemental analysis, have been extensively used to distinguish between terrestrial and algal sources for organic matter in sediments (*45*). However, in most cases, their downcore variations shall mostly reflect diagenetic activity (*46*). Indeed, an increase in C/N ratio is generally observed when comparing either samples of increasing depths within a given water column (*47-49*) or surficial sediments under water columns of increasing depths (*50, 51*). Within the first few centimeters of sediment cores, an increase in this ratio is also often noted. Such an increase is usually associated with a preferential loss of amino acids due to the bacterial degradation of proteins (*17, 52-56*). In addition, planktonic constituents, which are characterized by relatively low (< 10) C/N ratios, are generally considered to be more sensitive to degradation than are higher plant constituents (C/N ratios > 20). Therefore, the faster degradation of planktonic material results in an increase in the C/N ratio (*16*). In kerogens, C/N values are typically in the 20-200 range whatever their source organisms. However, in some cases, only little changes in C/N ratios or even decreases are observed with depth in Recent sediments (*16, 18, 53, 57-59*). When such a decrease was observed, it was suggested that the organic nitrogen is protected by adsorption onto minerals (*60*). Another process which can be proposed for explaining this decrease in C/N ratios with depth is nitrogen incorporation into geopolymers and subsequent protection from degradation during diagenesis. This is supported by recent results reported by Quayle et al. (*61*) from Lake Pollen where an increase in the residual nitrogen (as a percentage of total nitrogen), corresponding to the insoluble organic matter, is observed downcore. Such differences in C/N variations with depth suggest that several factors control the preservation of sedimentary nitrogen. Whatever the observed trend in C/N variations with depth, parallel changes are often found between this ratio and TOC (*59, 62*).

Amino acids

Amino acids represent the largest pool of organic nitrogen identified to date in Recent sediments (*60*). Indeed, a number of studies reveal the occurrence in sediments of amino acids in many forms ranging from free monomers to polymers. In addition to free amino acids, hydrolysed amino acids are usually quantified so as to assess protein concentrations. Direct measurement of protein amount in sediments has been shown to afford variable results due to differences in the analytical techniques used for such measurements (*63*). However, it was recently stated that, depending on the procedure employed, adsorption or condensation reactions may also occur during release and isolation of hydrolysed amino acids. Hence, due to such potential losses and underestimation of protein contents, a new procedure was developed so as to increase the amino acid recovery (*64*). The total amount of free and hydrolyzed amino acids usually only comprises a relatively minor fraction of total organic nitrogen in Recent sediments, although higher levels are noted in samples from upwelling regions (up to 80 % of total nitrogen) (*18, 60, 65, 66, 67*) and the bulk of sedimentary organic nitrogen remains unidentified in most cases.

As stressed in the C/N section, total extractable amino acid concentration rapidly declines with increasing depth in the water column and in sediments. Despite this lowering, amino acids have been detected in sediment traps and surface sediments even at abyssal depths (*65, 68-71*). This resistance is not only observed with amino acids but also with whole proteins. Proteins have been detected in dissolved (DOM) and particulate (POM) organic matter from Atlantic and Pacific Oceans (*72-76*). In these cases, a limited number of proteins account for most of the DOM and POM proteins and different types of proteins dominate when comparing near surface samples with samples from bottom waters. These specific proteins are supposed to accumulate because of their structure which would inhibit degradation (*76*). In addition, it was also suggested that some proteinaceous amino acids undergo abiotic transformations such as irreversible adsorption or geopolymerization in the upper layer of the sediments, producing more refractory material, still detectable by hydrolysis (*18, 67, 77*).

Since an overall decrease in the total amount of amino acids with depth was generally observed, the variations in amino acid distribution were also investigated so as to reveal potential selective degradation. However, only minor changes are generally noted within amino acid distributions (*63, 66, 78, 79*). Moreover, similar rates of degradation were recently observed during laboratory decay of phytoplankton for two non-protein amino acids and for the total hydrolysable amino acids (*63*).

As discussed below, several pathways can be invoked to explain the preservation of some proteinaceous amides. Whatever the protection process, it appears that (i) some amino acids can be preserved in Recent sediments, (ii) their precise amount is difficult to determine and (iii) the residual (non hydrolyzable) nitrogen is still to be characterized and often predominates.

Nitrogen in macromolecular structures

Several attempts to elucidate the nature of macromolecular nitrogen in sediments were previously carried out using a variety of methods. Upon pyrolysis, upper sediments from Peru upwelling area released typical products related to proteins in agreement with results from acid hydrolysis (*17, 80*). However, an X-ray photoelectron spectroscopy (XPS) study of these sediments revealed that they contained four different types of organic nitrogen functional groups, namely pyridine, pyrrole, amide and (tentatively) quaternary nitrogen. The heterocyclic structures (pyrrole and pyridine) were always predominant with pyrroles more abundant than pyridines. The relative amide content was shown to decline with increasing sample depth. For the shallowest sample in the core, an early Holocene one (25 to 35 cm depth), the amides represent, at most, 40% of the total nitrogen, suggesting that changes in N functionality already began at or above the sediment/water interface.

Recent results obtained by solid state ^{15}N NMR on a Holocene sediment (9.7 m depth, *ca.* 4000 years old) from Mangrove Lake, Bermuda (*19, 81*), are in contrast with these findings from sediments of the Peru upwelling region. The ^{15}N NMR spectrum of the Mangrove Lake sediment is dominated by a peak corresponding to amides and also shows a minor peak for amines. No significant signal could be detected in the region of pyrroles and pyridines, indicating that the relative abundance of these groups is very low. On the basis of these data, it was suggested that the main nitrogen constituents of this sample were proteinaceous although a previous study, based on amino acid hydrolysis, indicated that only *ca.* 10% of the organic nitrogen should correspond to proteins (*82*). Nevertheless, after the above hydrolysis, the ^{15}N NMR spectrum of the sediment appears very similar to the starting material and tetramethylammonium hydroxide (TMAH) thermochemolysis products indicated a protein origin for the nitrogen in this sediment (*81*). This study showed that some amide groups, likely derived from proteins, can withstand both the first steps of fossilization and acid hydrolysis, in contradiction with the general belief that proteins are rapidly degraded during early diagenesis.

When comparing the ^{15}N NMR spectrum of an algal compost (algal mixture incubated for two months with a natural compost) with that of the starting algal biomass, no substantial modification could be detected (*19*): both spectra are sharply dominated by the amide peak which represents more than 80% of the total nitrogen. The only other nitrogen function which could be detected corresponded to amines. These results indicated that amides are able to survive microbial digestion, consistent with the data from Mangrove Lake showing the preservation of some proteins and amide-containing, protein-derived, products.

The resistance of amide functions towards acid hydrolysis was also observed in the case of insoluble, non-hydrolysable biomacrolecules termed algaenans (*19, 83*). Algaenans can be isolated from the algal biomass via successive extractions, drastic base and acid hydrolyses. They were shown to occur in the outer walls of numerous microalgal species including *Scenedesmus communis* and *Botryococcus braunii* (reviewed in *84*) and their geochemical importance was established (*85, 86*). In previous studies, the algaenans isolated from ^{15}N-enriched *S. quadricauda* (recently renamed *S. communis*) biomass and from a mixed algal biomass, comprising *Scenedesmus* species, were examined by ^{15}N NMR and showed a predominance of amide functions (*19, 83*). Due to the low sensitivity of ^{15}N NMR, most of the previous studies using this technique were carried out on enriched materials. However, it was recently shown with soil samples that some spectra can be obtained at natural abundance levels (*87*).

As a result, we recorded the ^{15}N NMR spectrum of the algaenan isolated from unenriched *S. communis*. This spectrum was obtained using high power dipolar decoupling, cross polarization (contact time 1 ms, pulse delay 0.15 s) and magic angle spinning at 4 kHz and 5.5 kHz with a Bruker MSL 300 and a Bruker DMX 4000 spectrometer operating at 30.137 MHz and 40.55 MHz for [15]N. It is the result of 365 000 scans. The spectrum is dominated by an intense peak at -260 ppm (reference nitromethane) assigned to amides and shows a minor peak at -340 ppm due to amines (Figure 1). Considering the isolation procedure used for the algaenans, this demonstrates that some amides can withstand drastic hydrolyses. *S. communis* is characterized by a thin (10-20 nm), algaenan-containing outer wall. Such walls were shown to be selectively preserved upon fossilization and thus to be the precursors of very thin lamellar structures observed by TEM in a number of oil shales and source rocks of various ages and origins (*86, 88, 89*). Such structures, termed ultralaminae, could not be detected by light microscopy due to their thinness and the corresponding kerogens were thus considered to be amorphous when observed only by light microscopy.

Following the above observations on the unenriched algaenan of *S. communis*, nitrogen functionality was therefore examined in two immature, ultralaminae-rich kerogens, the Göynük oil shale, Oligocene, Turkey and the Rundle oil shale, Eocene, Australia. These samples were previously analysed in detail by pyrolysis and GC/MS identification of the released products (*86, 89*). Their ^{15}N NMR spectra are similar and show a large peak ranging from -220 to -260 ppm, corresponding to pyrroles (Figure 2, same experimental conditions as Figure 1 but with a number of scans of ca. 10^6). A small signal may be detectable at -95 ppm, which may result from pyridinic-N. However, for interpretation of this signal, one has to consider the low signal-to-noise ratio of this spectrum. Accordingly, the comparison between the algaenan from *S. communis* and the immature ultralaminae-rich kerogens shows that upon diagenesis, the amides which were present in the algaenan are either degraded or underwent chemical transformations yielding heterocyclic pyrrole-like structures.

So as to test the general character of the above change in nitrogen functionality upon fossilization, a second set of samples, derived from the microalga *B. braunii* was examined by ^{15}N NMR. *B. braunii* is a freshwater colonial microalga with thick (1-2 μm) algaenan-composed outer walls. Although nitrogen is even less abundant in this algaenan when compared to *S. communis*, its ^{15}N NMR spectrum could be obtained at natural abundance levels. As in the case of *S. communis*, it is strongly dominated by a peak at -259 ppm, corresponding to amides and it possibly exhibits a weak signal at -330 ppm due to amines, but the signal to noise is too low to conclude.

The selective preservation of *Botryococcus* algaenan is known to be responsible for organic matter accumulation in Torbanites (*85*). These extremely organic-rich sedimentary rocks are chiefly composed of fossil remains of *Botryococcus*. An immature Permian Torbanite from South Africa was analysed along with a Pliocene alginite from Pula, Hungary. A Recent material, Balkashite, was also examined in this study. This material is formed, after *B. braunii* blooms, on the shores of Lake Balkash, Kazakhstan. The floating *B. braunii* biomass is pushed ashore by wind and the resulting accumulations, upon drying and air and light exposure, are transformed into a rubbery material. Balkashite was previously studied using a wide range of techniques such as FTIR, solid state ^{13}C NMR and GC/MS identification of its pyrolysis products (*90*). The selective preservation of *B. braunii* algaenan was thus shown to have played an important role in Balkashite formation. However, the occurrence of some specific pyrolysis products indicated that oxygen cross-linking of the alkadienes and alkatrienes, abundantly produced by the A race of *B. braunii* growing in this lake, also took place during the very early stages of diagenesis under oxic conditions. The ^{15}N NMR spectrum of Balkashite reveals that the only nitrogen function corresponds to amide groups. Taking into account the highly oxic conditions under which Balkashite is formed, the survival of amides in this sample demonstrates

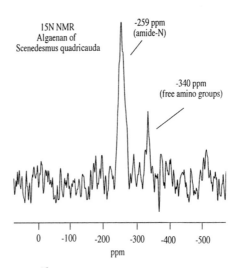

Figure 1. Solid-state ^{15}N NMR spectrum of the algaenan of *Scenedesmus quadricauda (S. communis)* .

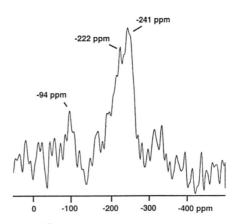

Figure 2. Solid-state ^{15}N NMR spectrum of an ultralaminae-rich kerogen (Göynük Oil Shale, Turkey).

that at least some of such functions exhibit a very high resistance and can survive early diagenesis under especially harsh conditions.

Afterwards, so as to estimate the potential of amide preservation in the sedimentary record, N functionality was investigated in two immature *Botryococcus*-derived kerogens. Pula alginite was previously extensively studied by electron microscopy, spectroscopic and pyrolytic methods and bitumen analysis (*91*). It was thus shown that two races of *B. braunii* (the alkadiene- and lycopadiene-producing ones, A and L, respectively) contributed to this immature, 5-6 millions years old, maar-type oil shale and that the morphological features of the algal colonies are fully retained. The selective preservation of the algaenan was the most important pathway in Pula kerogen formation along with condensation of high molecular weight lipids from the A race of *B. braunii*. Although the depositional conditions are very different for Pula alginite and Balkashite, similarities are therefore observed in the mechanisms involved in their formation. However, as shown by the ^{15}N NMR spectrum which shows a peak that maximizes at -240 ppm, we conclude that the nitrogen in Pula alginite is mainly located in pyrrole groups. Taking into account the above observations on Balkashite, the very low maturity of the Pula sample and the exceptionally good preservation of the morphology of the colonies, this change in functionality from amide-containing structures to pyrrole-containing structures was highly unexpected. A similar observation is made in the case of the immature Permian Torbanite. This Torbanite was previously shown to be derived from the selective preservation of *B. braunii* algaenan and the bulk chemical structure of the algaenan did not undergo any substantial alteration upon fossilization (*85*). In contrast, when nitrogen functionality is considered, a drastic change is observed between the algaenan isolated from extant *B. braunii* and this immature kerogen whose nitrogen chiefly corresponds to pyrrole units.

Conclusions and pending questions

Amino acid and protein analysis indicated that such constituents can be partly preserved in sediments. This behaviour is rather unexpected if one considers the fact that the sensitivity of amide functions towards bacterial degradation is rather high. However, some amide groups were shown to exhibit a high resistance against chemical and microbial attack upon algaenan isolation, composting of plant material, Mangrove Lake sediment and Balkashite formation.

Several processes can be invoked to explain such a resistance. The adsorption of proteins or, more generally speaking, of organic matter, on mineral surfaces, especially clays, and the incorporation into mineral pores as has been suggested recently (*60*). Such a physical protection has traditionally been used to explain the preservation of organic matter in soils and it has been recently recognized in Recent sediments with relatively low organic carbon content (*92*). However, other mechanisms must be considered in the cases of the algaenans, of the algal compost, of laboratory decay of phytoplankton and also for Balkashite and immature oil shales. In all these cases, the amount of organic matter is high in comparison with the mineral matter. Protein sorption onto or encapsulation into a refractory organic matrix has been recently suggested to explain the survival of some proteins in sediments and upon composting (*19, 63, 81*). In addition, long chain amides were shown to occur in some algaenans (*83*). Such functions were supposed to be sterically protected within the highly aliphatic macromolecular network to which they belong. Specific proteins with structures resistant to degradation were also reported in ocean DOM and POM (*76*).

Nevertheless, such relatively resistant amides do not survive into fossil materials, even when immature samples are considered. Indeed, a conspicuous change in N functionality occurs during diagenesis and the dominant nitrogen functions then

250

correspond to heterocyclic structures like pyrroles, as observed here in immature
Torbanites (Pula and South Africa) or ultralaminae-rich oil shales (Göynük oil shale,
Rundle oil shale). However, the precise timing, the controlling environmental factors
and mechanisms of this change are still to be elucidated. Indeed, a study of Peru
upwelling upper sediments indicates that, in that case, the above change may have
begun to occur at or above the sediment/water interface, *i. e.* during early diagenesis
(*17, 80*). In sharp contrast, in Mangrove Lake sediment, *ca.* 4000 years old, amides
remain the major nitrogen function (*19, 81*).

Concerning the mechanism(s) of such a change in nitrogen functionality, amides
may either be selectively eliminated hence emphasizing minor preexisting aromatic
nitrogen-containing units or, as suggested for proteinaceous amino acids, amides
undergo transformations producing more refractory material likely comprising
heterocycles as dominant nitrogen functions (*18, 67, 77*). To discriminate between
these two mechanisms one would need to study a series of samples comprising both
amides and heterocyclic moieties along with a precise quantitation of nitrogen.

Nitrogen in fossil fuels is known to occur in heterocyclic structures. In oils, it
mostly corresponds to pyrroles and pyrrole-containing constituents (*5, 20-23*). Coals
were shown to chiefly comprise pyrrole moieties (*93, 94*) or pyrrole and pyridine
ones (*24-27*). These pyrrole units are generally thought to result from cyclization and
aromatization taking place during catagenesis. However, the presence of pyrroles in
immature kerogens leads us to reconsider such an origin in fossil fuels and to
examine whether they are derived from thermal processes and/or are already present
as diagenetic products. Further studies such as artificial coalification are needed to
discriminate between these two pathways.

Acknowledgments - One of us (SD) thanks the Petroleum Research Fund for travel
support to attend the ACS meeting.

Literature cited

1. Frankenfeld, J. W.; Taylor, W. F. *Am. Chem. Soc. Div. Fuel Chem., Preprints,* **1978**, *23*, 205.
2. Ford, C. D.; Holmes, S. A.; Thompson, L. F.; Latham, D. R. *Anal. Chem.,* **1981**, *53*, 831.
3. Snape, C. E.; Bartle, K. D. *Fuel,* **1985**, *64*, 427.
4. Schmitter, J. M.; Vajta, Z.; Arpino, P. J. In *Adv. Org. Geochem 1979.,* 1980, 67.
5. Dorbon, M.; Ignatiadis, I.; Schmitter, J. M.; Arpino, P.; Guiochon, G.; Toulhoat, H.; Huc, A. *Fuel,* **1984**, *63*, 565.
6. Theodore, L.; Buonicore, A. J. *Energy and the Environment: Interactions*; CRC Press Inc.: Boca Raton, FL, 1986; Vol. I, Part A.
7. Crocker, M. E.; Marchin, L. M. *J. Petrol. Technol.,* **1988**, 470.
8. Jenden, P. D.; Kaplan, I. R.; Poreda, R. J.; Craig, H. *Geochim. Cosmochim. Acta,* **1988**, *52*, 851.
9. Whiticar, M. J. *Org. Geochem.,* **1990**, *16*, 531.
10. Krooss, B. M.; Leythaeuser, D.; Lillack, H. *Erdöl und Kohle-Edrgas-Petrochemie,* **1993**, *46*, 271.
11. Krooss, B. M.; Littke, R.; Müller, B.; Frieglingsdorf, J.; Schwochau, K.; Idiz, E. F. *Chem. Geol.,* **1995**, *126*, 291.
12. Littke, R.; Krooss, B. M.; Idiz, E. F. ; Frieglingsdorf, J. *AAPG Bull.* , **1995**, *79*, 410.
13. Billen, G. In *Heterotrophic Activity in the Sea*; Hobbie, J. E.; Williams, P. J., Eds.; Plenum Press, 1984; pp. 313-355.
14. Romankevitch, E. A. *Geochemistry of Organic Matter in the Ocean*; Springer-Verlag: 1984.

15. de Leeuw, J.W.; Largeau, C. In *Organic Geochemistry principles and applications*; Engel M.H.; Macko S.A., Eds.; Plenum Publishing Corp.: New-York, 1993; pp. 23-72.
16. Henrichs, S. M.; Farrington, J. W. *Geochim. Cosmochim. Acta*, **1987**, *51*, 1.
17. Patience, R.L.; Clayton, C. J.; Kearsley, A. T.; Rowland, S. J.; Bishop, A. N.; Rees, A. W. G.; Bibby, K. G.; Hopper, A. C. *Proc. ODP, Sci. Results*; Suess, E.; von Huene, R., Eds.; College Station (Ocean Driling Program): TX, 1990; pp. 135-153.
18. Haugen, J. E.; Lichtentaler, R. *Geochim. Cosmochim. Acta*, **1991**, *55*, 1649.
19. Knicker, H.; Scaroni, A. W.; Hatcher, P. G. *Org. Geochem.*, **1996**, *24*, 661
20. Schmitter, J. M.; Ignatiadis, I.; Arpino, P. *Geochim. Cosmochim. Acta*, **1983**, *47*, 1975.
21. Schmitter , J. M.; Garrigues, P.; Ignatiadis, I.; de Vazelhes, R.; Perin, F.; Ewald, M.; Arpino, P. *Org. Geochem.*, **1984**, 6, 579.
22. Ignatiadis, I.; Dorbon, M.; Arpino, P. *Rev.Inst. Fr. Pet.*, **1986**, *41*, 551.
23. Li, M.; Larter S. R. In *Organic Geochemistry*; Øygard, K., Ed.; Falch Hurtigtrykk, Oslo, 1993; pp. 576-579.
24. Bartle, K. D.; Perry, D. L.; Wallace, S. *Fuel Proc. Technol.*, **1987**, *15*, 351.
25. Burchill, P.; Welch, L. S. *Fuel*, **1989**, *68*, 100.
26. Wallace, S.; Bartle, K. D.; Perry, D. L. *Fuel*, **1989**, *68*, 1450.
27. Kirtly, S. M.; Mullins, O. C.; van Elp, J.; Cramer, S. P. *Fuel*, **1993**, *72*, 133.
28. Treibs, A. *Angew. Chem.*, **1936**, *49*, 682.
29. Ocampo, R.; Callot, H. J.; Albrecht, P. J. *Chem. Soc. Chem. Commun.* **1985**, 200.
30. Ocampo, R.; Callot, H. J.; Albrecht, P. J. In *Metal complexes in fossil fuels*, Filby, R. H.; Branthaver, J. F., Eds.; ACS Symp. Ser.; 1987; *344*, pp. 68-73.
31. Prowse, W. G.; Maxwell, J. R. *Geochim. Cosmochim. Acta*, **1989**, *53*, 3081.
32. Callot, H.J.; Ocampo, R.; Albrecht, P. *Energy Fuels*, **1990**, *4*, 635.
33. Keely, B. J.; Maxwell, J. R. *Org. Geochem.*, **1993**, *20*, 1217.
34. Keely, B. J.; Harris, P. G.; Popp, B. N.;Hayes, J. M.; Meischner, D.; Maxwell, J. R. *Geochim. Cosmochim. Acta*, **1994**, *58*, 3691.
35. Keely, B.J.; Blake S. R.; Schaeffer, P.; Maxwell, J. R. *Org. Geochem.*, **1995**, *23*, 527.
36. Keely, B. J.; Prowse W. G.; Maxwell, J. R. *Energy Fuels*, **1990**, *4*, 628.
37. Huseby, B.; Ocampo, R.; Bauder, C.; Callot, H. J.; Rist, K.; Barth, T. *Org. Geochem.*, **1996**, *24*, 691.
38. Sinninghe Damsté J. S.; Eglinton T. I.; De Leeuw J. W. *Geochim. Cosmochim. Acta*, **1992**, *56*, 1743.
39. Sinninghe Damsté J. S.; Wakeham, S. G.; Kohnen, M. E. L.; Hayes, J. M.; de Leeuw, J. W. *Nature*, **1993**, *362*, 827.
40. Schaeffer, P.; Ocampo, R.; Callot, H. J.; Albrecht, P. J. *Nature*, **1993**, *364*, 133.
41. Schaeffer, P.; Ocampo, R.; Callot, H. J.; Albrecht, P. J. *Geochim. Cosmochim. Acta*, **1994**, *58*, 4247.
42. Baxby, M.; Patience, R. L.; Bartle K. D. *J. Pet. Geol.*, **1994**, *17*, 211.
43. Dungworth, G.; Thijssen, M.; Zuurveld, J.; van deer Velden, V.; Schwartz, A. *Chem. Geol.*, **1977**, *19*, 295.
44. Shimoyama, A.; Hayishita, S.; Harada, K. *Geochem. Journ.*, **1988**, *22*, 143.
45. Tyson, R.V. *Sedimentary Organic Matter*; Chapman and Hall: London, 1995; pp. 384.
46. Muller, P.J.; Suess E. *Deep-Sea Res.*, **1979**, *26A*, 1347.
47. Gordon, D.C. *Deep-Sea Res.*, **1971**, *18*, 1127.
48. Suess, E.; Muller, P. J. In *Biogéochimie de la Matière Organique à l'Interface Eau-Sédiment Marin*; Daumas, R., Ed.; Colloques Internationaux du CNRS; Paris, 1980, 293; pp. 17-26.
49. Valiela, I. *Marine Ecological Processes*; Springer-Verlag: New-York, 1984.

252

50. Blackburn, T.H.; Henriksen, K. *Limnol. Oceanogr.*, **1983**, *28*, 477.
51. Lancelot, C.; Billen, G. *Adv. Aquatic Microbiol.*, **1985**, *3*, 263.
52. Stevenson, F.J.; Cheng, C.N. *Geochim. Cosmochim. Acta*, **1972**, *36, 653.
53. Rosenfeld, J. K. *Limnol. Oceanogr.*, **1979**, *24*, 1014.
54. Tanoue, E.; Handa, N. *J. Oceanogr. Soc. Jap.*, **1980**, *36*, 1.
55. Balzer, W.; Erlenkeuzer, H.; Hartmann, M.; Müller, P. J.; Pollehne, F. In *Seawater-Sediment Interactions in Coastal Waters*; Rumohr, J.; Walger, E. Zeitschel B., Eds.; Lecture Notes on Coastal and Estuarine Studies; 1987; pp. 111-161.
56. Silliman, J. E.; Meyers, P. A.; Bourbonniere, R. A. *Org. Geochem.*, **1996**, *24*, 463.
57. Macko, S. A.; Pereira, C. P. G. In *Initial Reports of the Ocean Drilling Program Leg 113*, Barker P. F.; Kennett, J. P., Eds.; Part B; College Station, TX, 1990; pp. 881-897.
58. Boudreau, B. P.; Canfield, D. E.; Mucci, A. *Limnol. Oceanogr.*, **1992**, *37*, 1738.
59. Meyers, P. A.; Silliman, J. E.; Shaw, T. J. *Org. Geochem.*, **1996**, *25*, 69.
60. Macko, S. A. ; Engel, M. H.; Parker, P. L. In *Organic Geochemistry. Principles and Applications*, Engel M.H.; Macko S.A., Eds.; Plenum Publishing Corp.: New-York, 1993; pp. 213
61. Quayle, W. C., Collins, M. J.; Farrimond, P. In *Organic Geochemistry: developments and apploations to energy,climate, environment and human history;* Grimalt, J. O.; Dorronsoro, C., Eds.; AIGOA: Donostia-San Sebastian, 1996; pp. 1044-1046.
62. Tyson, R.V. *Sedimentary Organic Matter*; Chapman and Hall: London, 1995; pp. 389.
63. Nguyen, R. T.; Harvey, H. R. *Org. Geochem.*, **1997**, in press.
64. Cowie, G. L.; Hedges, J. I. *Mar. Chem.*, **1992**, *37*, 223.
65. Mayer, L. M.; Macko, S. A.; Cammen, L; *Mar. Chem.*, **1988**, *25*, 291.
66. Rosenfeld, J. K. *Am. J. Sci.*, **1981**, *281*, 436.
67. Burdige, D. J.; Martens, C. S. *Geochim. Cosmochim. Acta*, **1988**, *52*, 1571.
68. Whelan, J. K. *Geochim. Cosmochim. Acta*, **1977**, *41*, 803.
69. Biggs, D.C.; Berkowitz, S. P.; Altabet, M. A.; Bidigare, R.R.; DeMaster, D. J.; Dunbar, R. B.; Leventer, A.; Macko, S. A., Nittrouer, C. A.; Ondrusek, M. E. In *Proc. ODP, Init. Repts.*; Barker, P. F.; Kennett, J. P., Eds.; 113; College Station (Ocean Drilling Program), TX, 1988; pp. 77-86.
70. Biggs, D.C.; Berkowitz, S. P.; Altabet, M. A.; Bidigare, R.R.; DeMaster, D. J.; Macko, S. A., Ondrusek, M. E.; Noh, I. In *Proc. ODP, Init. Repts.*; Barron, J.; Larsen, B., Eds.;119; College Station (Ocean Drilling Program), TX, 1989; pp. 109-120.
71. Horsfall, I. M.; Wolff, G. A. *Org. Geochem.*, **1997**, *26*, 311.
72. Tanoue, E. *Deep-Sea Res.*, **1992**, *39*, 743.
73. Tanoue, E. *Mar. Chem.*, **1995**, *51*, 239.
74. Tanoue, E. *J. Mar. Res.*, **1996**, *54*, 967.
75. Tanoue, E.; Nishiyama, S.; Kamo, M.; Tsugita, A. *Geochim. Cosmochim. Acta*, **1995**, *59*, 2643.
76. Tanoue, E.; Ishii, M.; Midorikawa, T. *Limnol. Oceanogr.*, **1996**, *41*, 1334.
77. Burdige, D. J.; Martens, C. S. *Geochim. Cosmochim. Acta*, **1990,** *54*, 3033.
78. Wakeham, S. G., Lee, C.; Farrington, J. W.; Gagosian, R. G. *Deep-Sea Res.*, **1984**, *31*, 509.
79. Müller, P. J.; Suess, E.; Ungerer, C. A. *Deep-Sea Res.*, **1986**, *33*, 819.
80. Patience, R.L.; Baxby, M.; Bartle, K.D.; Perry, D.L.; Rees, A.G.W.; Rowland, S.J. *Org. Geochem.*, **1992,** *18*, 161.
81. Knicker, H.; Hatcher, P. G., *Naturwissenschaften*, **1997**, *84*, 231.
82. Hatcher, P. G. *NOAA Prof. Paper*, **1978**, *10*, 90.

83. Derenne, S.; Largeau, C.; Taulelle, F. *Geochim. Cosmochim. Acta.* **1993**, *57*, 851.
84. Derenne, S.; Largeau, C.; Berkaloff, C.; Rousseau, B.; Wilhelm, C.; Hatcher, P. *Phytochem.***1992**, *31*, 1923.
85. Largeau, C.; Derenne, S.; Casadevall, E.; Kadouri, A.; Sellier, N. *Org. Geochem.*, **1986**, *10*, 1023.
86. Derenne, S.; Largeau, C.; Casadevall, E.; Berkaloff, C.; Rousseau, B. *Geochim. Cosmochim. Acta* **1991**, *55*, 1041.
87. Knicker, H.; Fründ, R.; Lüdemann, H. D., *Naturwissenschaften*, **1993**, *80*, 219.
88. Largeau, C.; Derenne, S.; Casadevall, E.; Berkaloff, C.; Corolleur, M.; Lugardon, B.; Raynaud, J.F.; Connan, J. *Org. Geochem.* **1990**, *16*, 889.
89. Gillaizeau, B.; Derenne, S.; Largeau, C.; Berkaloff, C.; Rousseau, B. *Org. Geochem.*, **1997**, *24*, 671.
90. Gatellier, J.P.; de Leeuw, J.W.; Sinninghe-Damsté, J.S.; Derenne, S.; Largeau, C.; Metzger, P. *Geochim. Cosmochim. Acta*, **1993**, *57*, 2053.
91. Derenne, S.; Largeau, C.; Hetényi, M.; Brukner-Wein, A.; Connan, J.; Lugardon, B. *Geochim. Cosmochim. Acta*, **1997**, *61*, 1879
92. Hedges, J. I.; Keil, R. G. *Mar. Chem.***1995**, *49*, 81.
93. Knicker, H.; Hatcher, P. G.; Scaroni, A. W. *Energy and Fuels*, **1995**, *9*, 999.
94. Buckley, A. N.; Riley, K. W.; Wilson, M. A. *Org. Geochem.*, **1996**, *24*, 389.

Chapter 15

Comparative Study of the Release of Molecular Nitrogen from Sedimentary Matter During Non-Isothermal Pyrolysis

B. M. Krooss[1], B. Müller[1], P. Gerling[2], and R. Littke[3]

[1]Institute of Petroleum and Organic Geochemistry, Forschungszentrum Jülich,
D-52425 Jülich, Germany
[2]Bundesanstalt für Geowissenschaften und Rohstoffe, Stilleweg 2, D-30655 Hannover,
Germany
[3]Lehrstuhl für Geologie, Geochemie und Lagerstätten des Erdöls und der Kohle Rheinisch-
Westfälische Technische Hochschule Aachen Lochnerstrasse 4-20,
D-52064 Aachen, Germany

The role of sedimentary organic matter (SOM) as a source of molecular nitrogen (N_2), occurring abundantly in natural gases of certain parts of the North German Basin, has been investigated by non-isothermal open-system pyrolysis of coals and of shales containing dispersed organic matter. Enrichment of N_2 in natural gases with respect to methane (CH_4) is attributed to fractional generation and trapping of these two compounds due to different thermal stabilities of the precursor substances. The release of N_2 and CH_4 was monitored as a function of temperature up to 1200°C. Nitrogen evolution from humic coals of different rank shows two generation maxima between 700 and 800°C, and above 1100°C, respectively. Shales with dispersed organic matter exhibit a strong variability in nitrogen generation characteristics. The TOC-normalized N_2 generation potential is mostly higher than for coals and frequently exceeds the methane generation potential. Nitrogen release from shales starts at significantly lower temperatures than from coals. Substantial amounts of this low-temperature N_2 can be attributed to mineral components (ammonium-bearing clays and feldspars) which presumably inherited the nitrogen from organic matter during diagenesis.

Nitrogen Sources in Sedimentary Basins

Molecular nitrogen (N_2) is a common and, in certain regions, major or even predominant component of natural gases. Various attempts have been made to explain the occurrence of nitrogen-rich natural gases and to elucidate the origin of the nitrogen gas in the lithosphere (*1, 2*). In two recent publications on this matter (*3, 4*) we have

254

reviewed the different hypotheses and attempted to validate them from a qualitative and quantitative point of view. Among the various sources and pathways that have been invoked to account for nitrogen-rich natural gases a significant contribution of a crustal radiogenic N_2 source can be practically ruled out (4). Also, the direct incorporation of atmospheric nitrogen e.g. by meteoric water in appreciable quantities is rather unrealistic because the Ar/N_2 ratio of nitrogen-rich gases is usually much lower than the atmospheric value.

Magmatism and volcanism are conceived to contribute regionally and episodically to the molecular nitrogen balance of the lithosphere. Thus, according to Gerling et al. (in preparation) high N_2 contents in some East German gases appear to be related to volcanitic rocks. For the occurrence of isotopically light nitrogen in the vicinity of the Bramsche massif, NW Germany, three tentative explanations have been proposed (5) involving either a direct (contribution of mantle-derived gases) or an indirect (structural inversion, heating event) influence of Cretaceous igneous intrusions. Natural gas reservoirs rich in N_2 are, however, not necessarily associated with volcanism or magmatism.

The flux of primordial nitrogen from the mantle would be considered as a long-term process. The average flux of primordial nitrogen, however, does not appear sufficient to give rise to nitrogen concentrations observed in some reservoirs so that either focusing from large drainage areas or regionally high fluxes must be invoked (4).

Sedimentary organic matter (SOM) is an important gateway for the incorporation of nitrogen into the lithosphere. It contains nitrogen, fixed by biochemical processes, in relatively high concentrations (usually 1-3 %) and occurs in appreciable quantities in a dispersed or concentrated form in all sedimentary basins. Therefore it appears to be the most likely, direct or indirect, source of N_2 in many natural gas reservoirs.

Nitrogen contents in coals of the Mahakam delta, from the peat stage (48 % daf TOC) up to bituminous coal (69 % daf TOC) vary between 0.6 and 2.7 wt % (daf) with an increasing trend up to the onset of the bituminous stage and then reversing (6). High rank coals (bituminous to anthracite) have nitrogen contents between 0.5 and 3.4 % decreasing with increasing maturity (3, 7). Figure 1 shows that a good correlation exists between total nitrogen content and TOC for sedimentary rocks with TOC values below 50 %. The larger scatter in the >50 % TOC region reflects the heterogeneity of coals with respect to nitrogen content which depends both on maceral composition and maturity (cf. 6).

Origin and Fate of Nitrogen in Sedimentary Organic Matter

Besides carbon, oxygen and hydrogen, nitrogen is one of the main constituents of organic matter. In living organisms it occurs overwhelmingly in the amide group of proteins and peptides. Other nitrogen compounds are nucleic acids which contain between 15 and 23 % nitrogen, amino sugars, alkaloids and tetrapyrroles. Chitin, the main structural element of arthropod skeletons is another nitrogen-rich macromolecular component of the biomass (references in this volume). The nitrogen content of marine plankton and bacteria ranges between 1 and 15 % whereas vascular

256

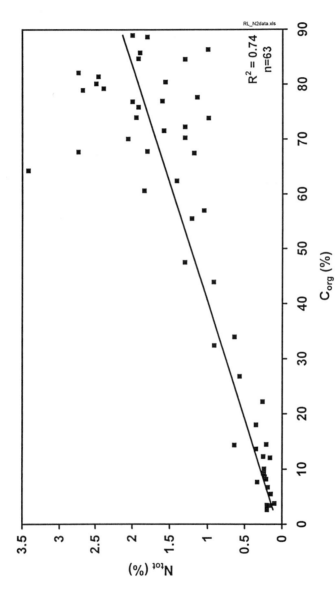

Figure 1. Total nitrogen content vs. C_{org} content of Paleozoic coals and sedimentary rocks (data from 3).

plants, consisting mainly of carbohydrates and lignin contain between 0.2 and 2 % nitrogen (*5*).

The hydrolyzable nitrogen components of sedimentary OM are largely recycled and the percentage of nitrogen preserved in the sedimentary column depends both on the type of organic matter and the depositional environment. According to Flaig (*8*), incorporation of nitrogen into coals occurs via phenolic species formed by microorganisms during lignite decomposition and only compounds, capable of forming quinones contribute to the formation of nitrogenous humic substances.

Preservation of nitrogen during coalification from lignite to subbituminous stages of maturation results in enrichment of this compound up to the bituminous stage. In a study of a Mahakam coal series Boudou et al. (*6*) found that the percentage of hydrolyzable nitrogen decreases sharply at the end of the peat stage and reduces to practically zero in the lignite stage (> 60 wt % C, daf). This indicates early decomposition of labile nitrogen-containing structures furnishing ammonia during early diagenesis.

Nitrogen in fossil fuels has been reported to occur mainly in pyrrolic and pyridinic functional groups. Investigations by Patience et al. (*9*) suggest that these compounds are present already in young marine sediments where they are selectively preserved while amino nitrogen and the total nitrogen content of these sediments declines with depth. The organic nitrogen in lignites and higher ranks is mainly heterocyclic, the structural units being either pyrrolic or pyridinic (*10*). Recent ^{15}N NMR studies (*11*) failed to detect, however, appreciable quantities of pyridinic nitrogen in sedimentary organic matter.

Inorganic Nitrogen in Sedimentary Minerals

While Littke et al. (*3*) found that practically all of the nitrogen encountered in a set of 63 Paleozoic coal samples was organic, Daniels and Altaner (*12*) report the occurrence of NH_4-rich illite in mineral assemblages associated with anthracite-rank coal from eastern Pennsylvania. Certain TOC-rich shales (oil shales) contain considerable amounts of non-organic nitrogen. Oh et al. (*13*) found N percentages between 0.59 and 0.75 for Green River shales with inorganic N representing between 22 and 52 % of the total N. The quantities of inorganic nitrogen in shales from the eastern parts of the US (New Albany shale) are usually lower. The inorganic nitrogen appears to reside in ammonium-bearing clays and feldspars. Thus, buddingtonite, an ammonium feldspar has been reported to occur in various oil shales (*14*). The ammonium is most likely derived from sedimentary organic matter. During diagenesis and catagenesis this organic nitrogen may be released and incorporated into clay minerals (*15*). This process is conceived to concur with the illitization of smectites.

Due to low concentrations the identification of ammonium-bearing clay minerals in shales by XRD is difficult and their occurrence is often only inferred. Everlien and Hoffmann (*16*) used non-isothermal pyrolysis to investigate the release of ammonia and molecular nitrogen from ammonium bearing clays and feldspars. The study was conducted with artificially prepared ammonium minerals because the gas yields from naturally occurring samples was below the detection limit of their analytical method. Both ammonia and nitrogen were liberated at different temperatures during heating of these materials up to 1100°C. The quantitative importance of this inorganic nitrogen, which is "organic" in origin, is difficult to assess and its relevance for the nitrogen balance of sedimentary basins remains to be investigated. Up to now no major occurrences of ammonium-bearing shales or feldspars have been reported for the North German Basin. In the course of the work reported here, however, evidence was obtained indicating significant amounts of inorganic nitrogen in Carboniferous black shales (see below).

Aims of the Study

Carboniferous coals are considered as the main source of natural gas in the North German Basin. In view of the large percentages of molecular nitrogen in some natural gases of the North German Basin the question arose whether this nitrogen gas is derived from sedimentary organic matter, what are the mechanisms of enrichment and how can the risk of encountering N_2 -rich gases be reduced. Pyrolysis experiments can provide information on the thermal stability of nitrogen-producing entities in sedimentary organic matter and the rates of gas generation as a function of temperature. Work performed by Klein and Jüntgen (17) on coals from the Ruhr area indicated differences in the reaction kinetics of methane and nitrogen release which, under geological conditions, could give rise to periods of preferential generation of molecular nitrogen. The nitrogen generation potential of coals is low compared to its hydrocarbon (methane) generation potential and therefore a mechanism for N_2 enrichment must exist during some phase of the evolution of reservoirs.

Enrichment of molecular nitrogen in natural gas can be envisaged to occur either by generation or migration fractionation processes. Various authors (18-21) have invoked migration-related fractionation to account for the observed enrichment (cf. discussion in 4). Their argumentation is mainly based on compositional and isotopic data and suspected migration pathways rather than on a qualitative and quantitative evaluation of fractionation mechanisms. In a review of potential transport-related fractionation mechanisms in the geosphere ("geochromatography") Krooss et al. (22) have shown how the occurrence and efficiency of these processes can be appraised by using chromatographic models. Experimental work by our group (23) on molecular transport (diffusion) of gases in sedimentary rocks documents the occurrence of compositional fractionation but indicates that due to the low transport efficiency of this process the formation of economic gas reservoirs by diffusion is highly unlikely and can be envisaged only under very special conditions (24).

One major prerequisite for the quantitative assessment of migration fractionation is information on source phase composition (as a function of geologic time), flow paths and flow regime (focused or pervasive) and interaction between stationary and mobile phases. In view of the poor constraints on most of these parameters in evolving geologic systems the suggestion of migration fractionation appears highly speculative.

The present study aimed at the elucidation of the source processes to elucidate the conditions of the release of methane and molecular nitrogen during thermal decomposition of sedimentary organic matter the reaction kinetics of the formation of these gases was studied by open-system non-isothermal pyrolysis.

Experimental

Figure 2 shows a flow diagram of the analytical set-up used in this study. Pyrolysis experiments were performed in a quartz reactor tube (Figure 2, 1) under a He current of approximately 30 ml/min. A NiCr-Ni thermocouple (2) placed 0.5 cm above the sample crucible was used for both temperature control and recording. Gas sampling and analysis was performed by automatically operated switching valves (3, 4, 5, 6). Via a 2 ml sample loop (4) the pyrolysis gases were transferred to a gas chromatograph for on-line quantitative analysis. Sampling intervals ranged between 3 and 8 minutes. Separation of nitrogen and methane was performed by means of a packed 5Å molecular sieve column (1/8" diameter, 1m length). Quantification of the gases was carried out by a thermal conductivity detector (TCD) or a mass spectrometer (MSD). Due to insufficient long-term signal stability this latter instrument was mainly used for compound identification and optimization of the analytical procedure. Gas mixtures containing certified quantities of the gases of interest were used for calibration.

Pyrolysis was carried out at different heating rates ranging from 0.1 - 2 K/min. The resulting generation curves for each gas component were evaluated in terms of reaction kinetics assuming a set of first-order parallel reactions with an Arrhenius temperature relationship and one common pre-exponential factor (cf. 4). This pseudo-kinetic approach is commonly used to extrapolate the gas generation reactions to the time and temperature scale of geologic systems. In combination with numerical basin modeling this procedure represents a tool to predict the timing of natural gas generation and the quantities and composition of gases in geologic time. A similar experimental approach has been used by other workers (15, 25).

Generation of Molecular Nitrogen from Coals

Experimental nitrogen generation curves for humic Carboniferous coals of different ranks are shown in Figure 3. All curves were measured at a heating rate of 0.5 K/min. They show, as a common feature, an onset of nitrogen generation around 600°C, a generation maximum between 700 and 800°C and a second rising slope indicating another generation maximum beyond the temperature range of the experimental set-up. The first generation maximum decreases in intensity with increasing rank and has vanished completely for the sample with highest maturity (6.1 % R_m). The generation curve of an algal coal (Torbanite) recorded at a heating rate of 0.1 K/min, which is included for comparison, shows the same overall structure as the humic coals. The nitrogen generation curves in Figure 3 are similar to those observed by other workers (17, 25). Boudou and Espitalié (25), by heating up to 1250°C, record the peak maximum of the second N_2 generation peak. Their results also document the disappearance of the first N_2 generation peak at maturity levels above 4 % R_m. These results indicate that the release of molecular nitrogen from coals occurs predominantly

260

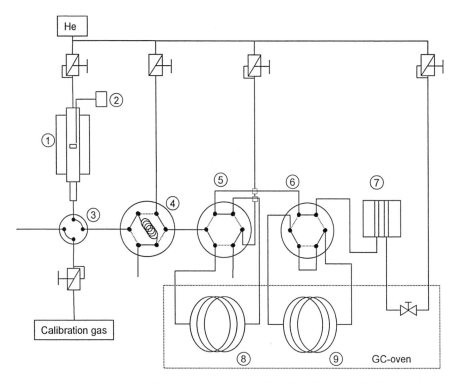

Figure 2. Flow scheme of open-system pyrolysis unit and gas analysis unit. 1. electrical furnace (Netzsch) with quartz reactor; 2. Eurotherm PID temperature controller; 3. 4-port switching valve (Valco) automatically operated; 4.-6. 6-port switching valves (Valco) automatically operated; 7. thermal conductivity detector (TCD); 8.-9. chromatographic columns (molecular sieve 5 Å).

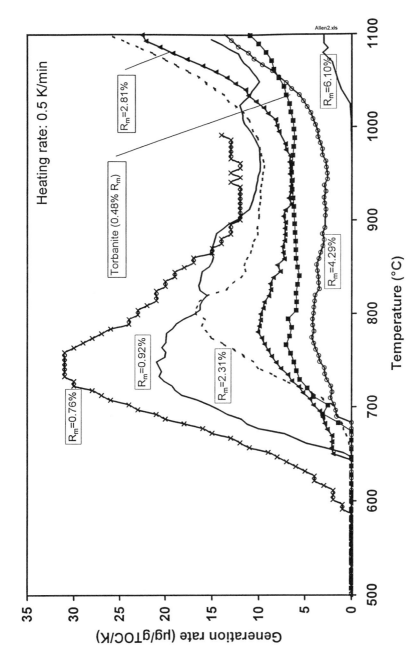

Figure 3. Selected nitrogen generation curves from a maturity series of Carboniferous humic coals of different rank and of an algal coal (Torbanite) at a heating rate of 0.5 K/min.

during the metagenetic stage of thermal maturation, subsequent to or partly coinciding with methane generation. Furthermore, the persistence of nitrogen release up to temperatures well above 1000°C testifies of the presence of extremely stable nitrogen-containing units in both humic and algal coals, the structure of which is essentially unknown.

Mass balance calculations were performed based on ash and nitrogen contents of the original samples and the residues (Table I) assuming that the quantity of ash was not affected by the pyrolysis procedure. The results show that the total nitrogen loss during pyrolysis up to 1100°C ranges between 13 and 96 % with an average value of 52 %. The fraction of original nitrogen liberated as N_2 under these conditions averages 32 %.

Table I. Nitrogen Mass Balance for Pyrolysis Experiments with Carboniferous Coals

	Original sample		Pyrolyzate				
R_m (%)	ash (wf) [wt. %]	N (wf) [wt. %]	ash (wf) [wt. %]	N (wf) [wt. %]	% loss of initial N	N_2 yield [kg/t]	N_2 yield % of initial N
0.7	4.99	1.53	8.22	1.18	53 %	11.3	74 %
0.75	79.80	0.51	89.26	0.30	47 %	0.5	10 %
0.75	4.82	1.94	8.44	1.48	56 %	7.7	40 %
0.76	16.32	1.59	24.70	1.29	46 %	9.5	60 %
0.79	18.04	1.58	27.98	0.21	91 %	5.8	36 %
0.92	15.17	1.31	22.44	1.04	46 %	2.5	19 %
1.26	0.76	1.91	4.32	1.44	87 %	6.3	33 %
1.58	19.18	1.30	24.21	1.44	13 %	3.2	24 %
2.09	10.73	1.56	18.24	1.32	50 %	5.1	33 %
2.3	2.93	2.04	3.50	0.09	96 %	4.3	21 %
2.31	1.34	1.96	1.76	1.37	47 %	6.0	31 %
2.81	2.36	1.76	2.43	1.20	33 %	4.5	26 %
4.29	5.73	1.32	5.92	0.90	34 %	3.7	28 %
6.1	4.28	1.03	4.19	0.70	30 %	0.8	7 %
			Average values:		**52 %**		**32 %**

Isotopic Composition of N_2 Generated During Open-System Pyrolysis of Coals.

The isotopic composition of the nitrogen in coals depends on the source material and on maturity. Nitrogen isotopes in Mahakam coals as well as in recent and living OM from this area have been reported by Boudou et al. (6). $\delta^{15}N$ values ranged between -1 and +2.5 with a slight trend to lighter composition towards the bituminous coal stage. The isotopic composition of $\delta^{15}N$ in Rotliegend reservoirs of the North German Basin, with N_2 contents from a few percent up to nearly 88 %, ranges between -3.2 and + 19.3 ‰ (5, 26).

Table II. Isotopic Composition of N_2 Generated during Closed System Isothermal Pyrolysis (27)

Sample	$\delta^{15}N$
Original HVB coal	+5.3 ‰
N_2 gas; 650°C (closed system, 3h)	+2.5 ‰
N_2 gas; 1000°C (closed system, 3h)	-6.5 ‰

In view of the characteristic nitrogen generation patterns of molecular nitrogen from coal the question arises as to potential differences in isotopic composition of the N_2 generated at different stages of the heating process. Stiehl and Lehmann (27) performed closed system pyrolysis of high volatile bituminous (HVB) coal (Gasflammkohle; 38.1 % vol. mat.) with an initial $\delta^{15}N$ of +5.3 ‰. After heating times of 3 hours at 650°C and 1000°C they found that the nitrogen produced in their experiments was depleted with respect to the nitrogen of the original sample. The experimental results are listed in Table II.

Table III: Maturity, TOC Content and Maceral Composition of a High Volatile Bituminous Carboniferous Coal used for Isotope Study of Pyrolytic Nitrogen.

R_m [%]	TOC [%]	Vitrinite [%]	Inertinite [%]	Liptinite [%]
0.70	72.8	75	11	14

Elemental analysis and mass balance

Component	Original sample wt %	Pyrolyzate wt %	Loss during pyrolysis wt %
Ash (wf)	4.99	8.22	
Total C (wf)	76.66	86.03	32 %
Total H (wf)	5.83	1.39	85 %
Total O (wf)	9.92	6.97	57 %
Total N (wf)	1.53	1.18	53 %
Total S (wf)	1.23	1.07	47 %
Sum (wf)	100.15	104.87	

For long-term closed system pyrolysis of coals (120 - 600 h) at 225 - 400°C (duration: 120 - 600 h) Stiehl and Lehmann (27) found that the $\delta^{15}N$ values of the residual material decreased slightly from +5.3 ‰ to values between +3.9 and +4.9 ‰. In the present work an attempt was made to assess the isotopic composition of the molecular nitrogen evolved during non-isothermal pyrolysis of a Carboniferous coal up to 1100°C at a heating rate of 2 K/min. The maturity, TOC and maceral composition of this coal are given in Table III. This table also summarizes the results of the elemental analysis of the coal and the residue after pyrolysis. Based on these data a mass balance of the losses of C, O, H, N, and S during pyrolysis was calculated. For this computation the ash content of the coal was used as an "internal standard" with the implicit assumption that the mass of the mineral components constituting the ash was not significantly affected by pyrolysis. According to this mass balance calculation the total nitrogen loss during pyrolysis amounted to 53 % of the initial nitrogen (8.2 mg/g out of 15.3 mg/g). The quantities of molecular nitrogen (N_2) released during several pyrolysis experiments with this coal at different heating rates average around 11 mg/g. The nitrogen content of the tar produced during these experiments was not measured. The results of this mass balance indicate the inherent problems of quantifying nitrogen as a minor component of sedimentary matter to a reasonable degree of consistency.

For the isotopic analysis of the nitrogen produced by pyrolysis gas fractions were collected at five different stages of the experiment. Figure 4 shows schematically the generation curves for methane and nitrogen and the corresponding temperature intervals for gas sampling which extend over the entire temperature range of N_2 generation. Sampling was performed by three different methods in order to detect and avoid potential isotopic fractionation effects arising from sample handling and transfer. In two instances N_2 was collected by cryo-trapping at liquid nitrogen temperature using molecular sieve (5A mesh 80/120) and Porapak Q as adsorbents. In a third experiment the pyrolysis gas was collected with the carrier gas in glass containers without adsorbent and then transferred to a stable isotope mass spectrometer (GC/IRMS, Finnigan MAT 252). Details of the experimental procedure will be presented elsewhere (Müller et al., in preparation).

The observed isotopic composition of the N_2 sampled during the pyrolysis experiment varies between -5 and +1 ‰. Apart from the first sampling interval the isotope values of all three experiments show the same overall trend (Figure 5): an enrichment in $\delta^{15}N$ up to the 750 - 890°C sampling interval, followed by a slight depletion. The inversion of the isotopic composition trend as a function of pyrolysis temperature resembles trends recently observed in the pyrolytic formation of methane and other hydrocarbon gases (28). These "nonlinearities" in the isotope fractionation effects occurring during pyrolysis of organic matter appear to be indicative of differences in the isotopic composition of different precursor structures. Detailed investigations of these effects using an on-line pyrolysis-GCI/IRMS method are presently under way.

Also indicated in Figure 5 are the isotopic compositions of the nitrogen in the original coal and in the residue. Considering the mass balance of nitrogen the strong depletion of the residue ($\delta^{15}N$ = -10.3 ‰) would require an average isotopic composition of $\delta^{15}N$ = +12 ‰ for the total product which is not in line with the gas data. In view of the reproducibility the gas isotope data appear to be reliable and the ^{15}N isotopic composition of the original coal is considered a realistic value. Due to the very low $\delta^{15}N$ value of the residue the results of this set of experiments are presently not conclusive and further work is required to solve this problem.

Closed system hydrous and anhydrous pyrolysis experiments (1 week at 365°C) with Carboniferous coals of different rank (85 - 95 wt % C, daf) yielded N_2 quantities between 0.1 and 1.1 mg/g (85 - 900 L/t) (5). The bulk isotopic

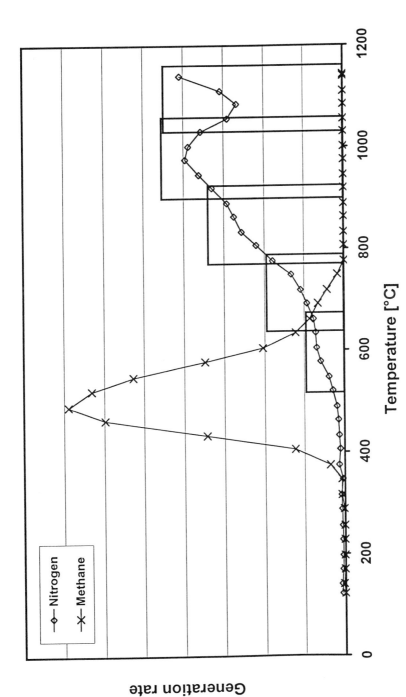

Figure 4. Temperature intervals of gas sampling for isotope analysis of nitrogen evolved during non-isothermal pyrolysis of a Carboniferous coal (generation rates: arbitrary units).

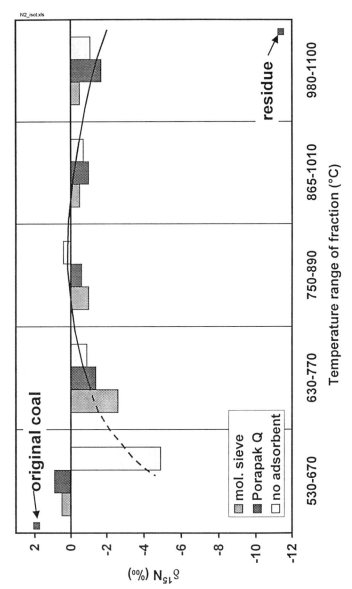

Figure 5. Isotopic composition of molecular nitrogen collected during non-isothermal pyrolysis of a Carboniferous coal.

compositions ($\delta^{15}N$) of the generated gases ranged from -7.7 ‰ up to values around +4 ‰ showing an increasing trend with maturity. The isotopic composition of the nitrogen in the parent material averaged +3 ‰.

Generation of Molecular Nitrogen from Shales

Sedimentary organic matter in dispersed form occurs in appreciable quantities in Paleozoic shales of the North German Basin. Depending on the geologic situation the amount of dispersed organic matter may exceed the quantities of OM concentrated in coal seams. To investigate the potential role of this dispersed OM in the formation of nitrogen-rich natural gases shale samples from different Paleozoic formations have been studied with respect to gas release during non-isothermal pyrolysis. Measurements were mainly performed on low-TOC shales (0.5 - 3 % TOC). Besides methane and nitrogen, the release of other gaseous products (H_2, CO_2, CO) was monitored in some experiments to obtain a more comprehensive image of the complex pyrolysis reactions. Experiments were performed at different heating rates (0.1 - 2°C/min) up to temperatures of 1150°C.

The two diagrams in Figure 6 show the evolution of methane, nitrogen and hydrogen from a Paleozoic shale with 2.9 % TOC at two different heating rates. The maturity of this sample ranged around 1.4 % R_m. Evidently, the total nitrogen generation potential of this shale (corresponding to the area under the generation curve) is significantly higher than the methane generation potential (16 mg N_2/g TOC vs. 5 mg CH_4/g TOC). The onset of nitrogen generation occurs between 500 and 600°C and the generation curve shows a distinctly different shape than the one observed for the coals. Nitrogen generation extends over a broad temperature range from 700 to 1100°C and the N_2 generation curve exhibits a spike in the 800°C region which is resolved as a double spike in the low heating-rate experiment. It is noteworthy that this feature is accompanied by a corresponding drop in the hydrogen generation curve indicating a process involving hydrogen consumption. Similar features were observed in other pyrolysis experiments with shale samples though the underlying reaction mechanisms could not yet be identified.

Although the exact quantification of H_2 poses some problems the experimental results show that considerable amounts of hydrogen are generated during the pyrolysis experiment and that hydrogen release continues even after methane generation has ceased completely.

It is evident that the conditions used in the pyrolysis experiments reported here deviate strongly from the conditions of thermal decomposition of organic matter in natural systems and that an extrapolation of the results to geologic systems is even more problematic than for the results obtained by other pyrolysis techniques. It is also clear that mineral reactions and mineral-organic matter interactions can contribute to the gas evolution patterns observed here. Furthermore it must be kept in mind that, due to the large temperature range used in these experiments the decomposition reactions – in particular those of the mineral components – may be controlled by thermodynamic stability rather than reaction kinetics.

268

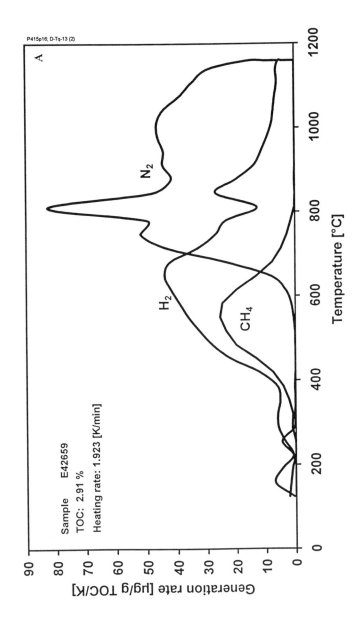

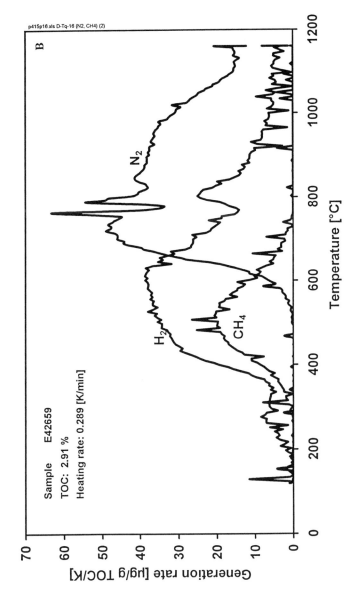

Figure 6. Generation curves obtained for methane, hydrogen and nitrogen during non-isothermal pyrolysis of a TOC-lean shale at different heating rates.

The open system pyrolysis approach with on-line gas component analysis should, at the present stage, be considered as an analytical tool to elucidate chemical reactions in complex systems and to collect compositional and stability information on both organic and inorganic components of sedimentary matter. We believe that the experimental results, although far from being understood, can contribute to an improvement of the still very limited understanding of gas generation processes in the lithosphere.

In order to demonstrate the qualitative differences of gas evolution from coals and dispersed organic matter a comparison of nitrogen generation curves from two Carboniferous shales and a Carboniferous coal is shown in Figure 7. The onset of nitrogen generation from the shales lies between 450 and 500°C, i.e. approximately 200 °C lower than for coals. The overall (TOC-normalized) nitrogen generation potentials of the shales are 21 and 33 mg/g TOC, respectively, and thus exceed the corresponding values observed for coals (max. 15.5 mg/g TOC). Finally, while the nitrogen generation curves of coals usually exhibit a relatively uniform pattern (see above) the corresponding curves differ significantly from each other and from the one of the coal. For one shale (R_r = 1.5 %) the generation maximum lies between 900 and 1000°C corresponding to the minimum of the N_2 generation curves for the coals. The second, overmature, sulfur-rich, shale (R_r = 3.8 %) shows a generation maximum between 500 and 600°C and several smaller maxima in the high temperature region.

In order to establish the potential contribution of inorganic nitrogen to the observed gas generation profiles shale samples were treated with a mixture of hydrochloric and hydrofluoric acid to remove the mineral phases. The resulting kerogen concentrates were then subjected to open system pyrolysis under the same conditions as the original shales. Figure 8 shows the resulting nitrogen evolution curve for the kerogen concentrate from one of the Carboniferous shales (R_r = 3.8%) in Figure 7. Comparison of the two generation curves shows that the bulk of the nitrogen released from this shale is associated with mineral matter removed during acid treatment. Only a high-temperature generation peak with a maximum around 990°C can be attributed to the organic matter in this shale. This example shows the importance of the aspect of inorganic nitrogen in shales addressed already by other workers (13-15). At the same time it documents for the first time the occurrence of significant quantities of inorganic nitrogen components in Carboniferous shales of the N German basin (see above). The pathways and interrelationships of organic and inorganic nitrogen in shales are, up to now, only marginally understood. Further systematic work is required to elucidate the origin and evolution of nitrogen-containing structures in both shales and coals, their stability and the transformation mechanisms resulting in the formation of N_2.

Implications for Nitrogen-Rich Natural Gas Accumulations

A formal kinetic evaluation of the pyrolytic gas generation curves using the pseudo-kinetic approach outlined above was performed for a selected Carboniferous coal and two Carboniferous shales. It is recognized that even the kinetic interpretation of N_2 evolution from coals based on high-temperature open-system pyrolysis experiments is,

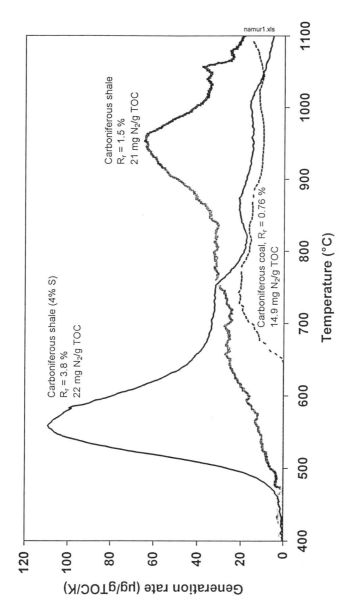

Figure 7. Comparison of nitrogen generation curves of two Carboniferous shales and a humic coal (experimental heating rate: 0.5 K/min).

272

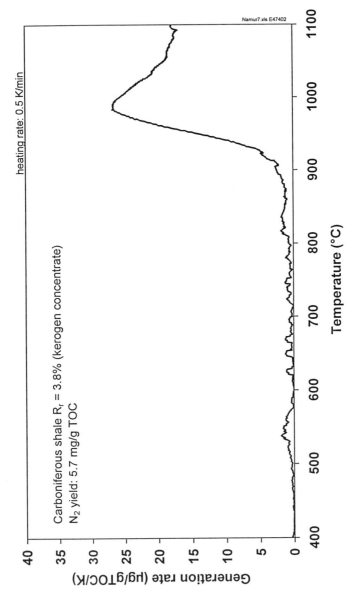

Figure 8. Nitrogen generation curve of kerogen concentrate prepared from an overmature Carboniferous shale (R_r = 3.8 %; cf. Figure 7).

at best, a rough approximation of the reaction processes in geological systems. This holds even more for the complex multicomponent system of shales involving organic and inorganic reactions like dehydration and mineral transformations. Therefore the following evaluation should be considered with due caution and as a tentative approach constrained by our still very limited understanding of geochemical reactions.

Based on the resulting kinetic data sets for methane and nitrogen the composition of gas generated from a coal and two shales at a geologic heating rate of 5.2 K/Ma was calculated. The computed nitrogen volume ratio of the gas is plotted as a function of temperature in Figure 9. It is evident from this diagram that for the coal a gas containing > 50 % nitrogen is expected to be formed only at temperatures in excess of 450°C. Although the top of the Carboniferous in certain parts of the N German basin has been buried to 10 - 13 km it is questionable whether considerable quantities of Carboniferous coals have been exposed to these temperatures. As a result of lower N_2 generation temperatures in laboratory experiments the onset of N_2-rich gas generation under geological heating conditions is predicted to occur at temperatures between 250 and 300°C which is substantially lower than for coals.

This result can be considered as a qualitative indication of shales acting as thermogenic N_2 sources at lower temperatures than coals. It disregards, for the time being, the aspects of thermodynamic rather than kinetic control on the decomposition of minerals (ammonium-clays/feldspars), mineral-organic interactions etc. and thus outlines the scope of research work still required on this issue.

Conclusions

The experimental studies using non-isothermal open-system pyrolysis have revealed significant differences in the thermal stability of nitrogen-bearing structures in coals and dispersed organic matter in a number of shale samples of Paleozoic age. The pyrolytic nitrogen (N_2) generation curves of shales exhibit a strong variability reflecting differences in organofacies, mineral composition, diagenetic evolution and mineral-organic matter interactions. Pyrolytic evolution of molecular nitrogen from shales (in laboratory experiments) occurs generally at lower temperatures than from coals. Furthermore, the total nitrogen generation potential of shales can exceed the methane generation potential. In some instances this observation could attributed unambiguously to the presence of nitrogen-containing mineral phases.

Generation of methane and nitrogen occurs at different temperatures. Kinetic evaluation of the experimental results suggests that the formation of nitrogen-dominated gas from shales with dispersed organic matter takes place at significantly lower temperatures than from coals.

Open-system pyrolysis with on-line analysis of evolved gas components is a valuable tool to monitor and compare gas generation reactions from different types of sedimentary matter and to investigate the chemical composition and structure of the precursor substances. On-line stable isotope analysis of gas components will add an additional dimension to this analytical approach. More work is required to collect and compare information on gas generation from various materials. Based on this work new concepts for the interpretation of the analytical data can be expected to arise.

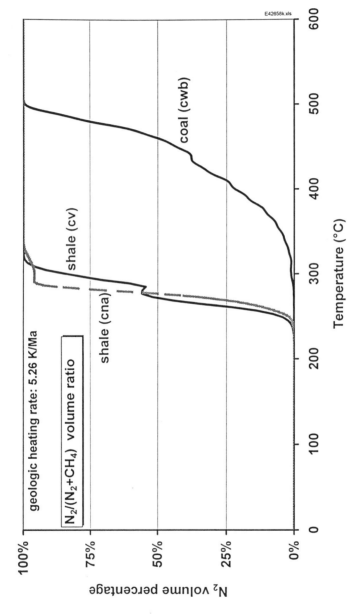

Figure 9.   Predicted evolution of nitrogen content (volume ratio) of gas generated from coal and shales at a constant geologic heating rate.

275

Acknowledgments

Major parts of this research work were funded jointly or individually by BEB Erdgas und Erdöl GmbH, Hannover and Mobil Erdgas und Erdöl GmbH, Celle. The authors thank A. Richter and D. Hanebeck for performing the shale pyrolysis experiments reported in this work. A PRF travel grant accorded to B. M. Krooss is gratefully acknowledged. Finally the authors wish to thank two anonymous reviewers for helpful comments and suggestions.

Literature cited

1. Maksimow, S. P.; Müller, E. P.; Botnewa, T. A.; Goldbecher, K.; Zor'kin, L. M.; Pankina, R. G. *Internat. Geology Rev.* **1975**, *18*, 551-556.
2. Jenden, P. D.; Kaplan, I. R.; Poreda, J. R.; Craig, H. *Geochim. Cosmochim. Acta* **1988**, *52*, 851-861.
3. Littke, R.; Krooss, B. M.; Idiz E.; Frielingsdorf, J. *AAPG Bulletin* **1995**, *79*, 410-430.
4. Krooss, B. M.; Littke, R.; Müller, B.; Frielingsdorf, J.; Schwochau, K.; Idiz, E. F. *Chem. Geol.* **1995**, *126*, 291-318.
5. Gerling, P.; Idiz, E.; Everlien, G.; Sohns, E. *Geol. Jb.* **1997**, *D103*, 65-84.
6. Boudou, J.-P.; Mariotti, A.; Oudin, J.-L. *Fuel* **1984**, *63*, 1508-1510.
7. Shapiro, N.; Gray, R. J. In *Coal Science;* Gould, R. F., Ed.; Advances in Chemistry Series 55; American Chemical Society: Washington, DC, United States. 1966; pp 196-210.
8. Flaig, W. *Chem. Geol.* **1968**, *3*, 161-187.
9. Patience, R. L.; Baxby, M.; Bartle, K. D.; Perry, D. L.; Rees, A. G. W.; Rowland, S. J. *Org. Geochem.* **1992**, *18*, 161-169.
10. Burchill, P.; Welch, L. S. *Fuel* **1989**, *68*, 100-104.
11. Knicker, H.; Hatcher, P. G.; Scaroni, A. W. *Int. J. Coal Geology* **1996**, *32*, 255-278.
12. Daniels, E. J.; Altaner S. P. *American Mineralogist* **1990**, *75*, 825-839.
13. Oh, M. S.; Taylor, R. W.; Coburn, T. T.; Crawford, R. W. *Energy & Fuels* **1988**, *2*, 100-105.
14. Oh, M. S.; Foster, K. G.; Alcaraz, A; Crawford, R. W.; Taylor, R. W.; Coburn, T. T. *Fuel*, **1993**, *72*, 517-523.
15. Scholten, S. O. *The distribution of nitrogen in sediments.* - PhD Thesis Rijksuniv. Utrecht, *Geologia Ultraiectina*, **1992**, *81*.
16. Everlien, G.; Hoffmann, U. *Erdöl & Kohle - Erdgas - Petrochemie/ Hydrocarbon Technology*, **1991**, *44*, 166-172.
17. Klein, J.; Jüntgen, H.; *Adv. in Org. Geochem. 1971*, **1972**, 647-656.
18. Boigk, H.; Stahl, W. *Erdöl und Kohle-Erdgas-Petrochemie* **1970**, *23*, 325-333.
19. Maksimow, S. P.; Sorkin, L. M.; Pankina, R. G. *Z. angew. Geol.* **1973**, *19*, 499-505.

20. Lutz, M.; Kaasschieter, J. P. H.; van Wijhe, D. H. *Proc. 9th World Petroleum Congress* **1975,** P.D. 2(3), 93-103.

21. Faber, E.; Schmitt, M.; Stahl, W. J. *Erdöl und Kohle-Erdgas-Petrochemie* **1979,** *32,* 65-70.

22. Krooss, B. M.; Brothers, L.; Engel, M. H. In *Petroleum Migration;* England, W. A., Fleet, A. J., Eds.; Geological Society, Special Publication, London,. 1991, No. 59, 149-163.

23. Schlömer, S.; Krooss B. M. *Marine and Petroleum Geology* **1997,** *14,* 565-580.

24. Krooss, B. M.; Leythaeuser, D. *AAPG Bulletin* **1997,** *81,* 155-161.

25. Boudou, J.-P.; Espitalié, J. *Chem. Geol.* **1995,** *126,* 319-333.

26. Sohns, E.; Gerling, P.; Faber, E. *Anal. Chem.* **1994,** *66,* 2614-2620.

27. Stiehl, G.; Lehmann, M. *Geochim. Cosmochim. Acta* **1980,** *44,* 1737-1746.

28. Berner, U.; Faber, E.; Scheeder, G.; Panten, D. *Chem. Geol.* **1995,** *126,* 233-245.

Role and Importance of Nitrogen in Crops, Waste, and Soil

Chapter 16

Utilization by Cattle of the Nitrogen in Forage Crops

Richard A. Kohn

Department of Animal and Avian Sciences, University of Maryland,
College Park, MD 20742

Domestic animals convert low quality forage and byproduct N into
protein for human consumption. The protein requirements of cattle
can be met in part from microbial protein synthesized in the rumen
from non-protein N and protein in forages. The same microbes also
destroy some amino acids consumed by ruminants. The utilization
efficiency for N consumed by cattle depends on the digestibility,
amino acid profile, and the propensity for destruction of amino acids
during storage and digestion. Transformations occur during harvest and
storage of forage crops that result in direct loss of some N to the field,
hydrolysis of protein and destruction of amino acids, and cross linkage
of protein. These changes alter the ruminal degradability and
digestibility of the N consumed by cattle. Diet formulation and feeding
further alters the N utilization in the herd. How crops and animals are
handled affects the manure N volume and composition (organic and
inorganic), and the subsequent losses from the farm to water resources.
Although most N pollution from animal agriculture occurs during
application of manure and fertilizer to crops, how the crops are
managed, harvested and how cattle are fed greatly affects subsequent
losses from farms.

Animal agriculture plays an important role in society by converting forage and feed
byproducts into foods available for human consumption. While some have argued
that more protein and energy would be supplied to humans if they directly consumed
the animal feed, much of the diets consumed by animals are comprised of fiber, non-
protein N, and low quality proteins that are not suitable for humans. Even with

supplementation of animal diets with grains to improve the efficiency of nutrient conversion, animal agriculture returns nearly as much or more digestible energy and protein for human consumption as it consumes, and the products of animal agriculture are of higher nutritive value than the grains consumed (*1*).

Increasing human population size and wealth have resulted in a greater demand for animal products and the agriculture industry has responded by intensifying production systems. As a result, animal agricultural production systems are becoming a major source of N pollution affecting ground and surface water resources. The N flow on a dairy farm is represented in Figure 1. A great deal of emphasis has been placed on soil and manure management as a means to reduce nutrient losses from farms. Recently, however, improved animal management and feeding have been identified as ways to reduce N contamination of water resources. For example, N losses to the environment from dairy farms were shown to be more sensitive to animal feeding and herd management than to manure or crop management. Improvements in animal diet and management to increase the conversion of feed N to animal product by 50% would decrease N losses by 36 to 40%, but improving manure availability to crops by 100% would only decrease N losses from the farm by 14%. Nitrogen losses could be further reduced by optimal crop selection which requires the integration of feeding and cropping systems (*2*).

The utilization efficiency for N consumed by cattle depends on digestibility, amino acid profile, and the propensity for destruction of amino acids during storage and digestion. This paper will focus on the N transformations that occur during the harvesting of crops, distribution and digestion of feed, and storage of manure. These changes affect the amount and form of manure N produced and the quantity of feed N required.

N Distribution in Crops

Nitrogen consumed by domestic animals is derived primarily from forages and the byproducts of grain processing (e.g. corn gluten feed), wet milling (e.g. brewer's grain, distiller's grain), and oil extraction (e.g. soybean meal), as well as from grains produced primarily for animal feeds (e.g. corn grain).

Herbivores also consume a large quantity of N from the leafy portion of plants. Legumes contain more N than non-legume crops, and cool season grasses typically contain more N than tropical species. Forages contain more N when immature because the N is diluted by storage carbohydrate and especially fiber as plants mature. In addition, some of the protein of mature plants is often bound and not digestible by animals. The leafy portion of plants contains more N than the stem portion, and whereas most forage protein is comprised of plant enzymes, the active parenchyma tissues contain more N than the less active sclerenchyma tissues (*3*).

Forage N is found in several different forms within different structures of plants. Fresh forages contain non-protein N including NO_3, NH_3, amino acids, nucleotides, and chlorophyll which comprise 10 to 25% of the total N while the rest of the N is contained in protein (*3*). Traditional Kjeldahl methods to determine N content underestimate NO_3 and nucleic acid N compared to combustion methods (*4*). About half of the true protein in fresh forages is soluble (*5*). The most abundant soluble protein is ribulose-bisphosphate carboxylase/oxygenase (RUBP c/o) which comprises 30% of the total soluble true protein of temperate forages and 10 to 25% of tropical forages (*3*). Other soluble proteins are also generally enzymatic and have diverse functions (*5*). Chloroplast membranes contain insoluble true proteins that are amphiphilic and comprise about 25% of the true protein (*3, 6-7*). These proteins include those involved with electron transport for photosynthesis (*6-7*). Similar proteins in cell mitochondria comprise an additional 5% of the true protein. Additional proteins are associated with plant cell wall (*8-9*) or are bound to tannins (*10-11*). The distribution of forage N depends on the plant species and maturity but most of the N is derived from plant enzymatic proteins.

Harvesting and Storage

In most climatic regions, crop production is seasonal so animal feeds must be harvested and stored for times when fresh forages are not available. Although only a small proportion of N is lost directly from crops during harvesting and storage, there are numerous transformations of the protein in crops that can reduce its digestibilty, destroy amino acid protein, and make proteins more susceptible to destruction in the rumen. These effects reduce the efficiency of N utilization by the animal, which ultimately increases feed N requirements, manure N production, and in turn increases N losses from the farm.

Crops are typically conserved for monogastric animals (e.g. swine, horses) by drying and for ruminants (e.g. cattle, sheep) by drying or ensiling. Herbivores (e.g. cattle, horses, sheep) also graze fresh forages or are fed green chopped material. The forms of N in forages begin changing the moment they are harvested, and these changes can affect the overall utilization of the material. Forage leaf material, which contains more N than the stem, may be lost in the field during the wilting (field drying) process. Because hay is dried more than silage, the leaf losses are generally greater for hay, and these losses increase due to rain damage. Wet silage may allow for runoff of soluble proteins and carbohydrates, but if properly managed runoff from silage can be avoided. During initial drying of forage crop, plant respiratory enzymes catalyze catabolism of soluble sugars which decreases the total dry matter of the crop. Additional respiration is catalyzed by plant and microbial enzymes during silage conservation. Respiration of forage dry matter causes apparent increases in crop N concentration. The expected changes in crop composition due to harvesting and storage conditions have been reviewed (*12*) and are shown in Table I.

The wilting process of forages is accompanied by proteolysis by plant enzymes (*13- 15*). Some soluble proteins are degraded to amino acids and NH_3. Ensiling of forages results in solubilization and degradation of plant proteins by microorganisms (*13- 14*). This process continues until the silage pH is reduced enough to inactivate plant and microbial enzymes (*16*). The only type of enzymes capable of activity at low pH, aspartic proteases, are not found in silage (*16*). However, clostridial bacteria

Table I. Typical dry matter and crude protein (N x 6.25) changes during harvesting, processing, and storage for hay and silage[1]

Crop and Process	Dry matter loss		Protein change	
	Legumes	Grasses	Legumes	Grasses
	------ % ------		--- % of DM ---	
Respiration	4.0	5.0	0.90	0.80
Rain Damage				
5 mm	5.0	2.0	-0.40	-0.20
25 mm	17.0	8.0	-1.70	-1.30
50 mm	31.0	15.0	-3.50	-2.70
Mowing and conditioning	2.0	1.0	-0.70	0.00
Tedding	3.0	1.0	0.50	-0.20
Swath inversion	1.0	1.0	0.00	0.00
Raking	5.0	5.0	-0.50	-0.30
Baling				
Small bales	4.0	4.0	-0.90	-0.50
Large bales	6.0	6.0	-1.70	-1.00
Chopping	3.0	3.0	0.00	0.00
Hay storage				
Inside	5.0	5.0	-0.70	-1.30
Outside	15.0	12.0	0.00	0.00
Silo				
Sealed	8.0	8.0	1.40	0.80
Stave	10.0	10.0	1.80	1.20
Bunker	12.0	12.0	2.30	1.50

[1]Adapted from Rotz and Muck (12).

actively degrade silage and hay proteins if the pH is not rapidly lowered as when anaerobic conditions are not maintained (13-14). Adding acids (e.g. formic acid) to quickly decrease silage pH reduces the proteolysis that occurs in the silage and spares amino acid protein (17).

High molecular mass phenolic compounds (tannins) bind to protein and carbohydrates reducing their availability to silage and rumen microbes and to decrease whole tract digestibility. This process occurs during crop growth especially when plants are stressed, and it occurs rapidly during wilting after harvesting (18). Tannins may form cross-links with proteins that are hydrolyzable or condensed (19). Both forms of tannins would theoretically inhibit protein degradation by preventing attachment of enzymes, but cross-linkage due to hydrolyzable tannins are reversed by the acid conditions of the abomasum (ruminants' true stomach). Thus these tannins decrease protein degradation in the rumen but may not affect the overall protein digestion (20). Other authors have correlated the concentration of tannins in sainfoin (20) and lotus with reduced ruminal degradation of protein (21). Tannins have also been associated with reduced protein digestion (22-24). Though much of the effect of tannins on digestion is likely to be caused by the cross-linkage, tannins may also be toxic to bacteria (25).

Heat can be produced during ensiling especially when air is allowed to penetrate into the silage. Heat can also result in moist hay due to respiration (26). While heat production generally has a negative effect on the quality of the forage, some heat production may decrease the susceptibility of the forage amino acids to degradation by ruminal microbes (27). Malliard product is formed by cross linkage of amine groups with the carbonyl groups of carbohydrates when forages are overheated which makes protein and carbohydrates indigestible (28). Generally heat production during conservation makes the forage protein and carbohydrate indigestible and may reduce the palatability of the forage (14, 26).

Harvesting and storage of forages does not appreciably affect the N content but results in numerous changes in proteins that may ultimately affect the digestion and utilization of the N by animals.

Feeding and Management

The potential losses from a farm animal are partitioned in Figure 2. Some feed N is apparently not digestible and appears in the feces. This fraction is composed of truly undigested feeds, undigested microbes that result from ruminal or hind gut fermentation, and an undigested fraction from animal origin including sloughed off cells or secreted enzymes from the digestion process. The largest and most controllable fraction is that of the undigested feed. If silage or hay is not preserved properly, heat damage to protein lowers its digestibility. Overly mature and lignified crops also may have less digestible protein. Both mature and heat-damaged forages are higher in acid detergent insoluble N, an indicator of protein crosslinked with carbohydrate to form Maillard product, and so suspicious feeds should be analyzed and the problem prevented (28).

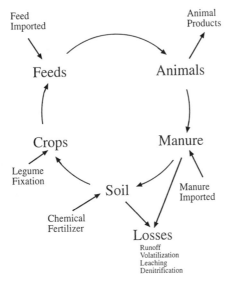

Figure 1. Nitrogen cycling in an animal production system.

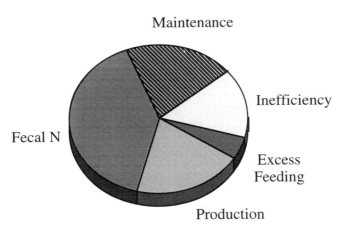

Figure 2. Partitioning of N losses from cattle. Fecal N includes undigested feed, and undigested microbial and animal protein. Maintenance, inefficiency of production and excess feed N digested will appear in urine.

The herd or flock nutritional efficiency can be improved simply by increasing production per animal. Maintenance costs are incurred to maintain animals and to grow replacement stock to a productive age. While maintenance losses per animal are fixed for each species (29), increasing production per animal lifetime can dilute maintenance and reduce the number of animals required for a given level of production. For example, if cows produce more milk or broilers grow faster due to a change in management style, their nutritional requirements increase more gradually than the production increases.   Improved management practices enable more efficient production in this way.

Some urinary N results from overfeeding of protein in general or overfeeding individual forms of protein. Diet formulation practices can reduce this overfeeding as they are used to select the right forms of protein and to determine the amounts required. Diet formulation is based on mathematical models that are derived from extensive empirical research. Consideration may be given to the forage maturity, potential heat damage evaluated by detergent insoluble N, and to the ruminal degradability estimated by protein solubility in different solvents (30) or calculated from tabular values for each feed ingredient (29).

The predictions made by these models are far from exact despite the complexity of the models. Because of the uncertainty of feeding animals, excess protein is often fed to insure that an adequate amount is provided. The greater the uncertainty, the greater the excess that is provided. Therefore, improving diet formulation models in ways that will decrease uncertainty will decrease the need for overfeeding, reduce the urinary N losses, and the associated farm N losses from manure and fertilized crops (31).

Digestion and Metabolism

Amino acids that can be synthesized by the body from common precursors are "non-essential", and amino acids that cannot be synthesized by the body from common precursors are "essential". "Limiting" amino acids are those which are absorbed from a given diet in the least concentration relative to requirements. That is, a decrease in the protein concentration of the diet would result in a decrease in production or growth due to the limiting amount of these amino acids. Proteins synthesized by body tissues have specific amino acid sequences. The non-essential amino acids can be synthesized *de novo* for the production of these proteins, but essential amino acids or their precursors must be supplied in the diet. If the body lacks an essential amino acid, protein synthesis will be arrested until that amino acid is supplied regardless of how much total protein is provided in the diet. In swine and poultry diets, one of the means to decrease the amount of N in the diet is to balance for specific amino acids rather than feed large excesses of the non-limiting amino acids just to provide enough of the limiting ones.

In ruminants, some feed protein is digested by microorganisms in the rumen (pre-stomach), hydrolyzed to amino acids and eventually to NH_3 and the respective ketoacids. Most non-protein N and the NH_3 released from protein degradation are consumed by the microorganisms and used to synthesize microbial proteins. The excess NH$_3$ released from protein degradation is absorbed from the gut. Some feed protein is not degraded by ruminal microorganisms. This undegraded feed protein and the microbial protein pass to the abomasum and small intestine and can be digested by animal enzymes. Peptides and amino acids may be absorbed from the small intestine and used to meet the animal's amino acid requirements. Some of the dietary protein is not digested and passes out of the animal in the feces. In addition, N in chlorophyll and other pigments are likely to pass from the animal in feces, as the hydrophobic nature of the molecule would inhibit degradation by gut microbes.

Some microbial protein synthesis occurs in the large intestine, but this protein is not available for absorption. Because this microbial protein cannot be absorbed it ends up in the feces. However, microbial growth may be important for digestion of fiber in the large intestine, and some end-products from this digestion may be absorbed in the large intestine and used for energy. Therefore, availability of N may be important to digestion in the large intestine, especially for post-gastric fermenters such as the horse. Nitrogen in the blood urea may be transported to the large intestine and used for microbial growth. Although this micribial activity can make energy available to the animal., the microbial protein synthesized in the large intestine ends up in feces.

Ammonia absorbed from the rumen has one of four possible fates. On high protein diets, most blood NH$_3$ is transported via the blood to the liver, converted to urea via the urea cycle, and excreted in urine. Some NH$_3$ is excreted directly from the blood by the kidney without first being converted to urea. Some urea and NH$_3$ are recycled back to the gut by passive transport. Urea and NH$_3$ diffuse into the saliva that enters the gut, and across the rumen and intestinal wall. Microbes are thus able to use this N for synthesis of microbial protein. Finally, NH$_3$ may be used for synthesis of non-essential amino acids from ketoacids..

The protein requirements of ruminants relate to the digestive process. As with non-ruminants, the animal must absorb essential amino acids from the small intestine. However, ruminal microorganisms are capable of synthesizing all of the common amino acids *de novo*. Therefore, ruminants have the unique ability to convert non-protein N to protein, and may maintain moderate production levels without consumption of true protein. None-the-less, microbial protein synthesis alone is not adequate to maintain high levels of production (*32-33*). Therefore, some feed protein that escapes ruminal degradation is required in the diet.

Adding sources of protein that are slowly degraded in the rumen has been shown to increase milk production in some cases (*34-37*), but not in other cases (*38-39*). A requirement of ruminally undegraded protein also has been demonstrated for maximal growth of beef steers (*40-42*). The amount of amino-acid protein absorbed in the small intestine was greater for cattle fed diets containing protein that is less degradable in the rumen. In addition, infusion of amino acid protein directly to the small intestine also increases production (*33*) and dry matter intake (*43*), thus demonstrating the importance of undegraded protein entering the small intestine.

Though providing for protein that escapes ruminal degradation is important to optimizing production of high-producing ruminants, feeding to increase microbial protein synthesis may also increase the flow of essential amino acids to the small intestine. Microbial growth requires an adequate supply of carbohydrate that is fermented in the rumen, and a rumen environment conducive to microbial growth. Microbial growth may be limited if carbohydrate degradation is too low to supply the energy for growth (44). Alternatively, on high starch diets, the pH of the rumen fluid will decrease which will lower microbial growth (45). The interactions of these factors result in a limit to the maximal amount of protein synthesis that can occur. Further research is needed to address ways to optimize microbial protein synthesis, but additional protein requirements must be met by providing ruminally undegraded protein.

Often times, microbial protein synthesis is limited by the amount and form of N available to rumen organisms (46-47). Inadequate rumen available N may result in a decrease in protein entering the small intestine due to a depression of microbial protein synthesis (47). If inadequate N is available to rumen microbes, fiber digestion may also be reduced (48-51). When microbes are not available to digest fiber in forages, fiber has a greater filling effect, and hence feed intake is reduced (35, 41). Therefore, while it is important to supply an adequate amount of rumen undegraded protein to help meet the needs of the host animal, there is also a requirement for rumen available N to provide for microbial growth to synthesize microbial protein, degrade fiber in the diet, and promote normal rumen function.

While most of the protein needs of rumen microorganisms can be met with NH_3 and carbohydrates, microbial growth is maximized by the inclusion of amino acids or peptides in the rumen (46, 52-53). In vitro studies involving incubation of rumen fluid with amino acid mixtures showed that maximal microbial growth was obtained by supplementation of urea with a mixture of leucine, methionine and histidine (54). Addition of other amino acids did not further increase microbial growth. Sulfide addition did not result in as high a microbial growth as either cysteine or methionine. Ruminal infusion of casein or supplementation with soybean meal increased microbial protein synthesis in the rumen above that observed from infusion of urea (47). Supplementation with soybean meal (49) or fishmeal (48-49) resulted in greater fiber digestion than supplementation with urea.

If N available to microorganisms limits microbial growth, then there may be an advantage to increasing rumen available N in the rumen. However, once the microbial requirements have been met, there is no additive effect of additional rumen N (55). The excess rumen available N is converted to NH_3 in the rumen and diffuses into the blood. The blood NH_3 is converted to urea in the liver and filtered from the blood by the kidney. Thereafter it ends up in urine. In the same way, if the rumen undegraded protein requirement has been met, and no amino acids are limiting, there is no advantage to further increasing the amount of rumen undegraded protein in the diet beyond the effect of providing additional energy. Excess amino acids are transported to the liver, amine groups are transaminated to glutamate or arginine and eventually urea is synthesized. This urea ends up in urine.

Understanding these possible effects and interactions of rumen protein availability makes it easier to understand why studies often fail to show positive effects of balancing diets for rumen degraded and undegraded protein. The requirements for each type of protein may depend on other feeds in the diet, and effects will not be seen unless the level of rumen available or undegraded protein is in fact limiting production. The difficulty of predicting protein requirements for ruminants should be clear from this discussion.

Effect of Protein Source on Ruminal Degradation

Forages provide a significant amount of the crude protein to high-producing ruminants. Though most forage proteins appear to be degraded in the rumen, they differ in their susceptibility to degradation. In general, proteins from hay are not as readily degraded in the rumen as those of silages, and grass proteins (including corn) are less readily degraded in the rumen than those of alfalfa (29, 56). The disappearance of protein from nylon bags suspended in the rumen was lower for orchard grass hay than for alfalfa hay (57). In a different study, sainfoin crude protein was degraded to a lessor extent in vitro in continuous culture than alfalfa, cicer milkvetch or birdsfoot trefoil (58). Silages of grasses were compared to clover silages incubated in situ (59). Initial disappearance of crude protein from the bags was greater for grasses, but clover silage degraded faster at later points in the incubation. Switchgrass (a tropical grass) crude protein disappeared from nylon bags more slowly than that of smooth bromegrass (60). A different study showed that protein disappeared faster from nylon bags in situ from the cool season grass species bromegrass and wheatgrass than for the tropical grasses big bluestem and switchgrass (61). Absolute values for protein degradation and comparison of crude protein degradation across studies is not possible due to the large variation in the techniques used to measure and to calculate protein degradation.

Conservation of forages results in changes in the protein available for degradation. Frozen grass and hay crude protein was estimated to have similar amounts of rumen available N, but silages of the same material were estimated to have higher amounts of N available to the rumen (15, 62). The same was true for comparison of alfalfa hay and silage (57). Dried perennial ryegrass provided more protein to the duodenum of sheep than the same forage that was fed fresh or had been frozen (63).

Some studies have suggested that advanced plant maturity may be associated with reduced ruminal degradation of forage protein. Disappearance of N from nylon bags was slower for silages made from more mature grasses and legumes (59). Initial N solubility was lower at advanced maturity of alfalfa, red clover, orchard grass, perennial rye grass, quack grass and timothy, but not different for birdsfoot trefoil and bromegrass (64). The rate of disappearance of N from nylon bags incubated *in situ* of the insoluble residue was slower with advancing maturity for most plant species (64). Protein disappearance from nylon bags was lower in mature alfalfa samples compared to early samples (65). Nitrogen disappeared more quickly from nylon bags incubated *in situ* for early bromegrass compared to late bromegrass, but bromegrass maturity had no effect on N retention in non-lactating Holstein cows (66). Maturity decreased protein degradation *in vitro* for alfalfa, bromegrass, and canarygrass (62).

Parenchymal tissues of plants are more rapidly degraded by rumen microbes than more lignified sclerenchymal tissues (67-68). The warm-season grass, green panic, was compared to the cool-season grass, Italian ryegrass, for *in situ* degradation studied by light and electron microscopy (69). Green panic leaves had a larger cross section, larger and more frequent vascular bundles, and less but more densely packed mesophyll. Due to the greater rigidity, green panic leaves were broken down more by chewing than ryegrass leaves. However, reduction of the chewed leaf particles by microbes was faster in ryegrass than in green panic because ryegrass epidermis and bundlesheath cells were easily split from the digestible mesophyll cells, but green panic mesophyll was protected from degradation by these lignified structures.

When isolated, RUBP c/o was readily degraded *in vitro* (70). This protein was readily degraded in silage of alfalfa, orchard grass, ryegrass and fescue, but not in vetch or pearl millet (71). The latter forages contain tannins which have been shown to bind to RUBP c/o (20), and thus could in turn reduce its susceptibility to degradation . Changing the diet from fresh to dried forages was shown to decreaseprotein solubility and *in vitro* proteolytic activity on RUBP c/o of rumen fluid. This suggests that bacterial populations may adjust to the presence of soluble protein by changing amount or type of enzymes produced, or by changing the species of organisms (72).

Manure Collection and Storage

Urine, feces, and discarded bedding material are collected and stored before application to crop land. During storage the manure N may undergo one of two major transformations: mineralization or volatilizaton. Mineralization is the microbial degradation of organic N to NH_3. Volatilization can result in losses of varying amounts of the manure NH_3 to the air. In addition, inorganic N (NH_3 or NO_3) is especially prone to runoff from the feed lot or storage facility, and leaching from the lot or manure storage area. The type of storage facility and the length of time in storage affect the extent of the losses.

Manure collected can be divided into two parts according to its rate of conversion to inorganic N. Urinary N from mammals and uric acid from birds is readily converted to NH_3, but the fecal N is 97% organic and is mineralized slowly during storage. Once mineralized the N is also a candidate for volatilization, leaching and runoff as is true of urine N. The first-order degradation of organic N to inorganic N is assumed in order to calculate the amount of organic N remaining at some point in time after storage:

$$N_t = N_0 \exp(-kt) + F/k \; x \; [1 - \exp(-kt)]$$

where N_t = organic N (kilograms) after t days of storage, N_0 = initial organic N, F = daily inflow of organic N into storage (kilograms per day), and k = rate constant (fraction per day). This equation was formulated as the antiderivative for the equation

$$dN/dt = F - kN_t$$

where dN/dt = change in organic N and other variables are as defined previously (73). Mineralization rate (k) depends upon temperature but is on the order of 0.2%/d (74).

A common rule of thumb is that N in manure is divided equally between that in inorganic and organic forms, but the truth is that the forms voided from animals depend on digestibility and utilizability of absorbed protein. Low N digestibility will lead to greater organic fecal N in manure, and poor utilization of absorbed protein and excess N feeding will increase urinary N that will become inorganic N in manure. Thus forage conservation and feeding affects the amount of manure N produced and its distribution to urine and feces. This distribution further affects the potential for volatilization and leaching, and timing of its availability to crops.

Conclusions

Nitrogen transformations occur when crops are harvested and stored, consumed by animals, and when manure is collected and stored. Nitrogen losses occur primarily from the manure storage facility or from manure or fertilizer applied to crops, however the way feeds are harvested and how they are fed to animals may have a bigger impact on the extent of those losses than manure management. Manure N output and composition (organic and inorganic forms of N in manure) are highly dependent on animal feeding and management practices and on crop handling to maintain nutritional quality of feeds.

The value of forage N varies according to forage species, crop management, harvesting methods, and what grain supplements are used. In general, the N in mature forages and those that are heat damaged during preservation may be bound and unavailable during digestion. This N will be excreted in feces. Forage N in young crops, especially alfalfa silage, is highly available to rumen microbes and can initially supply N and amino acid protein to support microbial protein production. However, feeding high concentrations of such forages results in poor N utilization because the rumen available N can exceed the capacity for use by rumen microbes. The goal of forage management is to preserve existing proteins, encourage reactions to optimally cross link proteins so that some amino acids will be protected from degradation in the rumen, and yet prevent irreversible crosslinkage of proteins that will make the proteins indigestible.

References

1. Oltjen, J. W.; Beckett, J. L. *J. Anim. Sci.* **1996**, 74: 1406-1409.
2. Kohn, R. A.; Dou, Z.; Ferguson, J. D.; Boston, R. C. *J. Environ. Manage.* **1997**, 50: 417-428
3. Lyttleton, J. W. In *Chemistry and Biochemistry of Herbage*; ed. G. W. Butter and R. W. Baily. Academic Press, New York, NY., **1973**, pp 63-103.

4. Simonne, A. H.; Simonne, E. H.; Eitenmiller, R. R.; Mills, H. A.; Cresman III, C. P. *J. Sci. Food Agric.* **1997**, 73: 39-45.

5. Mangan, J. L. 1982. In *Forage Protein in Ruminant Anim. Prod.*; ed. Thomson, D. J. et al., Br. Soc. Anim. Prod., London, **1982**, Occasional Publication No. 6, pp 25-39.

6. Askerlund, P.; Larsson, C.; Widell, S. W. *Phyiologia Plantarum* **1989**, 76: 123-134.

7. Marder, J. B.; Barber, J. *Plant, Cell and Environ.* **1989**, 12: 595-614.

8. Lamport, D.T.A. In *The Biochemistry of Plants*, Academic Press Inc., New York, **1980**; Vol. 3; 501-541.

9. Lee, K. B.; Loganathan, D.; Merchant, Z. M.; Linhardt, R. J. *Appl. Biochem. Biotech.* **1990**, 23: 53-79.

10. Sanderson, M. A.; Wedin, W. F. *Grass Forage Sci.* **1989**, 44: 151-158.

11. Sanderson, M. A.; Wedin, W. F. *Grass Forage Sci.* **1989**, 44: 159-168.

12. Rotz, C. A.; Muck, R. E. In *Forage Quality, Evaluation, and Utilization.* M. Collins, G. C. Fahey, L. E. Moser, and D. R. Mertens, ed. Agron. Soc. Am., Crop Sci. Soc. Am., and Soil Sci. Soc. Am., Madison, WI., **1994**, pp 828-836.

13. Ohshima, M.; McDonald, P. *J. Food Agric.* **1978**, 29:497-505.

14. McDonald, P. In Forage Protein in Ruminant Anim. Prod. Br. Soc. of Anim. Prod. **1982**, Occasional Publication No. 6; pp 41-49.

15. Petit, H. V.; Tremblay, G. F. *J. Dairy Sci.* **1992**, 75: 774-781.

16. McKersie, B. D. *Can. J. Plant Sci.* **1981**, 61: 53-59.

17. Vagnoni, D. B.; Broderick, G. A.; Muck, R. E. *Grass and Forage Sci.* **1997**, 52: 5-11.

18. Kohn, R. A.; Allen, M. S. *J. Sci. Food Agric.* **1992**, 58: 215-220.

19. Wong, E. In *Chemistry and Biochemistry of Herbage.* ed. G. W. Butler and R. W. Bailey. Academic Press, New York, NY., **1973**, pp 265-322.

20. Jones, W. T.; Mangan, J. L. 1977. *J. Sci. Food Agric.* **1977**, 28: 126-136.

21. Barry, T. N.; Manley, T. W. *Br. J. Nutr.* **1984**, 51: 493-504.

22. McNabb, W. C.; Waghorn, G. C.; Peters, J. S.; Barry, T. N. *Br. J. Nutr.* **1996**, 76: 535-549.

23. Hagerman, A. E.; Robbins, C. T.; Weerasuriya, Y.; Wilson, T.C.; Mcarthur, C. *J. Range Manag.* **1992**, 45: 57-62.

24. Hanley, T. A.; Robbins, C. T.; Hagerman, A. E.; Mcarthur, C. *Ecology* **1992**, 73: 537-541.

25. Hartley, R. D.; Akin, D. E. *J. Sci. Food Agric.* **1989**, 49: 405-411.

26. Ruxton, I. B.; McDonald, P. *J. Sci. Food Agric.* **1974**, 12: 706-714.

27. Broderick, G. A.; Yang, J. H.; Koegel, R. G. *J. Dairy Sci.* **1993**, 76: 165-174.

28. Goering, H. K.; Gordon, C. H.; Hemken, R. W.; Waldo, D. R.; Van Soest, P. J.; Smith, L. W. *J. Dairy Sci.* **1972**, 55: 1275-1280.

29. National Research Council. ed. Natl. Acad. Sci., Washington, DC. **1989**, 6th Rev.

30. Sniffen, C. J.; O'Connor, J. D.; Van Soest, P. J.; Fox, D. G.; Russell, J. B. *J. Anim. Sci.* **1992**, 70: 3562-3577.

31. Dunlap, T. F.; Kohn, R. A.; Kalscheur, K. F. *J. Dairy Sci.* **1997**.
32. Ørskov, E. R.; Hughes-Jones, M.; McDonald, I. In *Recent Advances in Animal Nutrition.* ed. W. Haresign, Butterworths, London, **1980**, pp 85-98
33. Schwab, C. G.; Bozak, C. K.; Whitehouse, N. L.; Mesbah, M.M.A. *J. Dairy Sci.* **1992**, 75: 3486-3491.
34. Kung, L. Jr.; Huber, J. T. *J. Dairy Sci.* **1983**, 66: 227-232.
35. Ciszuk P.; Lindberg, J. E. *Acta Agric. Scand.* **1988**, 38: 381-395.
36. Voss, V. L.; Stehr, D.; Satter, L. D.; Broderick, G. A. *J. Dairy Sci.* **1988**, 71: 2428-2439.
37. Hamilton, B. A.; Ashes, J. R.; Carmichael, A. W. *Austr. J. Agric. Res.* **1992**, 43: 379-387.
38. Folman, Y.; Neumark, H.; Kaim, M.; Kaufmann, W. *J. Dairy Sci.* **1981**, 64: 759-768.
39. Robinson, P. H.; Kennelly, J. J. *J. Dairy Sci.* **1988**, 71: 2135-2142.
40. Lindberg, J. E. and I. Olsson. In IVth *Int. Symp. Protein Metabolism and Nutrition*, Clermont-Ferrand (France) Sept. 5-9. INRA Publ. 1983. II (les colloques de INRA no 16), Clermont, France.
41. Hunter, R. A; Siebert, B. D. *Aust. J. Agric. Res.* **1987**, 38: 209-218.
42. Newbold, J. R.; Garnsworthy, P. C.; Buttery, P. J.; Cole, D. J. A.; Haresign, W. *Anim. Prod.* **1987**, 45: 383-394.
43. Meissner, H. H.; Todtenhöfer, U. *S. Afr. J. Anim. Sci.* **1989**, 19: 43-53.
44. National Research Council. National Academy Press. Washington, D. C. **1985**.
45. Staples, C. R.; Lough, D. S. *Anim. Feed Sci. Technol.* **1989**, 23: 277-303.
46. Polan, C. E. *J. Nutr.* **1988**, 118: 242-248.
47. Rooke, J. A.; Armstrong, D. G. *Br. J. Nutri.* **1989**, 61: 113-121.
48. McAllan, A. B.; Smith, R. H. In IVth *Int. Symp. Protein Metabolism and Nutrition*, Clermont-Ferrand (France) Sept. 5-9. INRA Publ. 1983. II (les colloques de INRA no 16)
49. McAllan, A. B.; Griffith, E. S. *Anim. Feed Sci. Technol.* **1987**, 17: 65-73.
50. Adamu, A. M.; Russell, J. R.; McGilliard, A. D.; Trenkle, A. *Anim. Feed Sci. Technol.* **1989**, 22: 227-236.
51. Zorrilla-Rios, J.; Horn, G. W. *Anim. Feed Sci. Technol.* **1989**, 22: 305-320.
52. Mathison, G. W.; Milligan, L. P. *Br. J. Nutr.* **1971**, 25: 351-366.
53. Thomsen, K. V. *Acta Agric. Scand. Suppl.* **1985**, 25: 125-131.
54. Fujimaki, T.; Kobayashi, Y.; Wakita, M.; Hoshino, S. *J. Anim. Physiol. Anim. Nutr.* -- Zeitschrift Fur Tierphysiologie **1992**, 67: 41-50.
55. Madsen, J.; Hvelplund, T. *Acta Agric. Scand.* **1988**, 38: 115-125.
56. Broderick, G. A., *J. Anim. Sci.* **1995**, 73: 2760-2773.
57. Janicki, F. J.; Stallings, C. C. *J. Dairy Sci.* **1988**, 71:2440-2448.
58. Dahlberg, E. M.; Stern, M. D.; Ehle, F. R. *J. Anim. Sci.* **1988**, 66: 2071-2083.
59. Vik-Mo, L. *Acta Agric. Scand.* **1989**, 39: 53-64.
60. Mullahey, J. J.; Waller, S. S.; Moore, K. J.; Moser, L. E.; Klopfenstein, T. J. *Agron. J.* **1992**, 84: 183-188.

61. Kephart, K. D.; Boe, A. In *Amer. Forage and Grassland Council, Interpretive Summaries.* Amway Grand Plaza Hotel, Grand Rapids, MI. **1991**, pp 24.

62. Kohn, R. A.; Allen, M. S. *J. Dairy Sci.* **1995**, 78: 1544-1551.

63. Beever, D. E.; Cammell, S. B.; Wallace, A. *Proc. Nutr.* **1974**, 33: 73A.

64. Hoffman, P. C.; Sievert, S. J.; Shaver, R. D.; Combs, D. K. *J. Dairy Sci.* **1992**, 75(Suppl. 1):210.

65. Griffin, T. S.; Hesterman, O. B.; Rust, S. R. In *American Forage and Grassland Council, Interpretive Summaries.* Amway Grand Plaza Hotel, Grand Rapids, MI. **1991**, pp 23.

66. Messman, M. A.; Weiss, W.P.; Koch, M. E. *J. Dairy Sci.* **1994**, 77: 492-500.

67. Akin, D. E. *J. Anim. Sci.* **1979**, 48: 701-709.

68. Grabber, J. H.; Jung, G. A. *J. Sci. Fd. Agric.* **1991**, 57: 315-323.

69. Wilson, J. R.; Akin, D. E.; McLeod, M. N.; Minson, D. J. *Grass Forage Sci.* **1989**, 44: 65-75.

70. Nugent, J. H.; Mangan, J. L. *Br. J. Nutr.* **1981**, 46: 39-59

71. Messman, M. A.; Weiss, W. P.; Erickson, D. O. *J. Anim. Sci.* **1992**, 70: 566-575

72. Hazlewood, G. P.; Orpin, C. G.; Greenwood, Y.; Black, M. E. *J. Dairy Sci.* **1996**, 45: 1780-1784.

73. Dou, Z.; Kohn, R. A.; Ferguson, J. D.; Boston, R. C.; Newbold, J. D. *J. Dairy Sci.* **1996**, 79:2071-2080.

74. Patni, N. K.; Jui, P. Y.; Anim. Res. Ctr., Ottawa, ON, Canada, **1986**.

Chapter 17

Characterization of Soil Organic Nitrogen after Addition of Biogenic Waste Composts by Means of NMR and GC–MS

Stefanie Siebert[1,4], Heike Knicker[2], Mark A. Hatcher[2,3], Jens Leifield[1], and Ingrid Kögel-Knabner[2]

[1]Soil Science Group, University of Bochum, 44780 Bochum, Germany
[2]Department of Soil Science, Technische Universität München, 85350 Freising-Weihenstephan, Germany
[3]Department of Chemistry, The Pennsylvania State University, University Park, PA 16802

Application of composts to soils leads to increases in soil organic carbon and nitrogen. The effect of biowaste compost on the chemical composition of the organic N fraction was examined in different soils. Sandy and loamy soils were mixed with biowaste composts, both fresh and mature, and incubated for eighteen months. NMR spectroscopic characterization of the bulk samples and their residues after hydrolysis with 6 N HCl revealed no major changes in the organic nitrogen functionality as a result of the compost application. These spectra show that more than 80% of the total organic nitrogen of all samples was derived from peptide-like material. This conclusion was supported by the results obtained from tetramethylammonium hydroxide (TMAH) thermochemolysis.

Biogenic waste composts are added to agricultural and reclaimed soils to reduce landfill volume for municipal waste. The application of biowaste composts is expected to increase soil organic carbon and nitrogen and therefore to improve the nutrient supply of soils. This treatment also leads to improvement of the physical structure of soils (1-5). The efficiency of such a procedure for recultivation is strongly influenced by the properties and the amount of newly formed humic material, but it also depends upon the availability of added nitrogen after recultivation.

The processes involved in the decomposition of organic biowaste appear to be comparable to those involved in the degradation of organic matter. The decomposition of organic wastes follows an exothermic process by biological oxidation. During degradation, materials are chemically and physically transformed into stable humified products (6-9). Biddlestone et al. (6) described the composting of

[4]Current address: In der Uhlenflucht 12, 44795 Bochum, Germany.

organic waste as a"building up" and "breaking down" process. Senesi (3) divided the degradation of organic materials into three stages. First, readily decomposable organic compounds, such as sugars, starch, hemicellulose, amino acids and some cellulose, are converted to CO_2 and other volatile compounds such as N_2O and ammonia. In the second stage, organic metabolites, biomass, and the remnants of cellulose, as well as parts of lignin, are incorporated into new biomass and metabolized to CO_2. The final stage is characterized by a gradual degradation of the more resistant compounds such as lignin, which are converted into humic substances.

However, while the soil organic carbon has been the subject of many investigations (10, 11), little is known about the fate of soil organic nitrogen (12). Characterization of the organic nitrogen pool is still incomplete and knowledge about the chemical structure and the linking of nitrogen compounds to humic substances is lacking (13).

Approximately 90% of the total nitrogen in soils is incorporated into the organic fraction. Generally, this organic nitrogen is analyzed by hot acid hydrolysis with 6 N HCl for 12 to 24 h. Applying this technique, 20 to 35% and 5 to 10% of the organic nitrogen can be found in amino acids and amino sugars, respectively. Twenty to 35% of the organic nitrogen is present as ammonium, while 10 to 20% of the organic nitrogen are hydrolyzed compounds not yet identified (hydrolyzable unknown nitrogen = HUN). Twenty to 35% of the organic nitrogen remains as acid insoluble nitrogen, which may be incorporated into refractory organic materials and protected against biological degradation (12-16). This resistance might be explained by the incorporation of nitrogen into heterocyclic compounds such as pyridines, indoles and pyrroles (16, 17). Recent evidence, on the other hand, indicated that most of the soil organic nitrogen is in amide functional groups, most probably of biogenic origin (18-20).

The intention of this study was to investigate the forms of soil organic nitrogen, found in soils after addition of fresh and mature biowaste composts by means of solid-state ^{15}N NMR and thermochemolysis with tetramethylammonium hydroxide (TMAH) (21-25). To learn more about the acid insoluble organic-N forms, these techniques were also applied to the residues obtained after acid hydrolysis of the composts and the soil/compost mixtures (26).

Materials and Methods

For incubation experiments soil substrate of a Haplic Luvisol under agricultural use near Witzenhausen, Germany (sand: 6.7 %, silt: 73.8 %, clay: 19.6 %) and a coal mine spoil of the Lusatian lignite mining district of Germany (sand: 93%, silt: 4.6%, clay: 2.4%) were each mixed with 70 t ha^{-1} of a fresh biowaste compost. In a second study, the substrates of the same soil and mine spoil were treated with 65 t ha^{-1} of a mature biowaste compost. The maturity of the composts was determined by the self-heating test (27). Under controlled conditions, at a temperature of 14°C and 50% maximum water holding capacity (WHC), the samples were incubated in a microcosm system, as described by Siebert et al. (28), for 18 months. Soil substrates without addition of compost were incubated as controls. Immediately after application of the composts and after 18 months of incubation, samples were taken, freeze-dried and ground. The nitrogen and carbon contents were determined for all samples using an CHN-1000 elemental analyzer (LECO).

Acid Hydrolysis. For the determination of the a-amino acid content, 2 g of each sample were hydrolyzed with 10 ml 6 N HCl for 12 h at 105°C (*29, 30*). The hydrolyzates were filtered (blue band filter No. 589) and washed with distilled water. The filtrates were lyophilized and freeze-dried. After addition of a sodium citrate solution and a solution containing a ninhydrin reagent, the supernatants were boiled for 20 minutes in a water bath. During this boiling time ninhydrin was allowed to react with α-amino acids to form purple-colored complexes. Their concentration was determined by measuring the optical density of the reacted solution at 570 nm. The amount of a-amino acid N was calculated by a calibration curve with a standard solution of 28 µg leucine ml^{-1}. The non-hydrolyzable nitrogen was obtained by measuring the nitrogen content of the hydrolytic residues.

^{13}C and ^{15}N NMR Spectroscopy. To improve the sensitivity of the NMR experiment, the samples and the hydrolytic residues were treated with 10 % hydrofluoric acid (HF), as described by Schmidt et al. (*31*). HF-treatment of soil samples was shown to increase the concentration of soil organic matter by removal of the mineral matter without any major alteration of the soil organic matter composition. Solid-state cross polarization magic angle (CPMAS) ^{13}C NMR spectra were taken on a Bruker MSL 100. A contact time of 1.0 ms and pulse delays between 100 and 600 ms were used. The magic angle spinning speed was 4.3 kHz. The ^{13}C chemical shifts are referred to external tetramethylsilane (= 0 ppm). Solid-state CPMAS ^{15}N NMR spectra were obtained on a Bruker MSL 300 (7.05T) with a contact time of 1 ms, a pulse delay of 100 ms, and a magic angle spinning speed of 4.3 kHz. The chemical shift assignment is referenced to the nitromethane scale (= 0 ppm) (*32*). A more detailed description of the applied parameters can be found in Knicker and Lüdemann (*20*).

Thermochemolysis with Tetramethylammonium Hydroxide (TMAH). Samples were subjected to thermochemolysis with TMAH in sealed tubes. Previously this technique was shown to cleave peptide-bonds in albumin and to lead to specific methylated derivatives of amino acids (*26*). For their identification, soil and compost samples and their residues (1 mg - 10 mg of sample) were heated with 100 µl TMAH (25% in methanol) for 30 minutes at 250°C. The thermochemolysis products were extracted with methylene chloride and concentrated to 50 µl. An aliquot of 1 µl was injected onto a J&W DB5 MS capillary column (30 m x 0.25 mm) of a combined gas chromatograph mass spectrometer (GC/MS) system (Fison, MD 800). The GC was programmed with an initial temperature of 40°C, a heating rate of 15°C min^{-1} to 100°C. The heating rate was then minimized to 6°C min^{-1} to a final temperature of 300°C. The methylated compounds were identified by their retention times and mass spectra.

Results

Characterization of SOM after Additon of Biogenic Composts to Soils. The biowaste composts are characterized by a low C to N ratio (C/N=13 fresh compost; C/N=12 mature compost). After addition of these composts to the loamy substrate of the Luvisol and to the sandy substrate of the mine spoil, the carbon and nitrogen content of the soil substrates increased. The carbon content increased from 13 mg C g^{-1} soil to 34 mg C g^{-1} soil in the Luvisol after compost addition. As well as the carbon content, the amount of nitrogen increased from 1.7 mg N g^{-1} soil to 3.7 mg N g^{-1} soil. The addition of biowaste composts to the mine spoil resulted in an enrichment of the C and N content from 0 to 17 mg C g^{-1} soil and to 1.8 mg N g^{-1} soil (Table I).

Table I: Chemical Characterization of Biowaste Composts and Soils

	C_t	N_t	C/N	N_{org}	N_{inorg}
	---- mg g^{-1} ----			% of total N	
Compost					
fresh compost	232.7 ± 3.6	18.5 ± 0.3	12.6 ± 0.3	99.9 ± 0.0	0.1 ± 0.0
mature compost	186.1 ± 1.9	15.9 ± 0.3	11.7 ± 0.3	99.5 ± 0.0	$0.5 \pm 0..0$
Mine spoil (0 months)					
soil	n.d.	n.d.	-	-	-
soil + fresh compost	19.1 ± 0.0	2.1 ± 0.0	9.1 ± 0.0	99.6 ± 0.0	0.4 ± 0.0
soil + mature compost	15.2 ± 0.0	1.5 ± 0.0	10.1 ± 0.0	99.0 ± 0.0	1.0 ± 0.0
Mine spoil (18 months)					
soil	n.d.	n.d.	-	-	-
soil + fresh compost	14.7 ± 0.0	1.6 ± 0.1	9.2 ± 0.4	91.2 ± 0.7	8.8 ± 0.7
soil + mature compost	11.9 ± 0.7	1.2 ± 0.1	9.9 ± 0.1	90.2 ± 1.0	9.8 ± 1.0
Luvisol (0 months)					
soil	13.0 ± 0.0	1.7 ± 0.0	7.7 ± 0.0	98.5 ± 0.0	1.5 ± 0.0
soil + fresh compost	33.7 ± 0.0	3.5 ± 0.0	9.6 ± 0.0	99.3 ± 0.0	0.7 ± 0.0
soil + mature compost	35.0 ± 0.0	3.8 ± 0.0	9.2 ± 0.0	99.0 ± 0.0	1.0 ± 0.0
Luvisol (18 months)					
soil	12.7 ± 0.0	1.7 ± 0.0	7.5 ± 0.2	94.6 ± 0.4	5.4 ± 0.4
soil + fresh compost	26.5 ± 0.1	3.3 ± 0.0	8.0 ± 0.1	88.6 ± 0.1	11.4 ± 0.1
soil + mature compost	27.0 ± 1.3	3.4 ± 0.1	8.0 ± 0.2	90.5 ± 0.0	9.5 ± 0.0

n.d = not detectable; Standard error (compost: n = 6; soil, soil + compost: n = 2)

After an incubation time of 529, days the total C content of the Luvisol treated with fresh compost decreased from 33.7 mg C g^{-1} soil to 26.5 mg C g^{-1} soil. A comparable decrease from 35 mg C g^{-1} soil to 27 mg C g^{-1} soil was determined for the Luvisol treated with mature compost. During the same time period, the nitrogen content of the Luvisol treated with fresh compost decreased only from 3.5 mg N g^{-1} soil to 3.3 mg N g^{-1} soil and in the Luvisol incubate mixed with mature compost from

3.8 mg N g^{-1} soil to 3.4 mg N g^{-1} soil. Consequently, the C/N ratios of the soil compost mixtures declined from approximately 9 and 10 at the beginning of the experiment to 8 after 18 months of incubation. In contrast to the Luvisol, the mine spoil, as a sandy sterile substrate, contains almost no organic material (Table I). Therefore, changes in organic matter composition of the mine spoil/compost mixture occurring during incubation are only related to the decomposition of the mature and fresh compost. During incubation, a decrease in the total C and N contents of approximately 23% and 24%, respectively, was observed in the mine spoil sample mixed with fresh compost. In the mine spoil substrate mixed with mature compost, the total C declined to 22% and the total N to 20% as result of 18 months incubation. This indicates that the changes in elemental composition occurring during incubation of the mine spoil/compost mixtures, but also in the Luvisol/compost mixtures, are mainly induced by the degradation of the compost materials. However, in both the incubates of the Luvisol and the mine spoil mixed with composts, the total nitrogen and carbon content is still higher than in the control samples (Table I). From these results it can be assumed that additon of biowaste compost to arable and reclaimed soils leads to an increase in the soil organic matter level.

For all samples, with the exception of the mine spoil substrate, the relative concentration of organic and inorganic nitrogen in relation to the total nitrogen of the samples was determined (Table I). Less than 2% of the total nitrogen was assigned to inorganic nitrogen in the pure composts and soil substrates. At the beginning of the incubation experiment, less than 1% of the total nitrogen originates from inorganic nitrogen. After 529 days of incubation, the inorganic N fraction increased in the soil and compost mixtures to 10% of total nitrogen, indicating that at least some mineralization of organic N occurred. The N mineralization in the soil without composts yielded only 4% of total nitrogen. However, at this time approximately 90% of the total nitrogen in all incubates were still found in the organic fraction, revealing that organic nitrogen of the composts was incorporated into the stable organic fraction of the soil.

Characterization of SON by Acid Hydrolysis. The characterization of SON by wet chemical analysis with 6 N HCl shows that 17 to 31% of the organic N is present as a-amino acids in the composts, Luvisol and in the soils-compost mixtures (Table II). The maturity of biowaste composts did not affect the yield in acid hydrolyzable amino acid content. After hydrolysis, a high N amount (>60% of organic N) remained in the hydrolytic residues of the composts, showing that only a small portion of the nitrogen pool is liberated by acid hydrolysis of compost materials. This agrees with the results of Bremner (33), who found that 20 to 60% of the nitrogen in humic acids is not solubilyzed by hydrolysis. Addition of biowaste composts to soils leads to a relative decrease of the N content in the hydrolytic residues. However, 28 to 43% of the organic N still remained in the refractory N fraction after compost addition to soil. Without compost addition the soils yielded the lowest amounts of acid insoluble N (<25% of organic N), while the a-amino N content in the soil was more or less the same as in the soil and compost mixture. After addition of composts to the soil, the hydrolyzable unknown N fraction yielded 36 to 47% of organic N. This fraction

298

consisting of amino sugars and ammonia from the partial destruction of amino acids, such as serine, threonine and tryphtophan during hydrolysis (13). Schnitzer (16) and others have pointed out that some of the unknown nitrogen may be present in heterocyclic structures. However, with the exception of a small amount of purines and pyrimidines, such heterocyclic compounds could not be identified in soils in relevant amounts (16).

Table II: Organic Nitrogen Fractions after Acid Hydrolysis

	a-amino N	HUN	Hydrolytic residue
		% of organic nitrogen	
Compost			
fresh compost	30.9 ± 0.7	5.2 ± 1.7	63.9 ± 1.1
mature compost	24.0 ± 0.3	0.0 ± 1.6	76.0 ± 1.4
Mine spoil (0 months)			
soil	n.d.	n.d.	n.d.
soil + fresh compost	21.8 ± 0.6	44.7 ± 1.8	33.5 ± 2.4
soil + mature compost	16.6 ± 0.0	43.1 ± 3.3	40.3 ± 3.3
Mine spoil (18 months)			
soil	n.d.	n.d.	n.d.
soil + fresh compost	34.8 ± 0.5	37.8 ± 0.6	27.4 ± 0.0
soil + mature compost	26.9 ± 0.2	36.1 ± 0.7	37.0 ± 0.5
Luvisol (0 months)			
soil	18.1 ± 3.3	58.1 ± 3.3	23.8 ± 0.0
soil + fresh compost	16.0 ± 0.7	40.8 ± 0.8	43.2 ± 1.4
soil + mature compost	19.8 ± 0.5	40.3 ± 1.8	39.9 ± 1.3
Luvisol (18 months)			
soil	21.8 ± 0.9	53.3 ± 1.3	24.9 ± 0.9
soil + fresh compost	19.5 ± 1.3	46.4 ± 1.2	34.1 ± 0.2
soil + mature compost	21.0 ± 0.3	46.5 ± 1.6	32.5 ± 1.5

HUN = hydrolyzable unknown nitrogen; n.d. = not detectable;
Standard error (compost: n = 6 ; 0 months: n = 2 ; 18 months: n = 4)

The results listed above in Table II indicate that the addition of composts to the soils leads to an increase in the nitrogen content of the hydrolytic residues. By this it can be concluded that, already in the compost material, some of the nitrogen is incorporated into the organic fraction. It has been further shown that detailed information about the linking of the organic nitrogen in the soils and composts cannot be obtained by acid hydrolysis alone.

Characterization of Soil Organic Matter by [13]**C NMR.** The ^{13}C NMR spectra of the composts and their hydrolytic residues are presented in Figures 1a and 1b. As revealed in Table III, the fresh and mature compost do not show major differences in the bulk chemical composition. Both ^{13}C NMR spectra are characterized by intense signals in the chemical shift region of 0-alkyl carbons (110 to 45 ppm), most tentatively assigned to carbohydrates. However, carbon bound in ether or alcoholic groups may also contribute to the signal intensity in this chemical shift region. It is also worthwhile to mention, that besides methoxyl carbon, amine-substituted carbon may add to the signal in the chemical shift region between 60 and 45 ppm. The peaks in the chemical shift region between 160 and 110 ppm are assigned to aromatic carbons. The signals between 220 and 160 ppm are attributed to carboxylic C and/or amide C, while those between 45 and 0 ppm originate from alkyl carbons, i.e. in aliphatic chains. The similarity between the ^{13}C NMR spectra of the fresh and mature composts suggests that most of the chemical alterations during degradation occurred during an earlier stage of the decomposition process. In general, such alterations are characterized by a loss of carbohydrates and a relative enrichment of aromatic, carboxylic and aliphatic carbons (*11, 20, 34, 35*). With compost maturity the degradation process slows down. Changes in the chemical composition of the degrading material become less detectable by ^{13}C NMR spectroscopy.

Table III: Relative Carbon Distribution of the Composts, the Luvisol, the Luvisol/Compost Mixtures after 18 Months of Incubation and their Hydrolytic Residues determined by Solid-State [13]**C NMR**

	Carboxyl C 220-160	Aromatic C 160-110	O-alkyl C 110-45	Aliphatic C 45 - -10
			ppm	
	% of total signal intensity			
Untreated materials				
Fresh compost	11	19	49	22
Mature Compost	9	22	54	15
Luvisol	10	23	49	18
Luvisol/fresh compost	10	25	45	20
Luvisol/mature compost	10	25	45	20
HCl - residues				
Fresh compost	6	30	40	23
Mature compost	7	32	42	19
Luvisol	7	34	36	23
Luvisol/fresh compost	7	35	36	22
Luvisol/mature compost	7	36	35	22

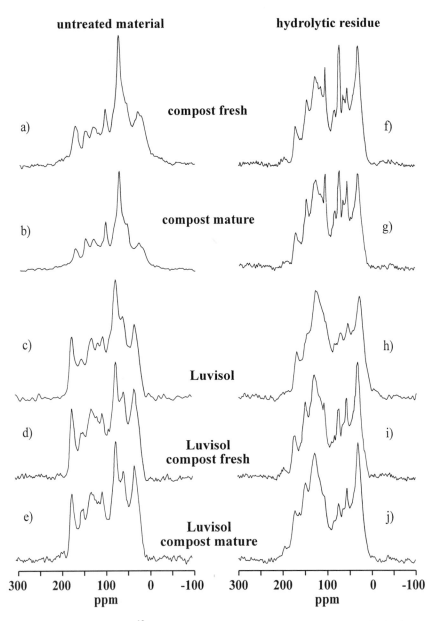

Figure 1.   Solid-state ^{13}C NMR spectra of the composts, the Luvisol, the Luvisol/compost mixtures and their hydrolytic residues.

The ^{13}C NMR spectrum of the untreated Luvisol (Figure 1c) still shows its major signal intensity in the chemical shift region assigned to carbohydrates (110 to 45 ppm). However, compared to the ^{13}C NMR spectra of the composts, the relative intensity in this region is diminished in favor of the relative intensity in the region assigned to aromatic carbons. The ^{13}C NMR spectra of the Luvisol, obtained 18 months after addition of the composts (Figure 1d, e) reveal a similar pattern to that of the natural soil. From this, it can be assumed that the composition of the organic material in the soil/compost mixture is still determined by the soil substrate. The application of biowaste composts to the soil, therefore, did not lead to major alterations in the soil organic matter of the Luvisol.

Acid hydrolysis with 6 N HCl of the composts and the soil substrate/compost mixtures resulted in a relative enrichment of aromatic compounds. This is revealed by comparison of the intensity distribution in the ^{13}C NMR spectra of the untreated samples and the hydrolytic residues (Table III; Figure 1a-j). The decrease of signal intensity in the chemical shift region between 110 and 45 ppm (O-alkyl carbon) and 220 to 160 ppm (carbonyl carbon, carboxyl- and amide carbons) as a result of the hydrolysis indicates the loss of labile compounds such as carbohydrates and amino acids. Compared to the ^{13}C NMR spectrum of the hydrolytic residue of the Luvisol (Figure 1h), a more intense signal at 72 ppm in the hydrolytic residues of the compost and soil/compost mixtures is observed. This can be explained by the presence of higher amounts of acid insoluble cellulosic material in the composts than in the more humified organic material of the soil.

Characterization of Soil Organic Nitrogen by ^{15}N NMR. The ^{15}N NMR spectra of the composts and their hydrolytic residues are presented in Figure 2a, b, f, g. The untreated composts and the hydrolytic residues of the composts show similar spectra. A broad signal is found in the chemical shift region between -220 and -285 ppm, most probably assigned to amide/peptide functional groups (20). No signals, distinguishable from the noise, are observed in the chemical shift region of pyridinic N (-40 to -145 ppm), indicating that such compounds were not formed in higher amounts during decomposition. However, nitrogen in acetylated amino sugars, lactams, unsubstituted pyrroles, indoles and carbazoles may contribute to the intensity of the main peak. The heterocyclic N in histidine, nucleic acid derivatives and substituted pyrroles is expected to contribute to the chemical shift region between -145 and -220 ppm, and may be hidden by the broad main signal at -260 ppm. Had such compounds been major contributors to the total nitrogen, the main peak at -260 ppm would be shifted towards lower field.

The ^{15}N NMR spectra of the Luvisol and the soil substrate/compost mixtures in Figure 2 c-e show a comparable pattern to that of the ^{15}N NMR spectra of the untreated composts. They are also dominated by a peak at -260 ppm, assigned to amide functional groups. The similarity of the ^{15}N NMR spectra of the composts, soil and soil substrate compost mixtures indicates that, in all of these samples, the nitrogen is bound in similar functional groups. Even after 18 months of incubation, no major alterations in N functionality are detected by ^{15}N NMR spectroscopy. In these spectra, no major signal intensity can be observed in the chemical shift region of

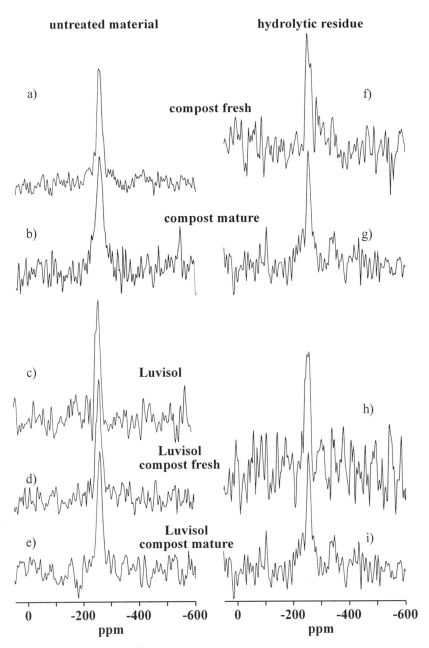

Figure 2. Solid-state ^{15}N NMR spectra of the composts, the Luvisol, the Luvisol/compost mixtures and their hydrolytic residues.

pyridinic N or pyrrolic N. From these results, it can be concluded that heterocyclic nitrogen is not a major form of organic nitrogen in these samples. The ^{15}N NMR spectra shown here rather support the assumption that some amide functional groups, presumably of biogenic origin, are more resistant to microbial degradation than previously thought (18-20, 26, 36, 37).

It is generally believed that most of the proteinaceous material is hydrolyzed by acid hydroylsis. In our experiments, on the other hand, amino acid N, liberated with acid hydrolysis, can only explain less than 30% of the total signal intensity observable in ^{15}N NMR spectra of the composts, Luvisol and soil substrate/compost mixtures. From this observation it can be concluded that some amides, identified with ^{15}N NMR spectroscopy, are not hydrolyzed by the commonly used acid hydrolysis.

To verify this conclusion, the acid insoluble residues obtained after HCl-hydrolysis of the composts and the soil substrate/compost mixtures were also subjected to ^{15}N NMR spectroscopy. In order to increase the sensitivity of the ^{15}N NMR experiment, the residues, depleted in organic material, were treated with HF. As revealed from Table IV, applying this technique increased the relative amount of organic nitrogen of the samples to a factor which allowed us to obtain ^{15}N NMR spectra with acceptable S/N ratios.

Table IV: Enrichment of C and N in the Hydrolytic Residue due to Treatment with 10% (v/v) HF

	Before treatment	After treatment[*]	Enrich-ment[1]	Before treatment	After treatment[*]	Enrich-ment[1]
	C mg g[-1]			N mg g[-1]		
Compost mature	189.4 ± 0.3	552.2	2.9	12.0 ± 0.1	33.8	2.8
Compost fresh	251.4 ± 0.6	571.2	2.3	11.8 ± 0.0	26.0	2.2
Luvisol/ compost mature	16.0 ± 0.1	467.4	29.0	1.2 ± 0.0	22.3	25.7
Luvisol/ compost fresh	17.0 ± 0.3	472.7	27.3	1.0 ± 0.0	25.7	22.3

[1] Enrichment = [content of C or N after HF treatment]/[content of C or N before HF treatment]. Standard error (n = 2); [*] - single analyzed.

The non-hydrolyzable residues (Figure 2 f-i) show the main intensity in the chemical shift region assigned to amide functional groups. Thus, hydrolysis with 6 N HCl failed to significantly alter the spectral signature of this refractory amide-like N. Similar to the ^{15}N NMR spectra of the bulk samples, no signals indicative of pyrrolic N or pyridinic N, can be observed in the ^{15}N NMR spectra of the residues. This result strongly contradicts common models, in which acid insoluble nitrogen in soils is explained by the formation of nitrogen heterocyclic aromatics during the humification process. The ^{15}N NMR spectra of the hydrolytic residues presented here rather support the above mentioned assumption that most of the N in the non-hydrolyzable residues of our samples is present in form of amides, which are protected from the intense chemical hydrolysis.

Characterization of SON by Thermochemolysis with Tetramethylammonium Hydroxide (TMAH). The procedure of thermochemolysis with tetramethyl-ammonium hydroxide (TMAH) has been demonstrated to cleave ester and ether bonds (*38*) and only recently was shown to efficiently cleave peptide bonds (*26*). Also, Schulten and Sorge (*39*) determined nitrogen compounds in the form of N,N-dimethylamides in soils applying a comparable technique, the pyrolysis methylation reaction with TMAH. If some of the amide nitrogen observed in the ^{15}N NMR spectra of the hydrolytic residues originate from proteinaceous or peptide-like material, such compounds should be cleaved during thermochemolysis with TMAH. Their methylated products were then identified with gas chromatography/mass spectrometry. Figure 3 represents the chromatogram of the TMAH/thermochemolysis products obtained from the hydrolytic residue of the mature compost. The numbered peaks (A-7 to A-17) indicate products, which were also identified after TMAH/thermochemolysis of bovine serum albumin (*26*). These products and their possible origin are listed in Table V. We did not find any peaks indicating higher amounts of TMAH/thermochemolysis products of pyridinic N in the hydrolytic residues of our samples.

Table V: Peak Assignments for Proteinaceous Material of Hydrolyzed Mature Compost by Thermochemolysis with TMAH

Peak	Compound	Possible origin
A-7	Dimethylalanine, methyl ester	Alanine
A-10	Benzaldehyde	Phenylalanine
A-11	Dimethylvaline, methyl ester	Valine
A-12	Benzene, (methoxymethyl)-	Thyrosine
A-14	Butanediocic acid, dimethyl ester	
A-15	N-methyl-, L-proline, methyl ester	Proline
A-17	N,N dimethyl, leucine (or isoleucine), methyl ester	Leucine or Isoleucine

Similar results were obtained for the soil substrate/compost mixtures. The identification of proteinaceous materials in the hydrolytic residues of our samples confirms the results of Knicker and Hatcher (*26*) that some peptide-like compounds in decaying organic material are protected against acid hydrolysis. Due to the low mineral matter content of the sediment, the resistance of proteinaceous material in the sapropel of Mangrove Lake was explained by the encapsulation of peptides into a refractory network of algal material. In soil with considerable mineral content, this mineral matrix may be involved in the stabilization of otherwise labile compounds. Peptide-like material can be adsorbed onto clay minerals and organic matter (*13, 40*).

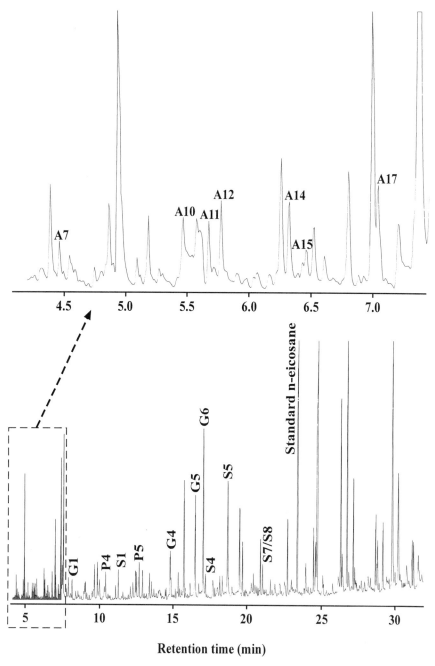

Figure 3. Gas chromatogram of the TMAH/thermochemolysis products of the hydrolytic residue of the mature compost.

Proteinaceous material protected in such a fashion may resist hydrolysis with 6 N HCl.

Summary and Conclusions

The objective of this study was to characterize the soil organic nitrogen pool after the amendment of biowaste compost to an agricultural soil and to a reclaimed sandy substrate of a mine spoil. For the characterization of SON, the composts, the soils and the soil/compost mixtures were analyzed by acid hydrolysis and by means of ^{13}C and ^{15}N NMR spectroscopy. Detailed information about the refractory nitrogen fraction was obtained by the examination of the acid hydrolysis residues from the different materials by ^{15}N NMR spectroscopy and by thermochemolysis with TMAH.

The addition of biogenic waste composts to soils led to an increase in soil organic carbon and nitrogen. After acid hydrolysis 17 to 31% of the organic N were identified as a-amino acids, 36 to 47% of the organic N was lost by hydrolysis as unknown nitrogen. 28 to 43% of organic N remained in the hydrolytic residue. ^{13}C NMR spectroscopic characterization of a Luvisol before and 18 months after application of biowaste composts indicated no major alterations of the soil organic matter composition. Acid hydrolysis resulted in a loss in intensity in the chemical shift region of O-alkyl carbons and carboxyl carbons of the ^{13}C NMR spectra of the composts, the substrate of a Luvisol, and mixtures of a Luvisol with biowaste composts. Increase in relative intensity was observed in the chemical shift region of aliphatic and aromatic carbons.

Characterization of the soil organic nitrogen pool was done by solid-state ^{15}N NMR spectroscopy. The organic nitrogen of the soils and composts is characterized by a broad signal in the chemical shift region between -220 and -285 ppm, which is tentatively assigned to amide/peptide functional groups. More than 80% of the total organic nitrogen is bound in peptide-like structures. To characterize the nature of the acid insoluble nitrogen, solid-state ^{15}N NMR spectroscopy was applied to the hydrolytic residues of the biowaste composts and the Luvisol/compost mixtures. Generally, it is assumed that peptide-like material in soils is quickly hydrolyzed with 6 N HCl. Only nitrogen bound in chemically- or physically-protected compounds should remain in the acid insoluble residue. In common models, it is assumed that this refractory nitrogen is incorporated into N-heterocyclic structures. However, in the studies presented here, such heterocyclic structures could not be identified by ^{15}N NMR spectroscopy. For detailed characterization we subjected the hydrolytic residues to a new technique of thermochemolysis with tetramethylammonium hydroxide (TMAH) and analyzed the products with gas chromatography/mass spectrometry. Several methylated products of amino acid derivatives were identified, but methylated compounds related to heterocyclic nitrogen compounds could not.

Our results confirm that refractory nitrogen in soil organic matter is probably composed of peptide-like material (*19, 20*). They further support the assumption of previous studies (*27*) that acid hydrolysis with 6 N HCl fails to attack all proteinaceous compounds in organic material of soils and sediments. The mechanisms, responsible for the resistance of such compounds towards acid hydrolysis may be similar to those involved in their protection against microbial degradation. Such mechanisms could be of physical or chemical nature involving the adsorption to clay minerals (*15*) or humic material incorporation into mineral sheets (*41*) but also as part or in connection with refractory biopolymers (*26, 36*). However, regardless of the mechanism involved in the protection of these generally thought labile compounds in soil, their presence in soils could partly explain the nature of the commonly-termed unidentified nitrogen.

Acknowledgements. Financial support was obtained from the Federal German Ministry of Science and Technology (Nr. 1460638 L). We thank PlanCoTec from Witzenhausen, Germany for providing the compost materials. We are grateful to H.-D. Lüdemann for providing the NMR facilities at the University of Regensburg.

Literature Cited

1. Almendros, G.J.; Leal, A.; Martin, F.; González-Vila, F.J. In: *Humic Substances in the Aquatic and Terrestrial Environment*; Allard, B.; Borén, H.; Grimvall, A., Ed.; Springer Verlag: Berlin, 1989, 205-216.
2. Grundmann, J. *Müll und Abfall* **1991**, 5, 268-273.
3. Senesi, N. *The Science of the Total Environment* **1989**, 81/82, 521-542.
4. Tester, C.F. *Soil Sci. Soc. Am. J.* **1990**, 54, 827-831.
5. Vogtmann, H.; Fricke, K. *Agriculture, Ecosystems and Environment* **1989**, 27, 471-475.
6. Biddlestone, A.J.; Gray, K.R.; Day, C.A. In: *Environmental Biotechnology*; Torser, C.F.; Wase, D.A.J., Ed.; John Wiley and Sons: New York, 1987, 135-175.
7. Hänninen, K.I.; Kovalainen, J.T.; Korvola, J. *Compost Science & Utilization* **1995**, 3, 51-68.
8. Witter, E.; Lopez-Real, J.M. In: *Compost: Production, Quality and Use*; De Bertoldi; M. et al., Ed.; Elsevier Applied Science: London, 1987, 351-358.
9. Zucconi, F.; De Bertoldi, M. *Global Bioconversions*; Wise, D.L., Ed.; CRC Press: Florida, 1987, 109-137.
10. Inbar, Y.; Chen; Y.; Hadar, Y.; Hoitink, H.A.J. *BioCycle* **1990**, 31, 64-69.
11. Kögel-Knabner, I.; Hatcher, P. G.; Tegelaar, E. W.; de Leeuw, J. W. *The Science of the Total Environment* **1992**, *113*, 89-106.
12. Kelley, K.R.; Stevenson, F.J. In: *Humic Substances in Terrestrial Ecosystems*; Piccolo, A., Ed.; Elsevier Science B. V.: Amsterdam, 1996, 407-427.
13. Stevenson, F.J. *Humic Chemistry*; John Wiley and Sons: New York, 1994; 496.
14. Greenfield, L.G. *Plant and Soil* **1972**, 36, 191-198.
15. Haider, K. *Biochemie des Bodens*; Enke: Stuttgart, Germany, 1996; pp 174.
16. Schnitzer, M. In: *Humic substances II*; Hayes, M.H.B.; Malcolm, R.I.; Swift, R.S., Ed.; John Wiley and Sons: New York, 303-325.
17. Bremner, J.M. *Journal of Agricultural Science* **1949**, 39, 183-193.

308

18. Clinton, P.W.; Newman, R.H.; Allen, R.B. *European Journal of Soil Science* **1995**, 46, 551-556.
19. Knicker, H.; Fründ, R.; Lüdemann, H.-D. *Naturwissenschaften* **1993**, 80, 219-221.
20. Knicker, H.; Lüdemann, H.-D. *Organic Geochemistry* **1995**, 23, 329-341.
21. Challinor, J.M. *Journal of Analytical and Applied Pyrolysis* **1995**, 35, 93-107.
22. del Rio, J.C.; Hatcher, P.G. In: *Humic and Fulvic Acids Isolation, Structure and Environmental Role*; Gaffrey, J.S.; Marley, N.A.; Clark, S.B., Ed.; ACS: Washington D.C., 1996, 82-95.
23. Hatcher, P.G.; Minard, R.D. *Organic Geochemistry* **1996**, 24, 593-600.
24. Hatcher, P.G.; Nanny, M.A.;. Minard, R.D.; Dible, S.D.; Carson, D.A. *Organic Geochemistry* **1995**, 23, 881-888.
25. Martin, F.; del Rio, J.C.; González-Vila, F.J.; Verdejo, T. *Journal of Analytical and Applied Pyrolysis* **1995**, 35, 1-13.
26. Knicker, H.; Hatcher, P.G. *Naturwissenschaften* **1997**, 84, 231-234.
27. *Methoden zur Analyse von Kompost*; BGK Bundesgütegemeinschaft Kompost e.V. Ed.; Abfall Now: Stuttgart, 1994.
28. Siebert, S.; Leifeld, J.; Kögel-Knabner, I. In: *The Science of Composting*; De Bertoldi, M.; Sequi, P.; Lemme, B.; Papi, T., Ed.; Black Academic & Professionell: London, 1996, 1335-1338.
29. Stevenson, F.J. In: *Methods of soil analysis*; Page, A.L.; Miller, R.H.; Keeney, D.R., Ed.; ASA-SSSA: Madison, 1982, Part 2; 625-641.
30. Kögel-Knabner, I. In: *Methods in applied soil microbiology and biochemistry*, Nannipieri, P.; Alef, K., Ed.; Academic Press: New York, 1995, 66-78.
31. Schmidt, M.W.I.; Knicker, H.; Hatcher, P.G.; Kögel-Knabner, I. *European Journal of Soil Science* **1997**, 48, 319-328.
32. Witanowski, M.; Stefaniak, L.; Webb, G.A. In: *Nitrogen NMR Spectroscopy*; Webb,G.A., Ed.; Academic Press: London, 1986.
33. Bremner, J.M. *Journal of Agricultural Science* **1954**, 44, 217-256.
34. Inbar, Y.; Chen, Y.; Hadar, Y. *Soil Science* **1991**, 152, 272-282.
35. Inbar, Y.; Chen, Y.; Hadar, Y. *Soil Sci. Soc. Am. J.* **1989**, 53, 1695-1701.
36. Knicker, H.; Almendros, G.; Gonzaléz-Vila, F.J.; Lüdemann, H.-D.; Martin, F. *Organic Geochemistry* **1995**, 23, 1023-1028.
37. Knicker, H.; Lüdemann, H.-D.; Haider, K. *European Journal of Soil Science* **1997**, 48 (3).
38. McKinney, D.E.; Carson, D.M.; Clifford, D.J.; Minard, R.D.; Hatcher, P.G. *Journal of Analytical and Applied Pyrolysis* **1995**, 34, 41-46.
39. Schulten, H.-R.; Sorge, C. *European Journal of Soil Science* **1995**, 46, 567-579.
40. Paul, E.A; Clark, F.E. *Soil microbiology and biochemistry*; Academic Press: London, 1989.
41. Theng, B.K.G.; Churchman, G.J.; Newman, R.H. *Soil Science* **1986**, 142, 262-266.

Chapter 18

Effects of Long-Term Fertilizer and Manure Treatments on the Distribution and ^{15}N Natural Abundance of Amino Acids in the Palace Leas Meadow Hay Plots: A Preliminary Study

R. Bol[1], J. M. Wilson[2], R. S. Shiel[2], K. J. Petzke[3], A. Watson[4], and J. Cockburn[4]

[1]Institute of Grassland and Environmental Research, North Wyke, Okehampton, Devon EX20 2SB, United Kingdom
[2]Department of Agricultural and Environmental Science, University of Newcastle upon Tyne, Newcastle upon Tyne NE1 7RU, United Kingdom
[3]German Institute of Human Nutrition, Potsdam-Rehbrucke, Bergholz-Rehbrucke, D-14858, Germany
[4]Institute of Grassland and Environmental Research, Plas Gogerddan, Aberystwyth, Ceredigion SY23 3EB, United Kingdom

The concentration and ^{15}N natural abundance of individual amino acids in soils treated annually, since 1897, with either farmyard manure (FYM), $(NH_4)_2SO_4$+PK, $(NH_4)_2SO_4$-only, or untreated, are presented. Total amino acid-N contents were substantially higher in the FYM treated soil and the O horizon of the $(NH_4)_2SO_4$-only treatment compared with the A horizon of the same treatment, and both the $(NH_4)_2SO_4$+PK treated and untreated soils. The percentage N content as glycine, isoleucine, glutamic acid, tyrosine and cysteic acid, were generally higher in the amended soils, whereas those of lysine, leucine and serine were lower, than in the untreated soil. The δ^{15}N values of phenylalanine, ornithine, serine, glycine and isoleucine were influenced by any N treatment, and those of leucine by soil pH.

Most soil nitrogen (N) is combined or closely associated with soil organic matter. Thus, a better quantification of the contribution of soil organic matter (SOM) to the mineralisable N pool would help to minimise N losses from agricultural systems to the environment and provide better scope for fertiliser advice (1). Nitrogen is undoubtedly important since it constitutes a major component of proteins, nucleic acids, porphyrins, and alkaloids. In SOM the majority of N is present as protein-N, peptide-N, amino acid-N, amino sugar-N, NH_3-N, N in purines and pyrimidines, and N in heterocyclic ring structures; many of which still remain to be identified (2, 3).

The difficulty of identifying and quantifying the forms and transformations of SOM-N residues remains acute. It is complicated by: the obscure role and fate of many individual N compounds of plant and microbial residues in the biogeochemical N cycle, and their role as exchangeable microbial and plant nutrient resources, and additionally by their participation in the renewal, formation and stabilisation of humic materials in both natural and agricultural soils (4). The presence of amino acids in soils has been recognised for a number of decades, and several hydrolysis studies have shown that they may comprise as much as 20-50 % of total soil N, and an even greater percentage of the annually mineralised organic-N (5, 2). Despite this, limited information is available on the effect(s) of N inputs and soil management on their composition or their transformations in agricultural soils.

Soil organic matter can be (arbitrarily) divided into relatively stable and active components, with the former being more closely associated with the physico-chemically stable soil matrix, and the latter more readily with the ability of the soil to cycle and supply nutrients such as N (6). Changes in amounts and forms of 'active' soil N compounds, such as amino acids, should therefore provide information on how management influences the ability of soils to sustain crop production. Since the majority of N immobilised in soil enters the acid-soluble fraction containing amino acids (AA) and amino sugars (AS), monitoring changes in hydrolysable AAs with Ion Chromatography (IC) should show management induced changes in soil organic N dynamics, before any such changes are observed in the total soil N (7, 8).

Many previous studies have successfully used stable ^{15}N isotope measurements to further clarify the effects of fertiliser, manure and other agricultural practices on the soil N cycle (see reviews: 9-11). If it is true, as suggested earlier, that changes in AA contents and distribution reflect those of the (sometimes undetectable) bulk soil N dynamics, then natural abundance ^{15}N isotope measurements of individual AAs would greatly enhance our prospects of detecting the (more subtle) effects of fertiliser and manure application on soil N dynamics. Although ^{15}N/^{14}N ratios of individual amino acids were reported as early as 1966 (12), the last few years have seen important advances in analytical equipment and methodology. Most recently, the introduction of Gas Chromatography-Combustion-Isotope Ratio Mass Spectrometry (GC-C-IRMS) has made it possible to measure the natural abundance level of ^{15}N/[14]N (δ^{15}N) in individual amino acids in both physiological and environmental samples (13-16, and references therein).

Tracer ^{15}N studies have revealed that: i) nitrate, nitrite and ammonia are assimilated in the same manner after sequential reduction of nitrate to nitrite and then to ammonia, ii) that the first amino acid that is produced by ammonia assimilation is glutamine with a rapid turnover of the amide nitrogen, and iii) that glutamate and other amino acids are produced by the transfer of the amide of glutamine to keto acids, by transamination, and by deamination and reassimilation of ammonia. Analysis of the $\delta^{15}N$ of amino acids and environmental effects on their ^{15}N composition are now required; so too is information on individual enzymatic reactions to further our understanding of isotopic variations in amino acids and other nitrogenous compounds (*17*). In our preliminary study, we therefore examined the effects of long-term fertiliser and manure treatments on soil amino acid-N dynamics using the approaches discussed earlier; that is we investigated changes in i) the distribution (using IC) and ii) the ^{15}N natural abundance signature of individual AAs (using GC-C-IRMS).

Materials and Methods

Study Site and Soil Characteristics. Since 1897, Palace Leas meadow hay plots at Cockle Park Experimental Farm, Northumberland, England (55°13'N 1°41'W) have been subject to virtually unchanged management. The plots, now in their centennial year were laid out in 1896 with the intention of demonstrating to farmers the potential for the improvement of old grassland by manuring (*18*). Even before the experiment began, the site had been described and classified as old grassland (*19*). Of the original 21 plots, only 13 now remain; these consist of parallelograms 120 x 15 m to which are applied a factorial combination of the presence/absence of N, P, K fertilisers, and five farmyard manure (FYM) treatments. The treatments are unreplicated, and lie within the same 2.5 ha, with no major topographical, climatic or significant hydrological differences. The only substantial change to the treatments occurred in 1976 when basic slag was replaced by triple superphosphate containing an equal amount of P. All phosphate and FYM treatments have since maintained an increased hay yield in contrast with the untreated plots (*20*). The soil is a pelo-stagnogley (Typic Ochraqualf) of the Hallsworth series with clay loam texture. The site can be classified as a grazed, cool temperate ecosystem, with an average annual temperature of 8.4 °C and rainfall of 670 mm, is located at 105 m altitude and has a slope of < 5°.

For the purposes of this investigation four plots were selected, providing soil which had been treated with FYM (plot 2), $(NH_4)_2SO_4$+PK (plot 13), $(NH_4)_2SO_4$-only (plot 7) and untreated (plot 6) since 1897 (Table I). The treatments have affected soil N content (*21*) and the nitrifier populations in these soils (*22*). Recent work has demonstrated differences in both the size and activity of the soil microbial communities between the experimental plots (23), which are reflected in differences in D- and L- amino acid metabolism (*24*). The upper 30 cm of soil in plots 2, 6 and 13 consists of an organic rich mineral Ah1 and underlying Ah (g)2 with gleyic features. As a result of the acidifying effect of the $(NH_4)_2SO_4$, plot 7 has a low pH and a substantial accumulation (5-6 cm) of a mor-type (Oh) organic matter at the surface, over a shallow, 2 cm thick, eluvial E below which the mineral Ah/B1 and Ah/B2 are situated (*25, 26*). Hay is cut from the plots and cattle graze the aftermath. The cattle are not confined to particular plots. This leads to additional N input in dung and faeces even on the control plot. In 1994 the highest hay yield in the selected plots (23) was on the FYM plot (7.4 t dry matter (DM) ha^{-1}) which is twice that of the (NH$_4$)$_2$SO$_4$+PK plot (3.7 t DM ha^{-1}), three times that of the plot receiving (NH$_4$)$_2$SO$_4$-only (2.5 t DM ha^{-1}) and nearly four times that of the untreated control (2.1 t DM ha^{-1}). The full treatment details and soil properties of the plots are presented in Table I.

Sample Collection. Soils were sampled on 14 January 1995, immediately after the annual FYM application and 3 months before the annual inorganic fertiliser application. The temperature during sampling was ca. 2 °C, but there had been ground frost overnight. Three samples were taken to a depth of 15 cm from the untreated plot, the FYM treatment and the (NH$_4$)$_2$SO$_4$+PK treatment, and were bulked to give a composite sample for each plot. In the case of the (NH$_4$)$_2$SO$_4$-only treatment three separate representative samples of both the O and A horizons were collected from depths of 0-6 cm and 6-15 cm, respectively. On return to the laboratory all soils were air-dried and then sieved (2 mm). The relative content and δ^{15}N of hydrolysable amino acids, total N content and ^{15}N natural abundance (further details below), and pH were determined for each soil. A cattle manure sample was collected on 30 November 1996 from the farm shed at Cockle Park Farm, Northumberland. The sample was returned to the laboratory and freeze-dried. An ammonium sulphate sample was also obtained. Subsamples of both nutrient sources were ground to a fine powder and analysed for total N content and ^{15}N natural abundance (Table I).

Amino acid composition. After an oxidation step, using performic acid (for 16 h, 0 - 4 °C), finely ground air-dried soil samples (0.25-0.50 g) of the < 2 mm fraction were subjected to conventional acid hydrolysis (6 M HCl, for 24 h at 110 °C) to release the constituent amino acids from the sample proteins. The initial oxidation step was necessary to avoid complete or partial losses of cysteine and methionine during the hydrolysis, by converting cysteine to cysteic acid and methionine to methioninesulphone. Unfortunately, tryptophan is always lost during the acid hydrolysis with or without the oxidation step. Also, the recovery of tyrosine is generally reduced as a result of the oxidation step. An aliquot of each acid hydrolysate was used for the amino acid determination, and further prepared for AA analysis using Ion Chromatography (IC). Another aliquot was taken for the determination of δ^{15}N for each amino acid (see next section). The analysis was carried out on a Hilger Chromospek amino acid analyser (Hilger Analytical, Margate, UK). The procedure was similar to that described in (27), but phenol was omitted at the oxidation step in the procedure to minimise the loss of tyrosine. The volume of hydrolysis reagent used was altered from 50 to 30 cm^3 to reduce the final salt content of the samples. For analyses of standard AAs concentrations were within 3% of the mean value, except for cysteic acid, methioninesulphone and tyrosine when values were within 5% and arginine where values were within 10%. For the soils, we expect a maximum possible error associated with the individual amino acid concentrations of 5% for all AAs except cysteic acid, methioninesulphone, and tyrosine with a maximum of 10% and arginine with a maximum error of 20%. Full details of the results are given in Table I.

Compound Specific Stable Isotope ($^{15}N/^{14}N$) Analysis. The acid hydrolysate (see previous section) was neutralised to pH 7.0. Each sample was then filtered under vacuum before passing through a chromatography column filled with cation exchange resin Dowex 50W (16-40 mesh size) allowing the extraction of the amino acids. Once all samples were loaded, the bound material was eluted with 2 M NH_4OH (pH 14). Samples were concentrated by rotary evaporation and stored at < 0 °C prior to analysis. Amino acids were derivatised to N-pivaloyl-i-propyl (NPP) amino acid esters through treatment with thionyl chloride solution in i-propanol and heated for 60 min at 100 °C. The product was dried and dissolved in pyrodine. After adding pivaloylchloride the solution was acylated for 30 min at 60 °C and dichloromethane was added following cooling. The mixture was then passed over a silica gel column and the eluate dried in a gentle nitrogen stream and redissolved in ethyl acetate for injection.

Compound Specific Stable Isotope ($^{15}N/^{14}N$) Analysis (CSIA) was carried out on a Finnigan delta S isotope ratio mass spectrometer (Finnigan MAT, Bremen, Germany), coupled on-line with an HP 5890 gas chromatograph (GC) (Hewlett-Packard, Waldbronn, Germany) via a combustion interface, generating and purifying N_2 gas from GC separated compounds introduced into the isotope mass spectrometer. The interface consisted of a combustion furnace reactor filled with copper oxide and platinum (980 °C) and a reduction furnace filled with elemental copper (600 °C). The introduction of a standard N_2 gas of known isotopic composition at particular reference points during the gas chromatographic run was used for the calibration of sample amino nitrogen. An Ultra 2 capillary column (50 m x 0.32 mm i.d., 0.5 μm film thickness; Hewlett-Packard) with a carrier gas (helium) stream of 1 ml min^{-1} (head pressure 18 psi) was used for amino acid separation. A volume of 0.5 μl was injected split-less by autosampler. The injector temperature was 280 °C and the following oven temperature gradient was used: 70 °C, held 1 min; 70 to 220 °C, ramp 3 °C min^{-1}; 220 to 300 °C, ramp 10 °C min^{-1}, held 8 min.

In the range of natural abundance (0.3626 to 0.3736 atom % ^{15}N), the $[^{15}N]/[^{14}N]$ ratio is conventionally expressed as:

$$\delta^{15}N \ (‰) = \{(R_{sample}/R_{standard}) -1\} \times 10^3, \tag{1}$$

where R is the $[^{15}N]/[^{14}N]$ ratio, derived from the respective ratios of the m/z 29 to m/z 28 ion current signals of the mass spectrometer. The international standard for N is air (0.3663 atom % ^{15}N); which is arbitrarily assigned a $\delta^{15}N$ value of 0 ‰. Further details of the sample preparation and analysis can be found in (15). Routine precision for the ^{15}N AA measurements are 0.3 ‰ or better for all AAs, except glycine with a precision of 0.5 ‰ and lysine with a precision of 1.3 ‰. The $\delta^{15}N$ values of individual AAs obtained from the GC-C-IRMS analysis for each soil are shown in Table III. Unfortunately, values for the Palace Leas manure sample are unavailable.

Total N Content and ^{15}N Natural Abundance. Subsamples of the finely-ground soil, manure and ammonium sulphate samples were put into pre-weighed tin capsules and analysed for percentage N content. In a second determination the sample size was adjusted in light of this to give 120 ± 3 µg N and then analysed in triplicate for ^{15}N natural abundance (*28*). The ^{15}N analyses were carried out on a Roboprep, a Dumas-type continuous-flow CHN analyser, coupled to a Tracermass isotope ratio mass spectrometer (both instruments from Europa Scientific Ltd. Crewe, UK). Values (Table I) are expressed according to equation 1 (above).

Results

Total Nitrogen and Amino Acid-Nitrogen Contents and Distribution. The differences in total (hydrolysable) amino acid-N (AA-N) in the soils of the four treatments were in line with their respective total N content, such that:

$$\text{AA-N} = 0.366 * \text{total N}; P < 0.001 \qquad (2)$$

In terms of total amino acid-N (AA-N) concentration, similar levels were found in the untreated soil and in the A horizon of both the $(NH_4)_2SO_4$-only and the $(NH_4)_2SO_4$+PK treated soils. The respective values were 1.6 mg g^{-1}, 1.5 mg g^{-1} and 1.4 mg g^{-1} air dry soil. Levels of total AA-N found in the FYM treated soil were *ca.* 1.5 times as great (2.4 mg g^{-1}) and for the O horizon of the $(NH_4)_2SO_4$-only treatment were *ca.* 4 times as great (6.3 mg g^{-1}) as in the untreated soil.

Of the eighteen individual amino acids measured only alanine and glycine accounted for > 10 % of total AA-N concentrations in all soils (Table II). Similar relative values between treatments were also found for ornithine which contributed ~1% towards the total AA-N of all the soils. In the untreated plot contributions of lysine, leucine, proline and aspartic acid were higher, whereas those of glycine and tyrosine were lower than in the amended plots. Levels of histidine and threonine were highest in the O horizon of the $(NH_4)_2SO_4$-only treatment, in contrast to levels of arginine which were greatly reduced relative to the A horizon and the other treatments. Interestingly, there were further marked differences between the O and A horizons of the $(NH_4)_2SO_4$-only treatment for several other amino acids. The percentage leucine content was at a minimum in the A horizon of the $(NH_4)_2SO_4$-only treatment whereas glutamic acid and phenylalanine showed a two-fold and a four-fold increase relative to the O horizon, and indeed, the other treatments. In the manure sample, alanine, glycine and glutamic acid each accounted for >10 % of the total AA content; isoleucine, aspartic acid, threonine and serine for > 5% but < 10%, whereas phenylalanine, lysine, histidine, tyrosine, methioninesulphone and arginine each contributed < 5% towards the total AA content (data not shown).

δ^{15}N of Total Soil N and Individual Amino Acids. The total soil N δ^{15}N values in the amended soils varied according to the type and δ^{15}N value of applied N (Table I). The application of FYM (with high δ^{15}N source value) to plot 2 was reflected in the more enriched soil δ^{15}N value (8.0 ± 0.4 ‰), whereas the application of inorganic N to plots 7 (A horizon, 4.5 ± 0.5 ‰) and 13 (4.4 ± 0.3 ‰) was associated with depleted δ^{15}N values relative to the untreated soil of plot 6. In fact, the δ^{15}N value of the O horizon of plot 7 ($(NH_4)_2SO_4$-only treatment of 1.8 ± 0.4 ‰ is very close to that

Table I. Soil properties of the Palace Leas plots and details of nutrient applications

	Depth	Applied N	Total N	Total Amino Acid-N	Total N $\delta^{15}N^a$ ± s.d.	pH[a]
	cm	*kg ha^{-1} yr^{-1}*	*mg g^{-1}*	*mg g^{-1}*	*‰*	
SOIL TREATMENT						
Untreated (plot 6)	0-15	0	3.6	1.6	5.2 ± 0.7	4.6
Manure (plot 2)	0-15	100	5.6	2.4	8.0 ± 0.4	5.1
$(NH_4)_2SO_4$ (plot 7)-O horizon	0-6	36	16.4	6.3	1.8 ± 0.4	3.6
$(NH_4)_2SO_4$ (plot 7)-A horizon	6-15	36	3.2	1.5	4.5 ± 0.5	3.9
$(NH_4)_2SO_4$+PK [b] (plot 13)	0-15	36	3.1	1.4	4.4 ± 0.3	5.0
NUTRIENT SOURCE						
Farmyard manure			24.0	n.a.[c]	13.6 ± 0.3	
Ammonium sulphate			207.0	0.0	1.3 ± 0.3	

[a] mean value of triplicate analyses of composite samples
[b] of 26 kg P and 59 kg K ha^{-1} yr^{-1}
[c] value not available

Table II. Individual amino acids as a percentage of total amino-N in the Palace Leas soils

Treatment	Untreated	FYM	$(NH_4)_2SO_4$-only O horizon	$(NH_4)_2SO_4$-only A horizon	$(NH_4)_2SO_4$+PK
Amino Acid			% of total amino acid-N		
Alanine	16.3	14.7	16.3	11.9	13.1
Glycine	11.2	17.4	19.2	16.9	15.8
Valine	5.9	3.5	7.5	7.0	6.4
Leucine	8.9	5.5	6.9	2.4	4.6
Isoleucine	2.6	4.4	4.3	2.8	3.8
Proline	6.5	5.8	6.1	5.1	5.3
Aspartic acid	7.2	4.2	3.8	3.6	2.7
Threonine	4.6	3.7	7.0	5.1	5.7
Serine	5.7	4.6	3.9	4.3	3.2
Glutamic acid	5.6	6.3	6.6	14.7	7.6
Phenylalanine	1.6	1.8	1.3	6.2	1.5
Ornithine	1.1	1.1	0.8	1.0	1.1
Lysine	13.7	7.4	6.9	7.1	7.6
Histidine	4.8	3.9	6.0	3.5	3.7
Tyrosine	0.1	0.5	0.6	0.8	0.9
Cysteic acid	0.1	1.6	0.6	0.4	0.7
Methioninesulphone	0.2	2.0	0.6	0.2	1.2
Arginine	4.0	11.6	1.8	7.0	14.9

of the applied $(NH_4)_2SO_4$ (1.3 ± 0.3 ‰). These distinct differences between total soil N $\delta^{15}N$ values and type of amendment were also generally reflected in the $\delta^{15}N$ of individual AAs (Table III).

The $\delta^{15}N$ values of the two 'source' AAs, i.e. glutamic acid and aspartic acid, together with alanine and proline were invariably relatively enriched (ca. + 4 to + 11 ‰) in the Palace Leas soils, whereas those of serine and histidine were relatively less enriched (ca. + 2 ‰, or lower). All other AAs generally had intermediate $\delta^{15}N$ values or the values differed widely among treatments. The largest range in $\delta^{15}N$ between treatments was observed for phenylalanine with values varying from − 8.0 to + 3.8 ‰, with the narrowest range being found for proline with values between + 5.7 and + 8.2 ‰. The two most distinct differences between treatments were firstly, that tyrosine was ca. 4 ‰ more enriched in the untreated soil and the A horizon of the $(NH_4)_2SO_4$-only treated soil than in the $(NH_4)_2SO_4$+PK treated soil, and was further enriched by ca. 4 ‰ in the manured soil, and secondly that glycine, aspartic acid, threonine, serine and phenylalanine were more depleted in ^{15}N in the amended soils than in the untreated soil, whereas isoleucine was invariably enriched by >1.4 ‰.

For the $(NH_4)_2SO_4$-only treated soil there were several notable differences between the O and A horizons of the plot. The majority of amino acids measured were depleted in the O horizon relative to the A horizon with the exception of isoleucine and phenylalanine which were more enriched. The $\delta^{15}N$ values of leucine and ornithine differed by > 4 ‰, with the A horizon showing the greater enrichment. By far the greatest difference between the two soil horizons was measured for phenylalanine with values of + 3.4 ‰ for the O horizon compared with − 6.5 ‰ for the A horizon.

Discussion

The data from our preliminary study of selected soils from the Palace Leas plots indicate that long term application of fertiliser and manure to the grasslands has led to substantial differences in the total and individual percentage amino acid-N contents, distribution and $\delta^{15}N$ signatures. Increased soil N and total (hydrolysable) amino acid-N as a consequence of fertiliser amendments and other agricultural practices was first described for the experimental Morrow plots at the University of Illinois (29), which started in 1876. Similar observations have also been reported for a number of other soil and crop studies, (for example, 7, 30, 31, 8, and 32). A recent pot experiment found a linear relationship between increased amino acid N and total N in plant dry matter, as a result of increased N applications (33).

The relative contribution of the individual AAs in our study were comparable with those of the aforementioned studies, and a further study on the distribution of N in Canadian soils (34). No consistent overall AA pattern emerged on how fertiliser and manure treatments affect soil AA composition. This could be a result of the fact that most of those studies were conducted in arable cropping systems, rather than a permanent grassland system or be due to the range of time-scales associated with the different investigations. Our observation that continuous applications of manure (relatively enriched in $\delta^{15}N$) leads to more enriched soil $\delta^{15}N$ has been reported

Table III. ^{15}N composition of individual amino acids in the Palace Leas soils

Treatment	Untreated	FYM	$(NH_4)_2SO_4$-only O horizon	$(NH_4)_2SO_4$-only A horizon	$(NH_4)_2SO_4$+PK
Amino Acid			$\delta\,^{15}N$ (‰)		
Alanine	9.1	10.9	5.1	6.9	6.7
Glycine	3.3	2.6	-0.8	0.4	0.2
Valine	4.9	4.9	1.6	3.6	1.9
Leucine	1.9	4.2	1.0	5.1	0.6
Isoleucine	1.4	5.5	4.9	2.9	4.0
Proline	7.3	8.2	5.7	6.0	6.7
Aspartic acid	8.5	7.6	4.1	6.1	6.2
Threonine	4.5	2.3	-0.2	1.2	2.4
Serine	1.8	1.2	-2.4	-0.3	0.0
Glutamic acid	8.5	8.9	5.4	6.5	6.6
Phenylalanine	3.8	-1.0	3.4	-6.5	-8.0
Ornithine	5.2	n.a.[a]	-1.5	3.2	2.5
Lysine	3.9	-3.1	0.1	2.0	-0.8
Histidine	1.7	2.2	-1.1	0.2	0.0
Tyrosine	3.8	8.0	2.6	4.0	-0.3
Cysteic acid	n.a.	n.a.	n.a.	n.a.	n.a.
Methioninesulphone	2.7	n.a.	3.1	n.a.	-1.3
Arginine	n.a.	n.a.	n.a.	n.a.	n.a.

[a] indicates value not available

elsewhere (reviewed in *11*). Our finding that the application of $(NH_4)_2SO_4$ to plots 7 and 13 has resulted in relatively low soil $\delta^{15}N$, is in agreement with the fact that the application of N fertiliser (low in $\delta^{15}N$) tends to deplete soil $\delta^{15}N$ in most ecosystems (*10*).

We are unaware of the existence of the literature necessary to corroborate our findings of differences in soil $\delta^{15}N$ values between the individual AAs. Some of the soil AAs $\delta^{15}N$ values were as expected from trends observed in plants and animal tissues, that the $\delta^{15}N$ values of serine, leucine and lysine were relatively low in comparison with glutamic acid (*17*). Serine, leucine and lysine are principally produced during transamination by the reaction of precursors, amino acids (glutamic acid being important) and keto acids, and are the end products of these reactions. For wheat hydrolysates, it has been shown that glycine, proline, serine, lysine and leucine had relatively low $\delta^{15}N$ values (< 3 ‰), whereas phenylalanine, tyrosine and aspartic acid had relatively more enriched $\delta^{15}N$ values (> 6 ‰), with intermediate $\delta^{15}N$ values being observed for alanine, valine and isoleucine (*14*). In contrast, for a wheat soil, $\delta^{15}N$ values of similar AAs were sometimes remarkably different, most notably the $\delta^{15}N$ value of phenylalanine was - 27 ‰ (Simpson, I.A.; Bol, R.; Dockrill, S.J.; Petzke, K.J.; Evershed, R.P. *Archaeol. Prosp.*, in press). This suggests that the soil environment and associated microbial population can substantially modify the $\delta^{15}N$ of a specific source AA. Indeed, large and compound specific effects of microbial alterations on stable nitrogen isotopic composition of individual AAs in soil organic matter have been reported (*35, 36*).

Differences in stable nitrogen isotope composition are known to occur at the molecular and intramolecular level, as a product of source values and kinetic isotope effects during glutamate synthesis, transamination and deamination reactions (*37, 17*) involving individual amino-N cycling in plants, soils and micro-organisms. The ^{15}N depletion of phenylalanine, aspartic acid, serine, threonine, and glycine and the enrichment of isoleucine found in the amended soils indicate that any N application influences their ^{15}N content. The observed difference between the A horizon of the $(NH_4)_2SO_4$-only treatment and the $(NH_4)_2SO_4$+PK amendment suggest an effect of pH (indirectly an effect of PK) on the ^{15}N content of leucine.

At this preliminary stage, it would be inappropriate to draw conclusions as to the most important factors operating within these soils which have induced the differences in the $\delta^{15}N$ values of the individual AAs. Previous work on these soils has shown that there are clear differences in terms of: i) the size and activity of microbial populations (*23*), ii) amino acid metabolism (*24*), iii) soil structure and texture (*25, 38, 21*) or iv) soil pH (*38*), as a result of the long-term treatments. All of these may yet prove to be determining factor(s).

Obtaining further, replicated data on the isotopic composition of the amino acids is a priority. More detailed investigation of the extent of variation within the plots, including both spatial (e.g. depth) and temporal scales is also necessary. Nevertheless, this study has shown that the pathways of transformation and fate of amino acid-N at the molecular scale are influenced by long-term N inputs, as reflected in the associated changes in concentrations and $\delta^{15}N$ values of individual amino acids.

319

Acknowledgments

We acknowledge the award of a PRF travel grant to RB, and Terra Industries (formerly ICI Fertilizers) for their contribution to the management of the Palace Leas plots. JMW is grateful to the Ministry of Agriculture, Fisheries and Food for financial support through a studentship, and the Natural Environmental Research Council for supporting the ^{15}N analyses of soil, fertiliser and manure samples at the Mass Spectrometer facility at the Institute of Terrestrial Ecology, Merlewood, Cumbria, UK. We also extend our sincere thanks to Chris Quarmby and Darren Sleep for performing those analyses. Finally, we acknowledge the constructive comments of an anonymous reviewer.

References

1. Jarvis, S.C.; Stockdale, E.A.; Shephard, M.A.; Powlson, D.S. In *Advances in Agronomy*; Sparks, D.L., Ed.; Academic Press: London, 1996; Vol. 57; pp 188-237.
2. Paul, E.A.; Clark, F.E. *Soil Microbiology and Biochemistry*; Academic Press: San Diego, CA, 1989; 273pp.
3. Schnitzer, M. *Soil Sci.* **1991**, *151*, 41-58.
4. Kuzyakov, Y.V. *Eur. J. Soil Sci.* **1997**, *48*, 121-130.
5. Stevenson, F.J. In *Nitrogen in Agricultural Soils*; Stevenson, F.J., Ed.; Agronomy Monograph 22; ASA,CSA and SSSA: Madison, USA, 1982; pp 67-122.
6. Campbell, C.A. In *Soil Organic Matter*; Schnitzer, M.; Kahn, S.U., Eds.; Developments in Soil Science 8; Elsevier: Amsterdam, 1978; pp 173-271.
7. Gupta, U.C.; Reuszer, H.W. *Soil Sci.* **1967**, *104*, 395-400.
8. Campbell, C.A., Schnitzer, M., Lafond, G.P., Zentner, R.P. and Knipfel, J.E. *Soil Sci. Soc. Am. J.* **1991**, *55*, 739-745.
9. Handley, L.L.; Raven, J.A. *Plant, Cell Environ.* **1992**, *15*, 965-985.
10. Kohl, D.H.; Shearer, G. In *Stable Isotopes in the Biosphere;* Wada, E.; *et al.*, Eds.; Kyoto University Press: Japan, 1995; pp 103-130.
11. Yoneyama, T. In *Mass Spectrometry of Soils;* Boutton, T.W.; Yamasaki, S-I., Eds.; Academic Press: London, 1996; pp 205-223.
12. Gaebler, O.H.; Vitti, T.G.; Vukmirovich, R. *Can. J. Biochem.* **1966**, *44*, 1249-1257.
13. Minagawa, M.; Egawa, S.; Kabaya, Y.; Karasawa-Tsuru, K. *Mass Spectroscopy,* **1992**, *40*, 47-56.
14. Hofmannn, D.; Jung, K.; Segschneider, H-J.; Gehre, M.; Schüürman, G. *Isotopes Environ. Health Stud.* **1995**, *31*, 367-375.
15. Metges, C.C.; Petzke, K.J.; Henning,U. *J. Mass Spectrometry.* **1996**, *31*, 367-376.
16. Metges, C.C.; Petzke, K.J. *Analytical Biochem.* **1997**, *247*, 158-164.
17. Yoneyama, T. In *Stable Isotopes in the Biosphere;* Wada, E. *et al.*, Eds.; Kyoto University Press: Japan, 1995; pp 92-102.
18. Pawson, H.C. *Cockle Park Farm*; Oxford University Press: London, 1960.
19. Shiel, R.S.; Hopkins, D.W. *N. England Soils Disc. Group Proc.* **1991**, *26*, 35-56.
20. Coleman, S.Y.; Shiel, R.S.; Evans, D.A. *Grass Forage Sci.* **1987**, *42*, 353-358.
21. Shiel, R.S. *J. Soil Sci.* **1986**, *37*, 249-257.
22. Hopkins, D.W.; O' Donnell, A.G.; Shiel, R.S. *Biol. Fertil. Soils.* **1988**, *5*, 344-349.

23. Hopkins, D.W.; Shiel, R.S. *Biol. Fertil. Soils*. **1996**, *22*, 66-70.
24. Hopkins, D.W.; O'Dowd, R.W.; Shiel, R.S. *Soil Biol. Biochem.* **1997**, *29*, 23-29.
25. Bragg, N.S. *Unpublished MSc dissertation*, University of Newcastle upon Tyne, UK, 1979.
26. Hopkins, D.W.; Chudek, J.A.; Shiel, R.S. *J. Soil Sci.* **1993**, 44, 147-158.
27. Mason, V.S.; Beck-Anderson, S.; Rudemo, M. *Tierphysiologie Tiernahrung und Lattermittlekde*. **1980**, *43*, 146-164.
28. Michelsen, A.; Schmidt, I.K.; Jonasson, S.; Quarmby, C.; Sleep, D. *Oecologia*, **1996**, *105*, 53-61.
29. Stevenson, F.J. *Soil Sci. Soc. Proc.* **1956**, *20*, 204-208.
30. Allison, F.E. *Soil Organic Matter and its Role in Crop Production*. Developments in Soil Science 3, Elsevier: Amsterdam and New York, 1973, 639pp.
31. Campbell, C.A., Schnitzer, M., Stewart, J.W.B., Biederbeck and Selles, F. *Can. J. Soil Sci.* **1986**, *66*, 601-613.
32. Tanacs, L.; Bartok, T.; Matuz, J.; Kovacs, K.; Gero, L.; Harmati, I.*Cereal Res Commun.* **1992,** *20*, 257-262.
33. Eppendorfer, W.H. and Bille, S.W. *J. Sci. Food Agri.* **1996**, *71*, 449-458.
34. Sowden, F.J. *Can. J. Soil Sci.* **1977**, *57*, 445-456.
35. Macko, S.A.; Estep, M.L.F. *Org. Geochem.* **1984**, *6*, 787-790.
36. Macko, S.A.; Fogel (Estep), M.L.; Hare, P.E.; Hoering, T.C. *Chem. Geol.* **1987**, *66*, 79-92.
37. Galimov, E.M. *The Biological Fractionation of Isotopes*; Academic Press: London, 1985.
38. Shiel, R.S.; Rimmer, D.L. In *Biological processes and soil fertility*; Tinsley, J.; Darbyshire, J.F., Eds.; Developments in plant and soil science, Vol. 11; Martinus Nijhoff/Dr W. Junk Publisher: The Hague, 1984; pp 349-356.

Chapter 19

Organic Geochemical Studies of Soils from Rothamsted Experimental Station: III Nitrogen-Containing Organic Matter in Soil from Geescroft Wilderness

Pim F. van Bergen[1,3], Matthew B. Flannery[1], Paul R. Poulton[2], and Richard P. Evershed[1]

[1]Organic Geochemistry Unit, School of Chemistry, University of Bristol, Cantock's Close, Bristol BS8 1TS, United Kingdom
[2]Soil Science Deparment, IACR-Rothamsted, Harpenden, Herts AL5 2JQ, United Kingdom

Three distinct soil horizons from a mature oak dominated woodland were studied in order to determine the changes in the molecular composition of nitrogen-containing organic matter down a soil profile. The total amount of nitrogen relative to soil organic carbon increased down the profile with most of the recognizable nitrogen-containing compounds in the leaf litter and humic horizon being either amino acid or amino sugar derived. In contrast, a significant proportion of the organic nitrogen moieties in the mineral horizon appeared to contain macromolecular-bound nitrogen which is believed to represent the so-called 'unknown' soil organic nitrogen and is not obviously related to known biomolecules. The increase in total amino acids in the humic and mineral horizons indicated contributions from sources other than the leaf litter. The increase in organic nitrogen-containing moieties, most probably amino acid derived, accounted for the less depleted $\delta^{13}C$ values observed in the mineral soil horizon.

Different forms of organic nitrogen have been recognized in soils including, proteins, amino acids, amino groups, ammonium ions, hexoamines and nucleic acids (*1-4*). However, it has long been known that substantial amounts, sometimes up to 50%, of the organic nitrogen occurring in the organic matter of soils are present as so-called 'unidentified' (*3*) or 'unknown' forms (*2-5*). The significance of these forms is that: (i) nitrogen potentially available for biological processes may become immobilized (*3, 6*), and (ii) that this organic nitrogen appears more difficult to assimilate (*1*). The nitrogen immobilization has been considered to affect soil fertility (*7, 8*) and to cause changes in the extent of litter decomposition, which in turn influences carbon dioxide

---

[3]Present address: Organic Geochemistry Group, Faculty of Earth Sciences, Utrecht University, P.O. Box 80021, 3508 TA Utrecht, The Netherlands.

emission to the atmosphere (6). Despite the importance of organic nitrogen, the origin and exact molecular composition of 'unidentified' or 'unknown' structural moieties as well as the detailed molecular composition of soil organic nitrogen in general, are still poorly understood (5).

The main problem with characterizing the 'unknown' or 'unidentified' organic nitrogen constituents of soils is that they derive from complex biological macromolecules the structures of which have been altered through a wide variety of poorly understood complex chemical and microbiological soil processes. Treatment of whole soils or soil fractions with 6 M HCl will yield amino acids which can be analyzed by HPLC. Whilst the profiles produced by such analyses are of value in assessing overall trends in the degradation of proteins and individual amino acids this approach is less useful in determining the nature of highly altered non-proteinaceous components (3, 4). Schnitzer was one of the first to attempt to identify components present in this nitrogen pool with 'unknown' N fractions being isolated using Sephadex gels of various sizes (9). The results obtained indicated that heterocyclic N-components are major contributors to the 'unknown' N-pool. More recently, two other techniques have been applied to study the 'unidentified' N, namely pyrolysis-gas chromatography-mass spectrometry (4, 5) and solid state ^{15}N NMR (10, 11). Interestingly, the pyrolysis data largely corroborated Schnitzer's interpretations (4, 5) whereas the NMR results revealed mainly amides and terminal amino groups with only minor signals resulting from heteroaromatic structures. The few studies that have been performed have highlighted the fact that the chemical nature of the 'unknown' N-containing components of soil is a potentially highly complex field worthy of further investigation. With this in mind we decided to initiate a systematic study of the nitrogenous components of soil using a hierarchical analytical approach which sought to relate the bulk chemical properties, e.g. elemental composition, of the soil to molecular and isotopic data obtained following various chemical treatments.

The objective of this paper is to study the organic nitrogen fraction present in solvent-insoluble organic matter from three soil horizons at Geescroft Wilderness, Rothamsted Experimental Station. The history of this site and the annual input of organic matter to it are well-documented (11-14). The solvent-insoluble material was treated with base and, subsequently, acid to determine the mode of occurrence of the organic nitrogen. Elemental analyses, bulk stable isotopes, amino acid analyses and flash pyrolysis-gas chromatography-mass spectrometry were used to investigate the molecular composition of nitrogen-containing soil organic matter down this soil profile.

Sample Description

Samples were taken from three soil horizons of Geescroft Wilderness at Rothamsted Experimental Station, Harpenden, Herts., UK, in May 1995. This site is located in a small area of once-arable land (1.3 ha). The land had long been cultivated as shown on a map dated 1623. The present site was part of an experimental field growing beans from 1847 to 1878, with frequent breaks towards the end of that period.

Subsequently it was fenced off and allowed to revert to natural woodland in 1883 (*12, 13*). The site has never been chalked, except for receipt of a little lime on occasions prior to reversion. This is reflected in the decreasing soil pH (in H_2O) 7.1, 6.1, 4.5 and 4.2 in 1883, 1904, 1965 and 1991 respectively. The soil organic carbon input has been estimated at 2.5 t ha^{-1} yr^{-1}, with leaf-fall being the main contributor (estimated 1.57 t ha^{-1} yr^{-1}; *13*). Dead roots, dead mycorrhizae, root exudates etc. have been suggested as minor contributors to the soil organic matter (*13*). For additional information about this site the reader is referred to references *12-14*.

The top horizon sampled was a leaf litter predominantly composed of *Quercus robur* leaves. The underlying soil was sampled using a 2 cm diameter auger to a depth of 23 cm. The soil sample was subsequently subdivided into a humic rich brown top horizon (ca. 5 cm) and a light brown mineral soil horizon (ca. 18 cm). Replicates of the underlying soil were taken at approximately 5 m from the first sampling location. The soil at the site is a silty clay loam also classified as Chromic Luvisol (*15*) or Aquic Paleudalf (*16*).

Experimental

Soil and litter samples were initially oven dried at 60°C. Samples were crushed with a pestle and mortar and sieved over 2 mm and 75 μm sieves; the litter sample was only sieved over 2 mm. The leaf litter was solvent extracted (dichloromethane/acetone, 9:1 v/v) by ultrasonication whereas the soils were Soxhlet extracted for 24 h using the same solvent mixture. The insoluble residues obtained were vacuum dried.

Extracted residues were base treated using a 1 M potassium hydroxide solution in 96% methanol. The suspensions, under N_2, were heated to reflux for one hour at 70°C. After cooling, the reaction mixtures were acidified to pH 3 using 2 M hydrochloric acid in methanol/water (1:1 v/v). After centrifugation (1 min, 3000 rpm), the residues were extracted with water (3x), methanol/water (1:1 v/v; 1x), methanol (2x) and dichloromethane (3x). After each extraction step the suspension was centrifuged and the supernatant was removed. The residues after base treatment were vacuum dried.

Acid treatment was performed by suspending the residue after base extraction in a 4 M hydrochloric acid solution in water. The samples were heated for six hours at 105°C. After cooling the samples were neutralized using a 16 M potassium hydroxide solution in water. Mixtures were centrifuged (1 min at 3000 rpm) and the residues obtained were again saponified and extracted as described above. These final residues were vacuum dried. All residues (insoluble, base and acid treated) were studied using elemental analysis (EA), bulk stable carbon isotope measurements (IRMS) and flash pyrolysis-gas chromatography/mass spectrometry (Py-GC/MS). The insoluble residues and residues after base treatment were also studied for their amino acid contents.

Individual free amino acids were obtained using an acid treatment in an evacuated reaction tube sealed with a PTFE lined cap (6 M HCl 1:500 w/v, 100°C, 24 hr) (*17*). Excess acid was evaporated at 40°C using a stream of dry N_2 and 30% of the hydrolysate was transferred to a reaction vial. The solvents were removed by evaporation. Free amino acids contained in the hydrolysate were derivatized to the phenylisothiocyanate (PITC) derivatives by means of the Pico-TAG method (18). The reagent mixture added to each sample consisted of 2.0 µl triethylamine, 2.0 µl double distilled H_2O, 2.0 µl phenylisothiocyanate, and 15 µl HPLC-grade methanol. The vial was sealed with a PTFE lined cap, the reagents agitated to mix and left to react for 10 minutes at room temperature. Excess reagent mixture was removed by evaporation. Separation of the amino acids was accomplished using a 30 cm Pico-Tag column with an elution profile as described by Suleiman (19). Amino acids were detected at 254 nm using a Waters 486 UV detector. Quantification of the amino acids was based on the peak areas of standard amino acids of known concentration and normalized to the dry weight of the samples.

Elemental analyses were performed using a Perkin Elmer 240C elemental analyzer. Values stated are based on duplicate analyses (Table I). IRMS analyses of the residues were performed in duplicate using a Carlo Erba NC2500 elemental analyzer coupled to a Finnigan MAT Delta S instrument. All $\delta^{13}C$ values have been corrected against a NIST sucrose standard and are relative to the PDB standard.

Py-GC/MS analyses were performed using a Carlo Erba 4130 gas chromatograph, equipped with a CDS 1000 Pyroprobe device (Chemical Data System, Oxford, Pennsylvania) connected to a Finnigan MAT 4500 mass spectrometer. Samples were loaded into quartz tubes into the CDS 1000 Pyroprobe. The temperature of the pyrolysis interface temperature and GC injector was set at 250°C. Pyrolysis time was 10 seconds at 610°C. The GC oven was programmed from 35°C (5 min) to 310°C (10 min) at a rate of 4°C min^{-1}. Separation was achieved using a fused-silica capillary column (50 m x 0.32 mm) coated with CPSil-5 CB (film thickness 0.4 µm). Helium was used as the carrier gas. The MS was operated at 70 eV scanning the range m/z 35-550 at a cycle time of 1 s. Compound identifications were based on mass spectral data and retention time comparisons with reference samples and data reported in the literature (20-24).

Results

Elemental and Isotope Measurements. The replicate soil samples of the mineral layer were similar in overall elemental composition whereas those of the humic layer showed a certain disparity in the amounts of $C_{org.}$ and N_{total} (Table I) which is most likely due to differences in the contributions of minerals (58.8% *vs.* 75.8% ash). The elemental compositions of all three soil horizons clearly revealed the reduction of total N and organic C content of the residues following the base and acid treatments. This is, however, not revealed in the C/N of the residues after base treatment due the greater loss of carbon relative to total N.

$\delta^{13}C$ values of insoluble residues of the humic rich top layer are very similar to those of the leaf litter. In contrast, the organic matter present in the mineral soil is relatively enriched in ^{13}C when compared with the overlying horizons. The $\delta^{13}C$ values of the various samples show no specific trend as a result of base treatment. In

Table I. Bulk elemental and stable carbon isotope data from the three soil horizons. N.D. = not determined. N/A = not applicable as N_{total} = 0. Values in parenthesis are corrected for the weight loss during the base and acid treatments.

Sample	$C_{org.}$ (%)	N_{total} (%)	Ash (%)	C/N	$\delta^{13}C$ (‰)
Leaf litter insoluble	46.08	1.88	0.63	28.7	-27.9
base treated	43.40 (20.83)	1.15 (0.55)	N.D.	34.0	-27.3
base and acid treated	48.55 (9.08)	0.55 (0.10)	N.D.	103.9	-29.4
Humic layer[1] insoluble	19.29 / 9.59	1.47 / 0.91	58.78 / 75.84	15.3 / 12.3	-27.8 / -28.0
base treated	14.77 / 6.33 (12.26) / (5.63)	1.32 / 0.82 (1.10) / (0.73)	N.D. / N.D.	13.1 / 9.1	-27.8 / -27.9
base and acid treated	9.80 / 4.08 (6.27) / (2.49)	0.23 / 0.14 (0.15) / (0.09)	N.D. / N.D.	49.7 / 35.3	-29.4 / -29.7
Mineral soil[1] insoluble	1.77 / 1.78	0.44 / 0.38	93.28 / 92.31	4.7 / 5.5	-26.4 / -26.6
base treated	1.20 / 1.28 (1.08) / (1.14)	0.49 / 0.26 (0.44) / (0.23)	N.D. / N.D.	2.8 / 5.4	-27.1 / -26.2
base and acid treated	0.70 / 0.52 (0.57) / (0.33)	0.06 / 0.00 (0.05) / (0.00)	N.D. / N.D.	13.6 / N/A	-27.9 / -27.3

[1] Values for the replicate soil samples are reported separately

contrast, the residues after both base and acid treatment were depleted in ^{13}C compared with the untreated soil (Table I).

Pyrolysis. The pyrolysis products of insoluble residue of the leaf litter (Fig. 1a) are primarily dominated by methoxyphenols, furan and pyran derivatives indicative of dicotyledonous angiosperm ligno-cellulose (*25*). Only relatively few nitrogen-containing products have been detected including, acetamide (**18**, note that this product is present as a hump rather than a single peak), pyrrole (**1**), benzeneacetonitrile (**16**), indole (**8**) and diketodipyrrole (**23**; Table II). Other relatively abundant pyrolysis products, without nitrogen, which may have been derived from nitrogen-containing moieties include toluene and phenol (*26*) although these may originate from other sources, including ligno-cellulose. The pyrolysate of insoluble residue of the humic layer (Fig. 1b) resembles that of the leaf litter revealing evidence of a significant ligno-cellulose component. A substantial number of nitrogen-containing products are also identified including, pyrrole (**1**), C_1-pyridine (**7**), indole (**8**) and diketodipyrrole (**23**; Table II). Toluene, styrene and phenol are also present. The most abundant N-containing products were detected in the pyrolysate of insoluble residue of the mineral soil (Fig. 1c; Table II) including pyrrole (**1**), pyridine (**6**), benzonitrile (**14**) and diketodipyrrole (**23**). The most abundant pyrolysis products are, toluene and phenols, as well as styrene and C_1 phenols. Notably, this sample also revealed the presence of a number of alkyl nitriles (C_{14}, C_{16}, C_{18} and C_{20}; **27-30**) which were not detected in any of the overlying layers.

The pyrolysates of the residue after base treatment (Fig. 2) were similar in composition to those of the extracted insoluble residues (Fig. 1). The most significant differences observed are the increase in the abundance of cyclopentenones (Fig. 2) and a significant reduction in the abundance of levoglucosan and 4-ethenylphenol in the leaf litter. The distribution of nitrogen-containing pyrolysis products changed little except for a decrease in relative abundance of the alkyl nitriles in the mineral soils (cf. Fig. 2c *vs.* Fig. 1c).

The principal changes in the macromolecular composition of the organic matter in the samples occur after the acid treatment. The pyrolysates of the leaf litter residue after acid treatment (Fig. 3a) still shows evidence of lignin based on the methoxyphenols but the characteristic polysaccharide component is now undetectable. However, it should be noted that the distribution of the specific lignin markers is drastically altered when compared with the original litter (cf. Fig. 1a). N-containing compounds are virtually absent with the exception of pyrrole (**1**), N-methylpyrrole (**2**), pyridine (**6**) and diketodipyrrole (**23**; not shown). The pyrolysis products of residue after acid treatment of the humic layer (Fig. 3b) showed the same chemical alteration of the lignin-derived markers (drastic change of distribution pattern) as observed in that of the leaf litter (Fig. 3a). Nitrogen compounds similar to those in the base treated and extracted residues were still detectable, although in much lower relative amounts. Important non-nitrogenous pyrolysis products, phenol, toluene and styrene, are still relatively abundant (*22, 26*). The dominant pyrolysis products of the mineral soil after base and acid treatment (Fig. 3c) are N-

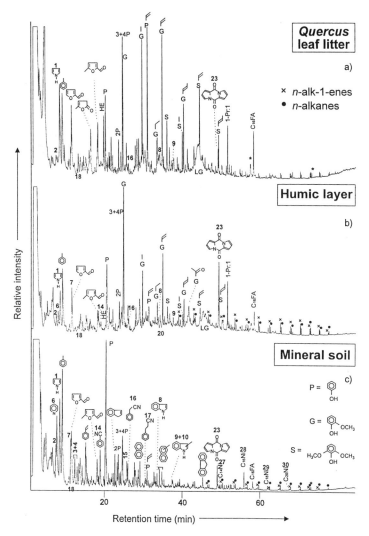

Figure 1. TICs of the pyrolysates of the solvent insoluble residues of a) *Quercus* leaf litter, b) humic rich top horizon and c) mineral soil horizon. Key: P = phenol; 3+4P = co-eluting 3- and 4-methylphenol; $C_{16}FA$ = hexadecanoic acid; 1-Pr:1 = Prist-1-ene; HE = Hemicellulose marker (4-hydroxy-5,6-dihydro-(2*H*)-pyran-2-one); LG = Levoglucosan; $C_{16}Ni$ = hexadecanenitrile; x= *n*-alk-1-enes; • = *n*-alkanes; * = contaminants. Numbers in bold refer to compounds listed in Table II. Side chains (attached at positions 4) of phenol- (P), 2-methoxyphenol- (guaiacyl; G) and 2,6-dimethoxyphenol- (syringyl-; S) components are indicated. For additional information and source of the material see text.

Table II. Nitrogen-containing pyrolysis products detected in the pyrolysates of the three different soil horizons.

	Pyrolysis products	Origin [a]	Mw	Leaf litter[b]			Humic layer[b]			Mineral soil[b]		
				1	2	3	1	2	3	1	2	3
1	Pyrrole	Pro, Hyp, Glu	67	+	+	+	++	++	+	+++	+++	++
2	N-methylpyrrole	AA / unknown	81	+	+	+	+	+	+	+	+	+++
3	2-Methylpyrrole	Pro, Hyp	81	-	+	-	+	+	+	+	+	+
4	3-Methylpyrrole	Pro, Hyp	81	-	+	-	+	+	+	+	+	+
5	C$_2$-pyrroles	Hyp	95	-	-	-	-	+	+	+	+	+
6	Pyridine	AS / Ala	79	+	+	+	+	+	+	++	++	+
7	C$_1$-pyridines	AS / Ala	93	-	-	-	++	+	+	+	+	+
8	Indole	Trp	117	+	+	-	+	++	+	++	++	+
9	3-Methylindole	Trp	131	+	+	-	+	+	+	+	+	-
10	C$_1$-Indole	unkown	131	-	-	-	-	-	+	+	+	-
11	Quinoline ?	unknown	129	-	-	-	-	+	+	+	+	-
12	Isoquinoline ?	unknown	129	-	-	-	-	+	+	+	+	-
13	Benzenamine	unknown	93	-	-	-	-	-	-	?	-	-
14	Benzonitrile	unknown	103	+	-	-	+	+	+	++	++	++
15	C$_1$-Benzonitriles	unknown	117	-	-	-	-	-	+	+	+	+
16	Benzenacetonitrile	Phe	117	+	+	-	+	+	+	+	+	+
17	Benzenepropanenitrile	Phe	131	+	-	-	+	+	+	+	+	-
18	Acetamide	AS	59	+	+	-	+	+	-	+	+	-
19	Acetylpyrrolidone	AS	127	-	-	-	-	+	-	-	-	-
20	3-Acetamido-5-methylfuran	AS	139	-	-	-	+	+	-	-	-	-
21	3-Acetamido-4-pyrone or 3-Acetamido-2-pyrone	AS	153	-	-	-	-	+	-	-	-	-
22	Oxazoline structures	AS	185	-	?	-	+	+	-	-	-	-
23	Diketodipyrrole	Hyp-Hyp	186	+	+	+	++	++	+	++	++	?
24	2,5-Diketopiperazines der.	Pro-Val, Pro-Arg	154	-	+	-	+	+	+	-	+	?
25	2,5-Diketopiperazines der.	Pro-Ala	168	-	-	-	+	+	-	-	-	-
26	2,5-Diketopiperazine	Pro-Pro	194	-	-	-	+	+	-	+	+	-
27	Tetradecanenitrile	unknown	209	-	-	-	-	-	-	+	+	-
28	Hexadecanenitrile	unknown	237	-	-	-	-	-	-	+	+	-
29	Octadecanenitrile	unknown	265	-	-	-	-	-	-	+	-	-
30	Eicosanenitrile	unknown	293	-	-	-	-	-	-	+	-	-

[a] AA = amino acid; AS = amino sugar; Pro = Proline; Hyp = Hydroxyproline; Glu = Glutamine; Ala = Alanine; Trp = Tryptophan; Phe = Phenylalanine; Val = Valine; Arg = Arginine

[b] - = not detected, + = present, ++ = abundant, +++ = very abundant.

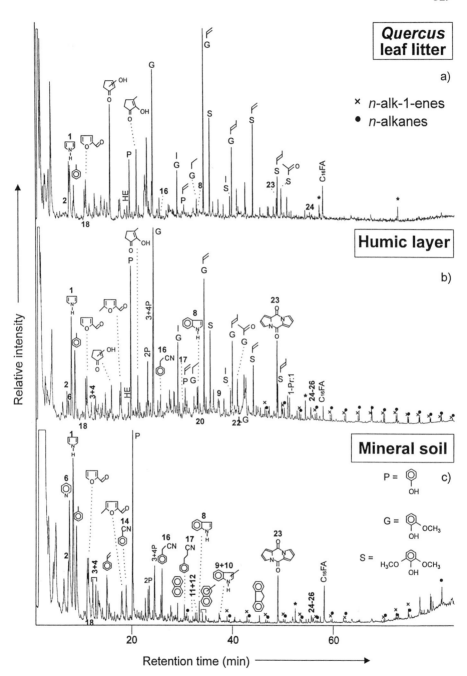

Figure 2. TICs of the pyrolysates of the residues after saponification of a) *Quercus* leaf litter, b) humic rich top horizon and c) mineral soil horizon. For key to symbols see caption of Figure 1.

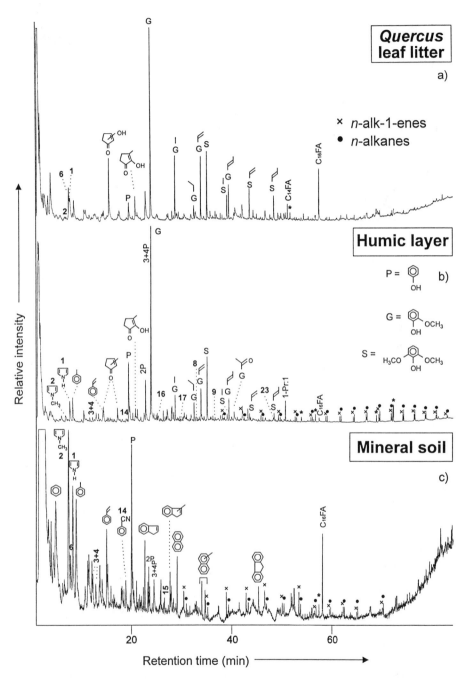

Figure 3. TICs of the pyrolysates of the residues after base and acid treatment of a) *Quercus* leaf litter, b) humic rich top horizon and c) mineral soil horizon. For key to symbols see caption of Figure 1.

methylpyrrole (**2**) and phenol. Other significant nitrogen products are pyrrole (**1**), pyridine (**6**) and benzonitrile (**14**).

Amino Acids. Based on total dry weight, the amino acid analyses of both the solvent insoluble (Fig. 4) and base treated residues (Fig. 5) revealed a steady decrease in concentration in going from the leaf litter to the mineral soil. The variation in absolute amounts present in the replicate soil samples of the humic layer (Fig. 4) most likely arises from the difference in mineral content of these two spatially separated samples. It should be noted that the amounts detected in residue after base treatment of mineral soil (Fig. 5) are close to the detection limit of the system. However, based on the total organic carbon content the results show no overall decrease, instead the absolute proportion of amino acids remained quite constant or showed a slight increase. The overall distribution of the different amino acids is most similar between the humic horizon and the mineral soil (Fig. 4). Compared with the litter horizon, and based on TOC, the most obvious differences are the absolute decrease in arginine and absolute increase in glycine, serine, alanine and proline, and the relative decrease in asparagine or aspartic acid (ASX), glutamine or glutamic acid (GLX). The most significant change after base treatment is the three fold reduction in concentration of the amino acids based on total organic carbon content (cf. Fig. 4d-f *vs* 5d-f). However, the susceptibility of the different amino acids to the base treatment appears to vary with proline being less affected while serine, glycine, and arginine are more affected.

Discussion

The overall chemical composition of the organic matter changed dramatically down the soil profile for all three residues as revealed by the elemental analyses (Table I) and pyrolysis data (Figs 1-3). The apparent increase in abundance of N-containing pyrolysis products down the soil profile (Figs 1-3) was mirrored by the relative increase in total N as revealed by the C/N ratios (Table I) and the proportion of amino acids based on total organic carbon (Fig. 4-5).

As expected the pyrolysate of insoluble organic matter of the leaf litter was dominated by products derived from ligno-cellulose (Fig. 1a). The contribution of N-containing products to the pyrolysate could be accounted for on the basis of either amino acid moieties, probably present as proteins or polypeptides, or from amino sugars, present as chitin. Although largely agreeing with the pyrolysis data, the results from the amino acid analyses revealed one significant discrepancy; diketodipyrrole (DK), which is known to be a characteristic pyrolysis product of free hydroxyproline (*18*), is a relatively abundant pyrolysis product. However, this amino acid was not detected in the HPLC analyses which is surprising as it is known to be present in the cell wall glycoproteins of dicotyledonous plants (*27*). Significantly, in an earlier study on the effects of land use on the amino acid composition of soils, hydroxyproline was similarly not detected in either manured or unmanured soils (*28*). A possible explanation could be that the hydroxyproline formed the corresponding sulfate ester during amino acid hydrolysis in the presence of sulfates originating from

332

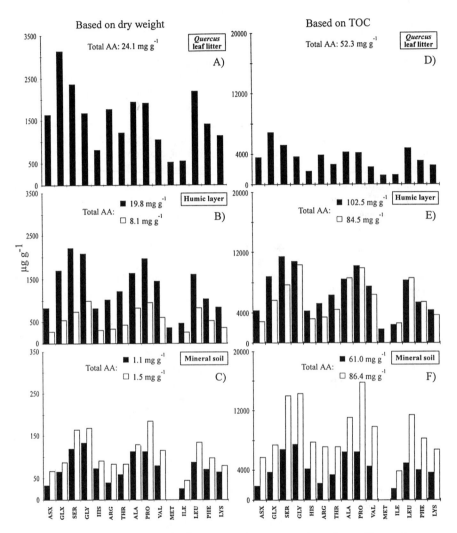

Figure 4. Amino acid distributions and concentrations in residues after solvent extraction of a and d) *Quercus* leaf litter, b and e) humic rich top horizon and c and f) mineral soil horizon based on dry weight (a, b, c) or total organic carbon (d, e, f). The black and white histograms in the humic and mineral horizons represent the replicate soil samples. Note that the different scales. Analyses performed by HPLC (see text for further details).

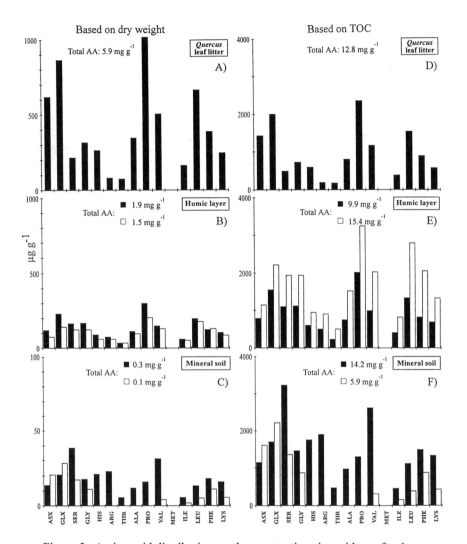

Figure 5. Amino acid distributions and concentrations in residues after base treatment of a and d) *Quercus* leaf litter, b and e) humic rich top horizon and c and f) mineral soil horizon based on dry weight (a, b, c) or total organic carbon (d, e, f). The black and white histograms in the humic and mineral horizons represent the replicate soil samples. Note that the different scales. Analyses performed by HPLC (see text for further details).

inorganic or organic components of the soil (*17*). However, this can only be a partial explanation since serine, which would also be affected, is still detected. Alternatively, diketodipyrrole may have originated from another combination of amino acids when present as peptides, since the original evidence for the formation of DK was based on the pyrolysis of pure hydroxyproline rather than peptides which included hydroxyproline. It should be noted that significant differences have been observed between pyrolysis products of free amino acids and peptides (cf. *29*).

The organic matter of the humic horizon was still dominated by ligno-cellulose which explains the similarity between bulk stable carbon isotope values of this layer and that of the leaf litter (both ca. -27.9 ‰). Like the organic matter in the leaf litter, most nitrogen-containing products can be explained as arising from amino acid moieties and amino sugars. The presence of chitin markers is to be expected since many soil mesofauna and fungi contain this structural amino sugar (*30*). Although similar in gross chemical terms, the amino acid composition of the humic horizon did reveal some changes, especially in ASX and GLX possibly because of depletion of aspartate and glutamate through enzymatic conversion to other amino acids (such as via the glutamate synthase cycle; *31-33*). In addition, the absolute amount of amino acids, based on total organic carbon, increased which indicates either a partial contribution to the amino acid pool from sources other than the leaf litter or alternatively, a faster loss of carbon with a 'stabilization' of the nitrogen. This effect is more noticable for serine, glycine and proline. The increased concentrations of serine and proline may be due to microbial production (*33*) whereas the increase in glycine is most probably due to the loss of the side chains of other amino acids, leaving a glycine remnant, in addition to a microbial contribution (*34*).

The most dramatic change in the chemical composition occurred between the humic and mineral horizons. The pyrolysate of the latter horizon was dominated by a range of products which could be directly related to amino acid moieties (Fig. 1c). The high relative abundance of amino acid moieties compared with the ligno-cellulose constituent may explain the more positive bulk $\delta^{13}C$ value of the residues from the mineral horizon, since proteins are generally relatively enriched in ^{13}C compared with lignin (*35*). The accumulation of amino acid moieties in soil has been seen previously; the associated N, presumably coming ultimately from the soil microbial biomass, either directly from resting organisms or indirectly by reaction between microbial N with other organic matter fractions and/or clay. The only possible pyrolysis product of amino sugars in this horizon is acetamide.

Apart from the amino sugar and amino acid moieties, the pyrolysate of this horizon also revealed evidence of N-containing products not obviously related to known biomolecules. These included quinoline, isoquinoline, one C_1-indole (**10**) other than 3-methylindole, and a series of alkyl nitriles. Although quinoline and isoquinoline have been identified previously in pyrolysates of soils (*4*), to date, no specific origin has been recognized. With respect to the C_1-indoles, pyrolysis of tryptophan releases specifically indole and 3-methylindole (*21, 22*), products which were identified in the pyrolysates of the litter and humic horizon. The isomeric C_1-indole was only detected in the mineral horizon soil suggesting that this product is soil specific. Although possibly derived from soil micro-organisms, it may also be a pyrolysis product released from chemically transformation product biomolecules formed in the soil. A chemically induced transformation product is supported by presence of this C_1 indole in the pyrolysate of the humic horizon after base and acid treatment. The long-chain alkyl nitriles also appear to be soil specific products (cf. 4). Such compounds have been recognized in the pyrolysates of algae (36) organisms also known to inhabit soils (1). More likely, however, they are formed in the soil during chemical transformations of biomolecules (such as alkaloids or reactions between fatty acids and ammonia or amines). Alternatively they are products formed during the pyrolysis process (cf. 21).

The most distinct change after base treatment is the substantial decrease in concentration of total amino acids (cf. Fig. 4 vs Fig. 5) resulting from the combined effects of solublization and partial hydrolysis during pH adjustment prior to the 6 M HCl hydrolysis for HPLC. During base treatment the more labile peptide linkages were hydrolyzed, reducing the recovery potential of the total amount of amino acid after the subsequent acid treatment by a factor of about 6 (cf. Fig. 4 and Fig. 5). Furthermore, the aqueous alkaline environment is conducive to the formation of Schiff bases through reaction of amino acid groups with sugar aldehyde groups. These then undergo Amadori rearrangement upon acidification of the reaction mixture (17). These amino acid/sugar derivatives are not detected in the HPLC analyses under the conditions used in this study. The pyrolysates of the residues after base treatment (Fig. 2) were similar to those of the insoluble residues (Fig. 1). The only significant changes were the decrease in the abundance of polysaccharide pyrolysis products, 4-ethenylphenol (Fig. 2a), and the alkyl nitriles (Fig. 2c). 4-Ethenylphenol is most likely derived from p-coumaric acid moieties (37) which are primarily ester-linked in the plant tissues and, hence, the decrease in the abundance of this component is not surprising. The reduction in alkyl nitriles indicates that these originate from moieties susceptible to base treatment. Another point worth noting is the detection of additional products indicative of amino sugars (acetylpyrrolidone and 3-acetamido-2/4-pyrone; Table II) in the base treated residue of the humic horizon. This observation indicates that due to the large amounts of certain pyrolysis products, in this case ligno-cellulose markers, other characteristic compounds may not have been observed, causing a bias against specific moieties such as amino sugars (cf. 29). This phenomenon could also explain the detection of quinoline and isoquinoline in this residue (Table II). Alternatively, the base treatment procedure may induce chemical alterations to the original organic material present. This could also explain the formation of cyclic hydroxypentenones recognized in the pyrolysates of the leaf litter and humic layer after saponification (cf. Fig. 2a, b) which most likely originate from modified polysaccharides.

The pyrolysates obtained after subsequent base and acid treatment revealed substantial chemical alteration to the organic matter present in the different soil horizons. Most notable was the virtually complete removal of the majority of the nitrogen-containing products derived from amino acid moieties and amino sugars. Such a removal of amino moieties upon hydrolysis would explain the more depleted $\delta^{13}C$ values of the final residues (Table II). The fact that the value of residue after acid treatment of the mineral soil is still less depleted then those of both the litter and humic horizons (c. -27.6 ‰ *vs* c. -29.4 ‰) indicates the presence of stable recalcitrant 'heavy' carbon; such a phenomenon has been observed previously by Lichtfouse and co-workers (*38*). Although some of the nitrogen-containing products detected in the pyrolysate of the mineral horizon could still derive from amino acid moieties and amino sugars, these products are assumed to represent the more recalcitrant soil organic nitrogen which may well constitute the so-called 'unknown' soil organic nitrogen pool. The form in which this nitrogen occurs is, however, as yet still unknown, but several mechanisms have been proposed for their formation in the soil. These include condensation products between polysaccharides and amino acid moieties (i.e. melanoidins; *1, 39*), phenols (including lignin) or quinones and amino groups or NH_3 (*1, 4, 11*), and enhanced cross-linking of proteins/ polypeptides (*40*). However, the absence of detailed molecular characterization of the nitrogenous transformation products in soil organic matter precludes any unequivocal interpretations regarding their origin. An additional aspect which should be considered is the extent to which both the base and acid treatments, as well as the actual pyrolysis affects the nitrogen-containing products released. One important observation in this context is that signals for pyrrolic N present in pyrolysates are seldom seen in ^{15}N solid state NMR spectra of soil organic matter (*10*).

Conclusions

Three soil horizons from a mature oak woodland at Rothamsted Experimental Station have been analyzed using elemental analyses, bulk stable carbon isotopes, flash pyrolysis-gas chromatography/mass spectrometry and amino acid analyses in order to determine the changes in the molecular composition of nitrogen-containing soil organic matter down a soil profile. The principal findings of the investigation are:

1) Based on elemental and pyrolysis data, the amount of nitrogen relative to soil organic carbon increases down profile.
2) The recognizable nitrogen-containing compounds in the leaf litter and humic horizon can be accounted for being either amino acid or amino sugar derived. In contrast, the nitrogen-containing pyrolysis products of the mineral horizon also reveal evidence of macromolecular-bound nitrogen not obviously related to known biomolecules. This may represent the so-called 'unknown' soil organic nitrogen.
3) The increase in the total amount of amino acids, based on total organic carbon, in the humic and mineral horizons indicates input from sources other than the leaf litter or a faster degradation of the carbon compared with the nitrogen.
4) The relative increase in organic nitrogen-moieties, most probably amino acid derived, is the most likely cause for the less depleted $\delta^{13}C$ values in the mineral soil.

Overall, these new data provide additional information concerning the molecular composition and fate of macromolecular bound organic nitrogen in soils. However, isotopic data on individual amino acids may provide further insights into the sources of these components in the lower horizons. Furthermore, it should be emphasized that more detailed molecular information is needed, in particular with respect to the molecular characterization of the 'unknown' nitrogen fraction, before we will fully understand the role and fate of soil organic nitrogen.

Acknowledgments

We thank J. Carter and A. Gledhill for their help with the GC/MS and IRMS analyses and D. Wallace and A. Moffet for their assistance with the HPLC analyses. C. Nott and Dr I. Bull are acknowlegde for their help with the sampling and some of the initial extractions. NERC provided Mass Spectrometry Facilities (Grant FG 6/36/01) and a research grant (GR3/9578) to RPE. IACR receives grant-aided support from the BBSRC.

References

1. *Soil Microbiology and Biochemistry (second edition)*; Paul, E. A.; Clark, F. E., Eds.; Academic Press: San Diego, **1996**.
2. Bremner, J. M. In *Soil Biochemistry*; McLaren, A. D.; Peterson, G. H., Eds.; Marcel Dekker: New York, **1967**; pp 19-66.
3. Anderson, H. A.; Bick, W.; Hepburn, A.; Stewart, M. In *Humic Substances II*; Hayes, M. H. B.; MacCarthy, P.; Malcolm, R. L.; Swift, R. S., Eds.; John Wiley & Sons: New York, **1989**; pp 223-253.
4. Schulten, H.-R.; Sorge, C.; Schnitzer, M. *Biol. Fertil. Soils* **1995**, *20*, 174-184.
5. Schulten, H.-R.; Sorge-Lewin, C.; Schnitzer, M. *Biol. Fertil. Soils* **1997**, *24*, 249-254.
6. Coûteaux, M.-M.; Bottner, P.; Berg, B. *TREE*. **1995**, *10*, 63-66.
7. Schulten, H.-R.; Leinweber, P. *J. Anal. Appl. Pyrolysis* **1996**, *38*, 1-53.
8. Olk, D. C.; Cassman, K. G.; Randall, E. W.; Kinchesh, P.; Sanger, L. J.; Anderson, J. M. *Euro. J. Soil Sci.* **1996**, *47*, 293-303.
9. Schnitzer, M. In *Humic Substances in Soil, Sediment and Water*; Aiken, G. R.; Mcknight, D. M.; Wershaw, R. L.; MacCarthy, P., Eds.; John Wiley & Sons: New York, **1985**; pp 303-325.
10. Knicker, H.; Fründ, R.; Lüdemann, H.-D. *Naturwissenschaften* **1993**, *80*, 219-221.
11. Knicker, H.; Kögel-Knabner, I. **1998**, This volume.
12. Jenkinson, D. S. *Rothamsted Experimental Station Report for 1970, Part 2.* **1971**, pp 113-137.
13. Jenkinson, D. S.; Harkness, D. D.; Vance, E. D.; Adams, D. E.; Harrison, A. F. *Soil Biol. Biochem.* **1992**, *24*, 295-308.
14. Poulton, P. R. In *Evaluation of Soil Organic Matter Models*; Powlson, D. S.; Smith P.; Smith, J. U., Eds.; NATO ASI Series 138; Springer-Verlag: Berlin, **1996**; pp 385-389.
15. *F.A.O.-UNESCO Soil map of the world: revised legend.* World soil resources report 60. Rome, **1990**.
16. *U.S.D.A. Soil Survey Staff. Key to soil taxonomy.* SMSS Technical monograph No. 19; 5th Edition, Pocahontas Press Inc., Blacksburg, Virginia, **1992**.

338

17. Hunt, S. In *Chemistry and Biochemistry of Amino Acids*; Barrett, G. C., Ed.; Chapman and Hall: London, **1985**; pp 415-425.
18. Cohen, S. A.; Meys, M.; Tarvin, T. L. *Pico-Tag advanced methods manual.* Millipore Corporation, Bedford, M. A., U.S.A., **1989**.
19. Suleiman, M. S.; Wallace, D.; Birkett, S.; Angelini, G. D. *Cardiovasc. Res.* **1995**, *30*, 747-754.
20. Stankiewicz, B. A.; van Bergen, P. F.; Duncan, I. J.; Carter, J. F.; Briggs, D. E. G.; Evershed, R. P. *Rapid Comm. Mass Spectrom.* **1996**, *10*, 1747-1757.
21. Chiavari, G; Galletti, G. C. *J. Anal. Appl. Pyrolysis* **1992**, *24*, 123-137.
22. Tsuge, S.; Matsubara, H. *J. Anal. Appl. Pyrolysis* **1985**, *8*, 49-64.
23. van Bergen, P. F.; Bull, I. D.; Poulton, P. R.; Evershed, R. P. *Org. Geochem.* **1997**, *26*, 117-135.
24. van Bergen, P. F.; Bland, H. A.; Horton, M. C.; Evershed, R. P. *Geochim. Cosmochim. Acta* **1997** *61*, 1919-1930.
25. Pouwels, A. D.; Tom, A.; Eijkel, G. B.; Boon, J. J. *J. Anal. Appl. Pyrolysis* **1987**, *11*, 417-437.
26. Bracewell, J. M.; Robertson, G. W. *J. Anal. Appl. Pyrolysis* **1984**, *6*, 19-29.
27. Dey, P. M.; Brinson, K. *Adv. Carbohyd. Chem. Biochem.* **1984,** *42*, 265-382.
28. Beavis, J.; Mott, C. J. B. *Geoderma* **1996**, 72, 259-270.
29. Reeves, J. B.; Francis, B.A. **1998**, This volume.
30. *Chitin*; Muzzarelli, R. A. A., Ed.; Pergamon Press, **1977**.
31. *Biology of Microorganisms*; Brock, T.: Madigan, M.T., Eds.; Prentice Hall: Englewood Cliff, N.J., **1991**.
32. Lea, P. J.; Wallsgrove, R. M.; Miflin, B.J. In *Chemistry and Biochemistry of Amino Acids*; Barrett, G. C., Ed.; Chapman and Hall: London, **1985**; pp 196-226.
33. Macko, S. A.; Engel, M. H.; Qian, Y. *Chem. Geol.* **1994**, *114*, 365-379.
34. Fogel, M. L.; Aguilar, C.; Bocherens, H.; Johnson, B. J.; Keil, R. G.; Teece, M. A.; Tuross, N. *Abstracts, 18th International Meeting on Organic Geochemistry* **1997**, 1, 23-24.
35. Deines, P. In *Handbook of Environmental Isotope Geochemistry*; Fritz, P.; Fontes, J. C., Eds.; Elsevier, Amsterdam, **1980**; pp 329-406.
36. Derenne, S.; Largeau, C.; Taulelle, F. *Geochim. Cosmochim Acta* **1993**, *57*, 851-857.
37. Tegelaar, E.W.; de Leeuw, J. W.; Holloway, P.J. *J. Anal. Appl. Pyrolysis* **1989**, *15*, 289-295.
38. Lichtfouse, E.; Dou, S.; Girardin, C.; Grably, M.; Balesdent, J.; Behar, F.; Vandenbroucke, M. *Org. Geochem.* **1995**, *23*, 865-868.
39. Maillard, L.C. *Ann. Chim. (Paris)* **1916**, *11/5*, 258-317.
40. Nguyen, R.; Harvey, R. **1998**, This volume.

Chapter 20

Soil Organic Nitrogen Formation Examined by Means of NMR Spectroscopy

Heike Knicker and Ingrid Kögel-Knabner

Department of Soil Science, Technische Universität München, 85350 Freising-Weihenstephan, Germany

NMR spectroscopy has been proven to be a valuable tool for the non-destructive examination of refractory geopolymers. Due to the low sensitivity of the ^{15}N isotope for an NMR experiment, ^{15}N NMR spectroscopy has been restricted to the examination of laboratory-derived ^{15}N enriched material. Recent developments have made it possible to obtain solid-state ^{15}N NMR spectra of natural soils at natural ^{15}N abundance. The results of recent ^{15}N NMR studies obtained from various soil and soil related systems are summarized. These studies indicate that amide functional groups represent the major nitrogen form in many soil systems. The formation of heterocyclic aromatic nitrogen, commonly proposed to be an important mechanism of nitrogen stabilization during humification, was not confirmed by ^{15}N NMR spectroscopic investigation of natural soils.

Nitrogen is a major nutrient element in the biosphere. Since nitrogen is considered a limiting factor for plant and microbial growth, the form and chemistry of the organic soil nitrogen is closely related to biological productivity in soils.

Nitrogen in soils is not a static entity, but takes part in a series of interconnected reactions which form the nitrogen cycle. The nitrogen of soil organic matter originates from living organisms in which it is mainly as protein and its constituent peptides and amino acids. Only small amounts can be assigned to amino sugars, nucleic acids, alkaloids and tetrapyrroles (1). Most of these nitrogen-containing compounds are known to be highly sensitive to microbial degradation and are therefore expected to be rapidly mineralized and/or re-utilized for biological production. Some of the nitrogen, on the other hand, is sequestered from the overall nitrogen cycle and becomes stabilized by incorporation into the stable soil organic matter pool. The significance of humic substances to the fertility of soil arises from the fact that much of their organic nitrogen resists further attack by microorganisms and is relatively unavailable to plants. The importance of refractory organic nitrogen in the overall nitrogen cycle is well recognized but little is known concerning its chemical composition or the mechanism(s) responsible for its resistance.

Greater knowledge of processes involved in the stabilization of nitrogen into the refractory soil organic matter pool will provide a better understanding of factors affecting the availability of soil nitrogen and may contribute to the development of improved management practices for the efficient and environmentally acceptable use of fertilizer N. The following chapter intends to review common models proposed for the explanation of refractory soil organic nitrogen formation. In addition, recent studies conducted to improve our knowledge about the nature and chemistry of organic nitrogen in soils by means of ^{13}C and ^{15}N nuclear magnetic resonance spectroscopy are summarized.

Soil Organic Nitrogen Identified by Wet Chemical Analysis

The most widely used approach for characterizing organic N in soils or humic substances is acid hydrolysis (2-4). In a typical procedure, the sample is heated with 3 N or 6 N HCl for 12 to 24 h, after which the hydrolyzed N is separated into several fractions. Applying this technique with subsequent colorimetric methods, approximately one third to one half of the total nitrogen can be identified as known biological compounds, mostly as amino acids (2). Approximately 1 to 2 % of the total nitrogen has been identified as amino sugars (3). Soils also contain trace quantities of nucleic acids and other known nitrogenous biochemicals such as degradation products of chlorophyll, uric acid and phospholipide amines (3,5,6). However, specialized techniques are required for their separation and identification.

After acid hydrolysis, about 20 to 25 % of the soil nitrogen of surface soils is recovered as NH_3-N by distillation with MgO (2). Some of this nitrogen derives from amino acid amides, such as asparagine and glutamine (7) while some is referred to as indigenous fixed NH_4^+. A part of this nitrogen may originate from partial destruction of amino sugars, but also amino acids such as serine, threonine or thryptophan. Threonine and serine are slowly degraded to ammonium and carbonyl. Thryptophan is stable if presence of air is avoided (9). However, it was shown that in the presence of iron (III) and copper (II) even degassing prior to hydrolysis offers no protection for thryptophan (8).

An accounting of all known potential sources of NH_3 in soil hydrolysates shows that about one half of the NH_3-N, equivalent to 10 to 12 % of the total organic N, is still obscure (2). Some of this unknown hydrolyzable nitrogen could originate from pseudoamines, such as iminoquinones, Schiff Bases and enamines, hydroxyamino acids, amino alcohols and sugars. Purines and pyrimidines were also suggested as a possible origin of this unidentified hydrolyzed nitrogen (7). After hydrolysis, a large portion of soil N, usually about 25 to 35 %, remains in the non-hydrolyzable residue. Because of its insolubility, this fraction is excluded from the investigation with common wet chemical approaches and is generally referred to as 'unknown soil organic nitrogen'. In search of the structure of this refractory organic nitrogen, most of the efforts were directed towards the identification of compounds which can resist microbial degradation and are inert against harsh chemical treatment. It was further concluded that such refractory organic nitrogen had to differ strongly from the labile nitrogen compounds of biogenic precursors. While the later are mostly characterized by amide functional groups, the inert nitrogen in soils was thought to occur mainly as heterocyclic aromatic nitrogen. Based on these assumptions, many models explaining the formation of such heterocyclic aromatic nitrogen were proposed (2,3). In the followings several of these models will be discussed.

Common Models Explaining the Formation of Refractory Soil Organic Nitrogen

Some of the models for the formation of refractory soil organic matter theorize that humic material represents the condensation products of partially degraded biopolymers. One of the early concepts (9) suggests that humic material is formed by condensation of C=O groups of lignin with NH_2 groups of proteinaceous material to Schiff Bases (ligno-protein model). More recently, Kuiters (10) assumed that soil protein-polyphenol complexes may derive from protein-tannin complexes formed during leaf senescence. Such protein-tannin complexes are known to resist degradation by most microorganisms. However, some ectomycorhizal fungi in coniferous forest soils are able to mineralize such compounds. They may be responsible for a competitive advantage of the plants associated with ectomycorhizal fungi (11).

Other models are based on the abiotic condensation of phenolic or quinonic structures with the side groups of N-containing compounds (recondensation models). Possible mechanisms involve covalent binding of the ε-lysylamino groups of proteins on quinones formed during lignin degradation (12) or the condensation via H-bridges between the hydroxyl groups of phenols and the carboxylic groups of proteins (13). Piper and Posner (14) suggested the formation of N-methylcarboxyl-quinoimine-complexes or structures which resemble N-(p-hydroxyphenyl)-glycine and N-(p-hydroxyphenyl)-glutamine. The presence of acid-persistent organic nitrogen in soils was explained with acid-insoluble indoles formed either as a condensation product of quinones and/or phenols with amino acids (15).

Another widely accepted mechanism for the stabilization of nitrogen in soil organic matter is represented by the formation of phenazine and phenoxazonederivatives via autopolymerization of phenols with ammonia (autopolymerization model) (16). The phenols necessary for this kind of reaction are thought to be delivered by the degradation of lignin or from secondary metabolites of fungi and microorganisms.

The concept of nitrogen stabilization via abiotic ammonia fixation on humic material was also the subject of recent ^{15}N NMR spectroscopic studies. Examining the structure of synthetic humic acids produced by air oxidation of benzoquinones with ammonium chloride, amino acids and peptides by means of ^{15}N NMR spectroscopy, Preston et al. (17) concluded that polymers are formed by coupling of semiquinone radicals. Amino compounds may be incorporated by formation of bonds between amino nitrogen and aromatic carbon. Zhuo and Wen (18), on the other hand, assumed from their ^{15}N NMR spectroscopic examination of synthetic humic acid that great differences exist between these synthetic compounds and natural humic acids. Investigating the reaction of [15]N -labeled ammonium hydroxide with the Suwannee River fulvic acid, the IHHS peat and a leonardite humic acids, Thorn and Mikita (19) recovered most of the incorporated nitrogen in indoles, pyrroles, followed by pyridine, pyrazine, amide and aminohydroquinones. Potthast et al. (20), subjecting lignin model compounds to oxidative ammonolysis, identified the incorporated nitrogen mostly as amides and benzonitriles. N-heterocyclic compounds were not found. From their results, they concluded that benzonitrile derivatives may also contribute to the active nitrogen-containing soil components that are released slowly from the soil.

An alternative mechanism explaining the stabilization of labile nitrogen in soils involves the Maillard reactions, which are well known to be responsible for the browning of food (melanoidin model) (21). During these reactions, ammonia or free amino groups of amino acids and amino sugars react with carbonyl groups of sugars to form Schiff Bases, which subsequently undergo the Amadori rearrangement to produce dark-colored melanoidins (21,22). Since these reactions occur independent from the availability of lignin and its degradation products, they are considered to be responsible for the formation of nitrogen-containing humic material in lignin-depleted ecological systems. Although comparative studies of soil organic matter and artificially produced melanoidins revealed high similarity concerning chemical properties ((22,23), this mechanism is still the subject of ongoing discussions. One of the main concerns involves the low reactivity of Schiff Bases at room temperature. The formation of melanoidins, therefore, may be in strong competition to microbial activity. On the other hand, the adsorption of the precursors onto clay minerals may create more favorable conditions for this reaction (24).

The consensus of all the models aforementioned is that during humification, biomacromolecules are microbiologically degraded to oligomeres or monomeres which, for the most part, are further degraded. A small fraction of these recombines with amino compounds (amino acids, NH_3) to heterocyclic compounds forming the brown-colored macromolecular network of humic material. All these models are based on laboratory studies. Schnitzer and Spiteller (25) were able to identify some possible heterocyclic products of the proposed mechanism in soils. However, it is still questionable if the concentrations of such compounds are high enough to account for most of the humified soil organic nitrogen.

Analytical Techniques for the Investigation of Refractory Organic Nitrogen in Soils

Principally, there are two approaches for the investigation of the chemical structure of refractory nitrogen in soils. The first involves thermolytic or chemolytic degradation of macromolecules into small fragments that are analyzed by gas chromatography or gas chromatography-mass spectrometry. It is assumed that the fragments are representative of the original larger macromolecules. Since secondary reactions (rearrangement, cracking, hydrogenation and polymerization) cannot be excluded, conclusions regarding the original structure of nitrogen in the macromolecular phase have to be drawn with caution. A more detailed introduction into the application of pyrolytic studies for the investigation of soil organic matter can be obtained from the literature (26).

Solid-state ^{13}C NMR spectroscopy. Alternative techniques for the examination of nitrogen in heterogeneous macromolecular mixtures are non-destructive spectroscopic methods, one of which is solid-state nuclear magnetic resonance (NMR) spectroscopy (27). The big advantage of this technique is that complex heterogeneous macromolecular structures such as soil organic matter can be examined as a whole, allowing conclusions on the chemical composition of the bulk sample. This is in contrast to most degradative and extractive techniques, in which only the extractable parts of the sample are analyzed, not allowing conclusions about the chemistry of the whole sample. Applying solid-state NMR, unexpected secondary reactions can be avoided, because harsh thermolytical and chemicolytical pretreatment or extended extraction procedures are not necessesary. Although the complexity of heterogeneous samples makes it difficult to obtain detailed compositional information, it is possible to identify various types of functional groups at the same time and to measure their average relative distribution. These advantages have made, in particular, ^{13}C NMR spectroscopy a well established technique for the characterization of soil organic matter (28) leading to some important findings concerning its chemical composition. Probably the most important of these is the discovery that phenolic carbons are not as dominant a structural feature of these materials as previously thought.

Table 1: Chemical shift assignment of various peaks in a ^{13}C NMR spectrum

ppm	Assignment
0-45	**paraffinic structures**
45-110	**carbohydrates**
45-60	aliphatic *C-N,* methoxyl
60-95	alkyl-O (carbohydrates, alcohols)
95-110	acetal and ketal carbon (carbohydrates)
110-160	**Aromatic-C**
110-140	aryl-H and aryl-C carbons
140-160	aryl-O and *aryl-N* carbons
160-220	**Carbonylic-C/carboxylic-C**
160-180	carboxyl and *amide carbons*
180-220	aldehyde and ketone carbons

Applying ^{13}C NMR for the investigation of nitrogen functionalities, one takes advantage of the fact that N-substituted carbon results in signals in distinct chemical shift regions. Table 1 shows a tentative assignment of differnt chemical shift regions of a ^{13}C NMR spectrum to certain N-containing functional groups. For example, the C_2 carbon of all amino acids except glycine contribute to the signal intensity in the chemical shift region between 51 and 61 ppm, N-substituted carbon of amino sugars contribute to the region between 51 and 58 ppm (29). González-Vila et al. (30) observed that the signals between 50 and 70 ppm in ^{13}C NMR spectra of three humic substances were strongest in the sample with the highest nitrogen-to-carbon ratio and weakest for the samples with the lowest nitrogen-to-carbon ratio. Based on these results, the authors assigned those signals to amino acids. However, one has to bear in mind that methoxyl-substituted carbons can add to the intensity in this specific chemical shift region. Another chemical shift region, indicative for nitrogen-substituted carbon, represents the chemical shift region between 180 to 160 ppm. Here, amide carbons can be observed but may be superposed by resonance lines of carboxyl groups of acids and esters. The fact that signals of oxygen-substituted carbons overlap signals of nitrogen-substituted carbons obviously makes it difficult to obtain information about the nature of nitrogen-containing compounds in soils and soil-related samples by means of ^{13}C NMR spectra alone. Correlation of the relative signal intensity of the specific chemical shift regions within the spectra and the nitrogen content of the samples may help to overcome some of the difficulties. In a recent study, the changes in chemical composition of the organic material of plant material incubated for two years were examined by ^{13}C NMR spectroscopy (27). The signal intensities of carbonyl/carboxyl/amide carbon (220 to 160 ppm) of the spectra obtained from plant incubates at different degradation stages were plotted against the total nitrogen content of the samples. A positive correlation was found, indicating that a major part of the intensities in this chemical shift region may be attributable to amide groups. A correlation with a slope of two was found between the relative intensities in the chemical shift region of alkyl carbon (45 to 0 ppm) and those of the region between 220 and 160 ppm. From these results, the authors concluded that a significant part of the intensity in the chemical shift region of aliphatic and carbonylic/carbonyl/amide carbon can be attributed to amino acids in proteins or proteinaceous material.

Solid-state ^{15}N NMR spectroscopy of biogenic precursors. The development of solid-state ^{15}N NMR of nitrogen-containing structures has been hampered by two factors. First, the most abundant nitrogen isotope ^{14}N has a large quadrupole moment from which it is impossible to obtain high resolution ^{14}N NMR spectra (32). Second, the dipolar ^{15}N nucleus, unfortunately, has a low natural abundance of 0.37 % and a low and negative gyromagnetic ratio. Therefore, the sensitivity of ^{15}N NMR is approximately 50 times lower than ^{13}C NMR. Consequently, ^{15}N NMR has not been used widely for studies of soil and related geopolymers, other than studies involving ^{15}N-enriched material (33-36). However, in previous studies ^{15}N NMR-relevant parameters were determined. The first application of these parameters to natural soils opened the door for the ^{15}N NMR spectroscopic analysis of the chemistry of soil organic nitrogen under natural conditions.

Table 2: Assignments for the various chemical shifts in a ^{15}N NMR spectrum (reproduced after 31)

	Assignment to biological material	**Other assignments**
25 to -25 ppm		Nitrates, nitrites, nitro groups
-25 to -90 ppm		Imines, phenazines, pyridines, Schiff-bases
-90 to -145 ppm	Purine (N-7)	Nitrile groups
-145 to -220 pm	Chlorophyll,purines/pyrimidines, indoles, imidazoles, pyrroles,	Maillard products
-220 to -285 ppm	Amides/peptides, N-acetylderivatives of aminosugars,tryptophanes, prolines	Lactame (pyrroles, carbazoles, indoles)
-285 to -325 ppm	NH in guanidines	Aniline derivatives
-325 to -375 ppm	NH$_4^+$, -NH$_3^+$, -NH$_2$, -NHR and -NR$_2$ groups, free amino groups in amino acids and amino sugars	Anilinium salts

As mentioned above, most of the nitrogen in soils derives from decaying organic material of biogenic origin, such as plants, fungi, bacteria but also algae. ^{15}N NMR spectra of some possible precursors of soil organic matter are shown in Figure 1. The most tentative chemical shift assignment is given in Table 2. Comparable spectra were obtained by Skokut et al. (37) for soybeans and for *Neurospora crassa* (38). In these spectra, the predominance of proteinaceous material in undegraded biological material is reflected by the main signal between -220 and -285 ppm and a smaller signal between -325 ppm and -350 ppm. The region between -220 to -285 ppm is most probably assigned to amide functional groups. Acetylated amino sugars and

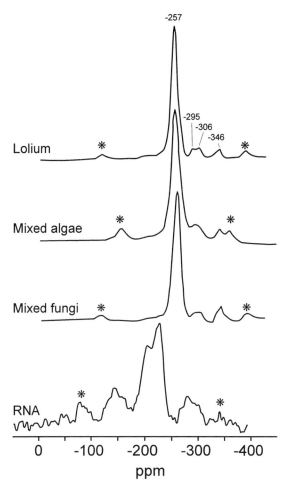

Figure 1. Solid-state ^{15}N NMR spectra of ^{15}N-labeled biogenic precursors of soil organic material in comparison to RNA at natural ^{15}N abundance (43). Asterisks indicate spinning side bands.

lactames, indoles (-245 ppm) and carbazoles (-262 ppm) may also contribute to the signal intensity in this chemical shift region. Unsubstituted pyrroles normally give a signal at -235 ppm, which may be overlapped by the amide resonance. Had pyrroles or indoles been major components, the broad peak would be shifted towards lower fields. The signal between -325 to -350 ppm is indicative for free amino groups of amino acids and/or amino sugars (32). The signals between -285 to -325 ppm can be assigned to primary and secondary amines not in α-position to carbonyl and may derive from citrulline, uric acids and nucleic acids but also from aniline derivatives. The later, however, are unlikely to occur in plant material in such amounts that they could be observed in their ^{15}N NMR spectra. The shoulder at the low field side of the main peak in the chemical shift region between -145 ppm and -220 ppm most tentatively originates from histidine, but some contribution of the ring nitrogen of nucleic acids may occur. For comparison, the solid-state ^{15}N NMR spectrum of ribonucleic acid is shown in Figure 1. Porphyrin structures in chlorophyll result in signals at -129 ppm, -169 ppm, -186 ppm and -188 ppm (39). The low concentration of chlorophyll in algae (approximately 0.2 pg cell $^{-1}$ for some algae) (40) and plant material (less than 1 % of the dry weight for plant leaves) (41) may be the reason why such signals cannot be discriminated from the noise in the NMR spectra. Signals, marked with asterisks (between -90 to -145 ppm and -390 to -420 ppm in the spectra of Lolium, the algae and the fungi and between -60 to -110 ppm and -320 to -375 ppm in the NMR spectrum of the RNA) are spinning side bands occurring due to application of the cross polarization magic angle spinning technique.

Solid-state ^{15}N NMR spectroscopy of ^{15}N-enriched plant incubates. Microbial degradation of plant and algal material incubated under laboratory conditions both at a sample moisture of 60 % of the maximal waterholding capacity and under water-saturation conditions resulted in solid-state ^{15}N NMR spectra showing a similar pattern to those obtained from the undegraded material (Figure 2, I) (27,42,43). The only changes observed were a slight increase in relative intensity of the amide signals with a simultaneous decrease of all other chemical shift regions. The signals for nitrate (-3.5 ppm) and ammonium (-356 ppm) reveal the accumulation of mineralization products after prolonged incubation. Additional signals, not observed in the spectra of the undegraded material and indicative for the formation of heterocyclic nitrogen, could not be identified. From these results, it was concluded that some peptide-like material of biogenic origin can survive microbial degradation. It was further assumed that the formation of aromatic heterocyclic condensation products did not represent a major pathway of nitrogen stabilization during the early stages of humification.

Comparable results were obtained from incubation studies, in which plant material with natural abundance of ^{15}N was incubated after addition of ^{15}N-labeled ammonium sulfate and a microbial suspension (Figure 2, II) (31). The solid-state ^{15}N NMR spectra obtained after one week of incubation indicated that most of the inorganic nitrogen was incorporated into amides and free amino groups. Also in this experiment, no indication for the formation of heterocyclic nitrogen was observed. In an additional experiment, organosolve lignin and sulfonated lignin were incorporated under the same conditions (Figure 2, III). It was assumed that depletion of carbohydrates and high concentration of aromatic compounds decreases the incorporation of nitrogen into microbial biomass in favor of nitrogen immobilization by condensation reactions. The solid-state ^{15}N NMR spectra of these incubates, however, showed no indication for the formation of heterocyclic nitrogen but for the immobilization of nitrogen into biomass. During humification in the presence of microorganism, the incorporation of nitrogen into biomass seems to be in strong

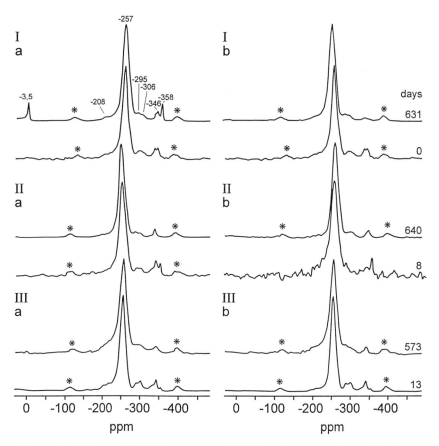

Figure 2. Solid-state ^{15}N NMR spectra of ^{15}N-labeled plant incubates (27,31) a) incubated at 60 % of the maximal waterholding capacity and b) under water saturation conditions. I) ^{15}N-enriched incubated wheat II) Beech saw dust incubated after addition of ^{15}N-labeled ammonium sulfate III) organosolve lignin incubated after addition of ^{15}N-labeled ammonium sulfate.

competition to nitrogen immobilization by abiotic condensation reactions. Similar conclusions can be drawn from experiments performed by Clinton et al. (44) examining the incorporation of ^{15}N-labeled ammonium into natural forest soils.

Solid-state ^{15}N NMR spectroscopy of natural soils. When examining processes under laboratory conditions, one has to bear in mind that such experiments may not reflect the natural situation. While a major fraction of the native soil organic matter is several hundred to several thousand years old (45), laboratory-produced material has only been incubated for a couple of years. Only the examination of natural soils, in which humic material was able to develop over an expanded period of time, enables us to draw conclusions about the structure of soil organic nitrogen and the processes involved in its formation. In order to obtain suitable ^{15}N NMR spectra of natural soils without artificial ^{15}N-enrichment, it was necessary to determine the optimal experimental conditions on soil organic matter related material providing ^{15}N NMR spectra with acceptable signal-to-noise ratios (27). ^{15}N-enriched plant incubates provided the strong signal for the optimization of instrumental parameters. Applying these optimized parameters to several natural soils under agricultural use (46,47), ^{15}N NMR spectra were obtained which resembled those of incubated plant material. Approximately 80 % of the total observable signal intensity of the spectra were detected in the region of amide functional groups followed by signals attributed to free amino groups, NH_2 groups and a shoulder in the area of possible indoles or pyrroles. No signals were observed in the chemical shift region of pyridinic N. Considering the low signal-to-noise ratio of the spectra, the broad lineshape of the resonance lines and the shoulder in the region of possible pyrrolic compounds, the authors concluded that heterocyclic aromatic N could not contribute more than 10 to 15 % to the total nitrogen signal intensity. These results are important to the origin and immobilization of organic nitrogen in soil systems and the chemical structure of nitrogen in refractory soil and sediment organic matter since they contradict the common view of the processes and chemistry responsible for the resistance of organic nitrogen. Rather than confirming rapid structural alterations via condensation reactions, they suggest that the nitrogen which survives microbial degradation remains as amide functional groups which mirror those in protein.

The dominance of the amide nitrogen can also be observed in ^{15}N NMR spectra obtained from sapropels (42) and some peats (43). This strongly indicates that the persistence of biogenic nitrogen against microbial degradation cannot be considered as a phenomenon limited to the soil environment but is also valuable for sedimental systems.

Possible Preservation Mechanism of Nitrogen in Soil Systems

Summarizing the results mentioned above, it can be concluded that nitrogen in amide functional groups, most probably deriving from peptide-like material of biogenic origin, is the major form of preserved organic nitrogen in soils. It can further be assumed that the postulated formation of N-containing aromatic heterocyclic structures (16,48) at most play a minor role in sequestering formerly labile nitrogen. If this refractory amide nitrogen derives - as indicated by the experiments mentioned above - from peptide-like or proteinaceous materials, then an explanation is needed to clearify how such labile compounds can survive microbial degradation.

One possibility to explain the high proportion of peptide material in soils could be a continuous de novo synthesis of microbial proteins from liberated nitrogen during the breakdown of decaying organic material. Such proteinaceous materials are commonly thought to be completely hydrolyzed after treatment with hot 6 N HCl. Applying this technique approximately 10 % to 60 % of the total organic nitrogen can be assigned to amino acids. In recent studies, this technique was applied to microbiologically degraded plant material (27,49) (Figure 3) and soils incubated with ^{15}N-labeled ammonium sulfate (50). Approximately 24 % of the total nitrogen remained in the insoluble residue of the Luvisol. Its ^{15}N NMR spectrum in Figure 3 reveals that at least some of the amides, but also some aliphatic amines, survive this harsh treatment. As shown recently (51), this is also true for biowaste composts and soils incubated after addition of biowaste composts. Subjecting soil organic matter (Figure 3b) and sedimental organic material (52) matured over a long time period to hot acid hydrolysis with subsequent ^{15}N NMR spectroscopic examination of the hydrolysis residue led to the same results. Obviously, HCl hydrolysis was not able to efficiently break down all amide functional groups. As shown by Siebert et al. (51), some of these amides can be assigned to proteinaceous material applying the new technique of thermochemolysis with tetramethylammonium hydroxide previously used for the identification of amino acids in sediments (52). From these data, it can be assumed that the chemical tests for the identification of nitrogenous compounds used in previous studies have failed to detect proteinaceous components, perhaps because they rely on extraction of protein into solution and the protein is tightly bound to a non-extractable phase.

Another explanation for the protection of proteinaceous material during humification may be its adsorption onto the mineral phase. Several reports have shown that clay minerals can protect organic molecules against microbial degradation (53) and that clays are the predominant sites of amino acid binding (54). It was shown that refractory organic matter in soils is intimately associated with the silt and clay fractions in soils (55), although some young organic material is present in these fractions (56). An understanding of the chemical structure and composition of the nitrogen functionality in these fractions may give insight in the mechanism involved in the stabilization of soil organic nitrogen. The high mineral content and therefore the low nitrogen content of such samples, however, may aggravate their ^{15}N NMR spectroscopic analysis. Treatment of soil samples with dilute hydrofluoric acids was shown to increase the relative amount of organic material by removal of mineral matter without major alterations of the chemical composition detectable by ^{13}C NMR spectroscopy (57). However, it cannot necessarily be assumed that this is also true for the behavior of the N-containing fraction, since solid-state ^{13}C NMR spectroscopy of complex organic mixtures such as they are in soil samples rarely allows direct conclusions about the chemical nature of their N-functional groups. In order to estimate possible HF-related alterations in the N-functionality of humic material, the solid-state ^{15}N NMR spectra of ^{15}N-enriched degraded biogenic material (i.e. algal material) obtained before (Figure 4a) and after (Figure 4b) treatment with 10 % HF were compared. After HF-treatment, only slight changes in the feature of the solid-state ^{15}N NMR spectrum of the degraded algal material were observed. The most apparent one is the decrease of the ammonium signal at -355 ppm. Considering the high solubility of ammonium in aqueous solution, it can be assumed that ammonium was removed with the aqueous HF-solution. In addition, a slight increase of the signal intensity of the free amino groups can be observed. This may be due to hydrolysis of very labile amide structures.

These results suggest that treatment with 10 % HF does not lead to major alteration of the bulk composition of the organic nitrogen fraction. Thus, enrichment of organic nitrogen in mineral rich soil samples by removal of mineral matter with 10 % HF may be considered a helpful tool for the improvement of the quality of solid-state ^{15}N NMR spectra.

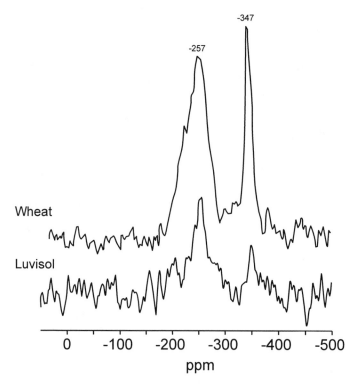

Figure 3. Solid-state ^{15}N NMR spectra of the residues after hydrolysis with 6 N HCl of a) a plant residues (wheat), incubated for 1 month (27) and b) a Luvisol (Germany) (53).

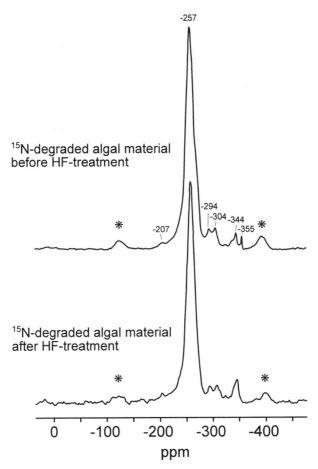

Figure 4. Solid-state ^{15}N NMR spectra of an ^{15}N-enriched degraded algal material a) before and b) after treatment with hydrofluoric acid (10%).

The solid-state ^{15}N NMR spectra of the fine particle size fractions of a Haplic Podzol after HF treatment in Figure 5 reveal the same features obtained for those of degraded plant material and bulk soils. They show the same pattern obtained for undegraded material of biogenic origin indicating a potential implication of mineral matter in the protection of originally labile organic compounds in soils. Since no major signals were detected in the chemical shift region assigned to pyrrolic or pyridinic nitrogen, pathways in which clay minerals are considered to catalyze the formation of Maillard products cannot be confirmed.

However, peptide-like material was also identified in solid-state ^{15}N NMR spectra of clay-free refractory biopolymers such as fungal melanins (58) or algaenan (36,42) and clay-free plant incubates (27). These studies strongly suggest the existence of additional processes other than mineral adsorption/protection which are responsible for the survival of peptide-like structures. A possible explanation was recently given by Nguyen and Harvey (59) suggesting an increasing stability of proteins by cross-links. Another possibility for the survival of the orginally thought labile proteinaceous material in organic-rich algal deposits was recently given by encapsulation of labile compounds in a hydrophobic network of the refractory algal biopolymer (52). It was also considered that parts of the algal cell walls are involved in the protection and that labile compounds may become sandwiched between algaenan layers. In most soils, algae do not present the major precursor of hydrophobic compounds. However, hydrophobic compounds of different origin, may participate in encapsulating labile organic soil components and may be responsible for the survival of at least some of the refractory soil amide nitrogen.

Summary and Concluding Remarks

The chemistry of soil organic nitrogen has been of major interest for many years. In particular, the processes involved in its stabilization and the mechanism responsible for its resistance against microbial and chemical degradation have been the research topics of scientist for generations. Most of our present knowledge of the formation and the structure of refractory soil organic nitrogen derives from laboratory experiments. From these results, it was concluded that most of the refractory organic nitrogen in soils is present in heterocyclic units, which are part of a highly aromatic macromolecular network. Conducting such experiments, one must be aware that under laboratory conditions, important parameters of complex natural systems may be ignored leading to results which do not reflect the events happening under natural conditions. The only way to prove the relevance of a mechanism discovered based on laboratory experiments, is its approval in a natural systems. However, in natural soils, N-containing heterocyclic units were not detected in such concentrations that they could be considered as a major form of refractory soil organic nitrogen.

The developments in ^{15}N NMR spectroscopy during the last years made it possible to apply this technique to natural soil systems, even to those with high mineral matter content. These studies suggest that amide nitrogen is the dominant form of soil organic nitrogen. It can further be assumed that these amides are most probably of biological origin. These results provide some major implication for the understanding of soil organic matter chemistry. One of these may be the conclusion that abiotic condensation reactions of organic material may not be as important in geological systems with high biological activity, as formerly thought. The mechanisms, involved in the stabilization of such amide groups are still obscure and have to be the subject of future research.

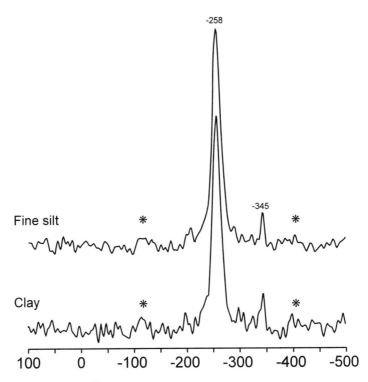

Figure 5. Solid-state ^{15}N NMR spectra of a) the fine silt fraction and b) the clay fraction of a Haplic Phaeozem (Germany) (57).

^{15}N NMR spectroscopy on soils, while still in its childhood shoes, has already been shown to be a valuable tool for the examination of soil organic nitrogen. Future efforts have to focus on improvements in spectra quality and the application of advanced NMR techniques in order to obtain more detailed information about the specific nature of the nitrogen, so far identified as amide functional groups. Obviously, this goal cannot be reached by means of ^{15}N NMR spectroscopy alone. In combination with other analytical techniques, ^{15}N NMR spectroscopy will add substantially to the understanding of the nature of soil organic nitrogen.

References

1) Paul, E. A.; Clark, F. E. *Soil Microbiology and Biochemistry*; Academic Press Inc.: San Diego, 1996.
2) Kelly, K. R.; Stevenson, F. J.: In Piccolo, A., Ed.; *Humic Substances in Terrestrial Ecosystems* Elsevier Science: Amsterdam, 1996, pp 407-427.
3) Schnitzer, M.: In Aiken, G. R.; Mcknight, D. M.; Wershaw, R. L.; MacCarthy, P., Eds.; *Humic Substances in Soil, Sediment and Water* John Wiley & Sons: New York, 1985, pp 303-325.
4) Anderson, H. A.; Bick, W.; Hepburn, A.; Stewart, M.: In Hayes, M. B. H.; MacCarthy, P.; Malcolm, R. L.; Swift, R. L., Eds.; *Humic Substances II* Wiley & Sons: Chichester, 1989, pp 223-253.
5) Cortez, J.; Schnitzer, M. *Canadian Journal of Soil Science* **1979**, *59*, 227-286.
6) Goodman, B. A.; Chshire, M. V. *Nature* **1973**, *244*, 158-159.
7) Kickuth, R.; Scheffer, F. *Agrochimica* **1976**, *20*, 376-386.
8) Stewart, M.; Nicholls, C. H. *Australian Journal of Chemistry* **1972**, *25*, 2139-2144.
9) Waksman, S. A.; Iyer, N. R. N. *Soil Science* **1932**, *34*, 43-69.
10) Kuiters, A. T. *Acta Botanica Neerlandica* **1990**, *39*, 329-348.
11) Northup, R. R.; Yu, Z.; Dahigren, R. A.; Vogt, K. A. *Nature* **1995**, *377*, 227-229.
12) Theis, E. R. *Journal Biol. Chem.* **1945**, *157*, 23-33.
13) Ladd, J. N.; Butler, J. H. A.: In Paul, E. A.; McLaren, A. D., Eds.; *Soil Biochemistry* Marcel Dekker: New York, **1975**; Vol. 4, pp 143-194.
14) Piper, T. J.; Posner, A. M. *Plant and Soils* **1972**, *36*, 595-598.
15) Rinderknecht, H.; Jurd, L. *Nature* **1958**, *181*, 1268-1269.
16) Flaig, W. J. A.; Beutelspacher, H.; Rietz, E. In Gieseking, J. E., Ed.; *Soil Components* Springer: New York, 1975; Vol. 1, pp 1-211.
17) Preston, C. M.; Rauthan, B. S.; Rodger, C. A.; Ripmeester, J. A. *Soil Science* **1982**, *134*, 277-293.
18) Zhuo, S. N.; Wen, Q. X. *Pedosphere* **1993**, *3*, 193-200.
19) Thorn, K. A.; Mikita, M. A. *Science of the Total Environment* **1992**, *113*, 67-87.
20) Potthast, A.; Schiene, R.; Fischer, K. *Holzforschung* **1996**, *50*, 554-562.
21) Maillard, L. *Ann. Chem. Phys.* **1916**, *5*, 258-317.
22) Ikan, R. *The Maillard Reaction; Consequences for the Chemical and Life Science*; John Wiley & Sons: Chichester, 1996.
23) Benzing-Purdie, L.; Ripmeester, J. A. *Soil Science Society of America Journal* **1983**, *47*, 56-61.
24) Hedges, J. I.: In Frimmel, F. H.; Christman, R. F., Eds.; *Humic Substances and their Role in the Environment* Wiley & Sons: New York, 1988, pp 45-58.
25) Schnitzer, M.; Spiteller, M.: In Schnitzer, M.; Spiteller, M., Eds.; *The chemistry of the "unknown" soil nitrogen* Hamburg, 1986; Vol. 2, pp 473-474.
26) Saiz-Jimenez, C.: In Piccolo, A., Ed.; *Humic Substances in Terrestrial Ecosystems* Elsevier: Amsterdam, 1996, pp 1-44.
27) Knicker, H.; Lüdemann, H.-D. *Organic Geochemistry* **1995**, *23*, 329-341.

28) Wilson, M. A. *NMR Techniques and Applications in Geochemistry and Soil Chemistry*, Pergamon Press, Oxford, 1987.

29) Kalinowski, H.-O.; Berger, S.; Braun, S. ^{13}C *NMR Spektroskopie*; Thieme: Stuttgart, 1984.

30) González-Vila, F. J.; Lentz, H.; Lüdemann, H.-D. *Biochemistry and Biophysics Research Communications* **1976**, *72*, 1063-1069.

31) Knicker, H.; Lüdemann, H.-D.; Haider, K. *European Journal of Soil Science* **1997**, *48*, 431-441.

32) Witanowski, M.; Stefaniak, L.; Webb, G. A. *Nitrogen NMR spectroscopy*; Aademic Press: London, 1993.

33) Benzing-Purdie, M., L.; Ripmeester, J. A.; Preston, C. M. *Journal of Agriculture and Food Chemistry* **1983**, *31*, 913-915.

34) Cheshire, M. V.; Williams, B. L.; Benzing-Purdie, L. M.; Ratcliffe, C. I.; Ripmeester, J. A. *Soil Use Management* **1990**, *6*, 90-92.

35) Almendros, G.; Fründ, R.; González-Vila, F. J.; Haider, K. M.; Knicker, H.; Lüdemann, H. D. *FEBS Letters* **1991**, *282*, 119-121.

36) Derenne, S.; Largeau, C.; Taulelle, F. *Geochimica Cosmochimica Acta* **1993**, *57*, 851-857.

37) Skokut, T. A.; Varner, J. E.; Schaefer, J.; Stejskal, E. O.; McKay, R. A. *Plant Physiology* **1982**, *69*, 314-316.

38) Jakob, G. S.; Schaefer, J.; Stejskal, E. O.; McKay, R. A. *Biochemical Biophysical Research Communications* **1980**, *97*, 1176-1182.

39) Martin, G. J.; Martin, M. L.; Gouesnard, J. P. ^{15}N-*NMR Spectroscopy*, Springer: Heidelberg, 1981.

40) Brown, M. R. *Journal of Exp. Mar. Biol Ecol.* **1991**, *145*, 79-99.

41) Greenfield, L. G. *Plant and Soil* **1972**, *36*, 191-198.

42) Knicker, H.; Scaroni, A. W.; Hatcher, P. G. *Organic Geochemistry* **1996**, *24*, 661-669.

43) Knicker, H.; Hatcher, P. G.; Scaroni, A. W. *International Journal of Coal Geology* **1996**, *32*, 255-278.

44) Clinton, P. W.; Newman, R. H.; Allen, R. B. *European Journal of Soil Science* **1996**, *46*, 551-556.

45) Hsieh, Y. P. *Soil Science Society of America Journal* **1992**, *56*, 460-464.

46) Knicker, H.; Fründ, R.; Lüdemann, H.-D. *Naturwissenschaften* **1993**, *80*, 219-221.

47) Knicker, H.; Fründ, R.; Lüdemann, H.-D.: In Nanny, M.; Minear, R. A.; Leenheer, J. A., Eds.; *Nuclear Magnetic Resonance in Environmental Chemistry*, Oxford University Press: New York, 1997.

48) Schulten, H.-R.; Sorge, C.; Schnitzer, M. *Biol. Fertil Soils* **1995**, *20*, 174-184.

49) Knicker, H.; Fründ, R.; Almendros, G.; González-Vila, F. J.; Martín, F.; Lüdemann, H.-D. *Finnish Humus News* **1991**, *3*, 313-315.

50) Zhuo, S. N.; Wen, Q. X.; Du, L. J.; Wu, S. L. *Chinese Science Bulletin* **1992**, *37*, 508-511.

51) Siebert, S.; Knicker, H.; Hatcher, M. A.; Leitfeld, J.; Kögel-Knabner, I.: In Stankiewicz, B. A.; van Bergen, P. F., Eds.; *Fate of Nitrogen Containing Macromolecules in the Biosphere and Geosphere* , ACS: Washington, DC, 1998.

52) Knicker, H.; Hatcher, P. G. *Naturwissenschaften* **1997**, *84*, 231-234.

53) Ensminger, L. E.; Gieseking, J. E. *Soil Science* **1942**, *50*, 205-209.

54) Rosenfeld, J. K. *Limnology and Oceanography.* **1979**, *24*, 1014-1021.

55) Christensen, B. T. *Advances in Soil Science* **1992**, *20*, 1-90.

56) Christensen, B. T.: In Carter, M. R.; Stewart, B. A., Eds.; *Structure and Organic Matter Storage in Agricultural Soils* CRC Lewis, 1996.

57) Schmidt, M. W. I.; Knicker, H.; Kögel-Knabner, I. *European Journal of Soil Science* **1997**, *48*, 319-328.
58) Knicker, H.; Almendros, G.; González-Vila, F. J.; Lüdemann, H.-D.; Martín, F. *Organic Geochemistry* **1995**, *23*, 1023-1028.
59) Nguyen, R. T.; Harvey, H. R.: In Stankiewiez, A.; van Bergen, P. F., Eds.; *Fate of N-Containing Macromolecules in the Biospere and Geosphere* ACS: Washington, DC, 1998.

INDEXES

Author Index

Subject Index

RETURN TO ➡

CHEMISTRY LIBRARY
100 Hildebrand Hall • 642-3753

LOAN PERIOD 1	2	3
4 **1 MONTH**	5	6

ALL BOOKS MAY BE RECALLED AFTER 7 DAYS
Renewable by telephone

DUE AS STAMPED BELOW

NON-CIRCULATING UNTIL: 7-30-99 MAY 1 9 2001		

FORM NO. DD5